The Science of Our Changing Climate

This compelling textbook provides a broad overview of the science underpinning our understanding of our climate, and how it is changing. Presented in clear and accessible language, and requiring only minimal algebra, it enables students to understand how our planet 'behaves' under 'normal conditions' and how human activity has moved us away from that normal. It walks the student comprehensively through the basic science, including how greenhouse gases absorb radiation and, crucially, a chapter on aerosols, major players in climate change. Diverse case studies and examples illuminate the impact and connections to real-world events while review questions and exercises consolidate knowledge. Including the latest results from the IPCC 6th Assessment Report, it concludes by exploring climate modelling, equipping students with an understanding of how to simulate both past climate changes and projections of future climate change. Online resources include lecture slides, solutions and Excel code.

Michael A. Box recently retired as Associate Professor in Physics at the University of New South Wales, Australia. He has taught Atmospheric Science and a number of advanced Physics courses. He has 100 publications, mostly in atmospheric science, with a particular focus on radiative transfer, remote sensing, and the radiative effects of aerosols. He is co-author of *Physics of Radiation and Climate* (CRC Press, 2016). He is a Fellow of the Australian Meteorological and Oceanographic Society, and a member of AGU. His former graduate students can be found in scientific establishments in Europe and Australia. He currently teaches Adult Education courses in Sydney.

Gail P. Box recently retired as Senior Lecturer in Physics at the University of New South Wales, Australia, where she taught a number of courses, including Atmospheric Science. She previously worked for BHP (Billiton). Her research focused on the environmental effects of atmospheric aerosols. In 2005 she founded the Australian (and New Zealand) Aerosol Workshop. She is co-author of *Physics of Radiation and Climate* (CRC Press, 2016). She is a member of AMOS and the Clean Air Society of Australia and New Zealand. Her former graduate students can be found in scientific establishments in Europe, Australia, and the USA.

The Science of Our Changing Climate

Michael A. Box
Retired from University of New South Wales

Gail P. Box
Retired from University of New South Wales

Shaftesbury Road, Cambridge CB2 8EA, United Kingdom

One Liberty Plaza, 20th Floor, New York, NY 10006, USA

477 Williamstown Road, Port Melbourne, VIC 3207, Australia

314–321, 3rd Floor, Plot 3, Splendor Forum, Jasola District Centre, New Delhi – 110025, India

103 Penang Road, #05–06/07, Visioncrest Commercial, Singapore 238467

Cambridge University Press is part of Cambridge University Press & Assessment, a department of the University of Cambridge.

We share the University's mission to contribute to society through the pursuit of education, learning and research at the highest international levels of excellence.

www.cambridge.org
Information on this title: www.cambridge.org/highereducation/isbn/9781009372343

DOI: 10.1017/9781009372305

First published 2024

A catalogue record for this publication is available from the British Library

A Cataloging-in-Publication data record for this book is available from the Library of Congress

ISBN 978-1-009-37234-3 Hardback
ISBN 978-1-009-37233-6 Paperback

Additional resources for this publication at www.cambridge.org/Box-Box

Cambridge University Press & Assessment has no responsibility for the persistence or accuracy of URLs for external or third-party internet websites referred to in this publication and does not guarantee that any content on such websites is, or will remain, accurate or appropriate.

Contents

Contents

Figures

Tables

Preface

Today, the words climate change, and global warming, are heard often; and much of the time with some degree of passion, one way or another. They are heard when a weather-related natural disaster strikes, which certainly seems to be happening more and more. Or is this just a recent 'media discovery'? We even hear 'this is not the time to talk about climate change' when such events occur, begging the obvious questions: Why not? When? We also hear these words whenever there is a big international meeting to discuss global action, such as the regular COP meetings, which often end with decidedly mixed results.

This book does not look in any detail at policy, much less politics, important though both are. Our focus is squarely on the branches of science that underpin the work of the many scientists who have devoted their careers to better understanding the challenges of our world, and its changing climate, and whose work is regularly assessed by the Intergovernmental Panel on Climate Change (IPCC).

There are many books that cover some of this material, and at many levels. Climate change is now taught in secondary schools, mainly in Geography/Earth Science classes. Books for such a course require readers to accept that the scientific consensus is rock solid: which we obviously agree is the case. At the other end of the spectrum, there is a growing number of university courses on climate, across a range of departments: the books for these can assume much more of their readers. This book aims to sit somewhere in between.

Our aim is to present all of the necessary information at a level that can be understood by any scientist, or science student, or indeed anyone with some basic science training, by keeping the technical details, and the mathematics (mostly algebra), to an absolute minimum. As such, it would be ideal for an introductory course for all science students, provided only that they have had a minimal exposure to physical science and reasonable competence in basic high-school mathematics. We are confident that an instructor can easily tune the level up or down to suit their particular student cohort.

The book is presented in **five parts**. The first three look at what we believe are the three basic branches of science that you need to know: chemistry, physics, and radiant energy. The last of these is, of course, a branch of physics, but is of such importance that we have chosen to treat it separately, if only for convenience. In Part IV we pull these subjects together, to see how they allow us to understand the climate system. In the final part we use that knowledge to focus on recent, and future, climate change. So, what do we cover?

Our climate is changing because we have been using the atmosphere as a dumping ground – for pollutants, carbon dioxide, CFCs, and so much else – changing its **Chemistry**. We devote three chapters to this branch of science, the first of which covers the basics. We then devote a chapter each to the greenhouse gases – their natural sources and sinks, and their recent anthropogenic sources – and one to aerosols, now recognised as major players in climate change. (This was our primary field of research, which you will see reflected in Chapter 4.)

In Part II we look at the atmosphere as a **Physical** system, starting with its 'static properties'. For pedagogic convenience, we devote a separate chapter to the role of water vapour, including a key focus on cloud droplet formation. The interactions between increasing aerosol levels, the seeds of all cloud droplets, and the reflectivity of clouds has assumed larger importance following AR6. The thermally driven circulation of the atmosphere is, along with ocean currents, the transporter of heat from low to high latitudes.

The inflow and outflow of **Radiant Energy** is the fundamental governor of our climate. So we start Part III by looking at the laws that govern radiation emission, and using these to compute the temperature of the Earth as a planet. But this is not the surface temperature, so we look to the greenhouse effect for an explanation, including very simple models that can help us to understand both climate and climate change. Why are some gases greenhouse gases, and others are not? This all-important question requires a short, qualitative journey into the wonders of quantum physics. With all of this knowledge under our belt, we append two applications chapters: one to cover the life-preserving ozone layer; and one to (satellite) remote sensing, and the vital data it provides for both weather and climate.

We now turn to the reason for writing this book in the first place: **Climate**, and more specifically, climate change. For convenience, we have split the last eight chapters into two parts. In Part IV, we pull together many of the necessary pieces. A major cause of climate variability lies in the ocean (e.g. El Niño): we need to understand this, so we can separate it from climate change. The climate is, in fact, a 'system', and we need to understand that it has many components which interact, sometimes leading to feedback phenomena. Finally, we should be in a position to build a global climate model, of appropriate sophistication, to answer the most crucial of questions. Climate has changed in the past: do we understand the causes, and the lessons to be learned?

How is our climate changing? Do we understand this? What does the future hold? The models just mentioned are the key to obtaining reliable answers to these, and other questions: something undertaken by thousands of scientists whose work underpins the IPCC's ongoing endeavours. These last four chapters draw heavily on the IPCC's 6th Assessment Report – AR6 – looking firstly at the data on climate change over the past ~100 years, and then at whether or not our models can replicate this. We also return to one of the issues introduced in the very first paragraph above: are the many reported extreme weather events really unusual? While future climate change will depend on the decisions humans make, our models are valuable tools in guiding that decision making, and predicting the likely consequences.

In Part IV, we introduce you to two interesting models. The first is an **Energy-Balance Climate Model**, which you can code up (an Excel version is available to instructors), which illuminates some of the key feedback ideas. The second is an online model that is freely available, via which you can play out many of the scenarios from Part V.

Most chapters contain suggestions for Further Reading, Review Questions, and Exercises. We also provide ~25–30 PowerPoint slides per chapter to assist instructors.

We have taught Physics at the University of New South Wales, Sydney, Australia, for many years, including an Atmospheric Physics elective. This is the book we wish we'd had then, but did not feel confident enough to write. Being an elective, we had the freedom to tailor the course to the interest and backgrounds of the students, which were varied.

In retirement, one of us has taught some of this material in adult education courses, firstly to maintain an interest, and secondly for the challenge of teaching a rather different cohort of students (mostly retired, many with a 'technical' background, but certainly not all). These students were in my class because they wanted to understand, and not to simply accept it all without question (or to pass an exam). It is the lessons learned from that experience, coupled with the University classes we taught, that impelled us to put finger to keyboard.

There are many people who deserve our gratitude for helping to make the book possible. Firstly, to our many old friends at the Climate Change Research Centre, UNSW, who were there to provide guidance and encouragement when needed. In particular, we thank Lisa Alexander, Gab Abramowitz, Alex Sen Gupta, and Laurie Menviel. Some great photographs and other inputs were provided by Pat McCormick, Dennys Angove, David Cohen, John Le Marshall, Ben McNeil, Sarah Fitzherbert, Majed Radhi, Maja Kuzmanoski, and Thomas Trautmann. Finally, an enormous vote of thanks to the team at CUP who guided (and, where necessary, goaded) us through the whole process, especially Emma Kiddle and Emma Collison. To all of you, we hope the end product does you the credit you deserve.

Abbreviations and Acronyms

AAO	Antarctic Oscillation (a.k.a. SAM)
AED	atmospheric evaporative demand
AMOC	Atlantic Meridional Overturning Circulation
ANSTO	Australian Nuclear Science and Technology Organization
AO	Arctic Oscillation
AOD	aerosol optical depth
AR6	Sixth Assessment Report of the IPCC
CAPE	convective available potential energy
CDR	carbon dioxide removal (a form of geoengineering)
CFC	chlorofluorocarbon
CINE	convective inhibition energy
EEA	extreme event attribution
ENSO	El Niño and the Southern Oscillation
ERF	effective radiative forcing
ERFari	ERF due to aerosol–radiation interactions
ERFaci	ERF due to aerosol–cloud interactions
ESM	Earth System Model
ET	evapotranspiration
FAR	fraction of attributable risk
GCM	general circulation model/global climate model
GHG	greenhouse gas
GMSL	global mean sea level
GMST	global mean surface temperature (LSAT + SST)
GSAT	global surface average temperature (LSAT + MAT)
IOD	Indian Ocean Dipole
IPCC	Intergovernmental Panel on Climate Change
IPO	Interdecadal Pacific Oscillation
IRF	instantaneous radiative forcing
ITCZ	Inter-tropical Convergence Zone
LCL	lifting condensation level
LSAT	land surface air temperature
MAT	marine air temperature
MSLP	mean sea-level pressure
NAO	North Atlantic Oscillation

NASA	National Aeronautics and Space Administration (USA)
NAT	nitric acid trihydrate [$HNO_3(H_2O)_3$]
NDC	Nationally Determined Contribution (under the Paris Agreement)
NH	Northern Hemisphere
NOAA	National Oceanic and Atmospheric Administration (USA)
NO_x	$NO + NO_2$
NPP	net primary production
ODS	ozone-depleting substance(s)
PAR	photosynthetically active radiation
PD	precipitation deficit
PDO	Pacific Decadal Oscillation
P-E	precipitation minus evaporation
QBO	quasi-biennial oscillation
RCP	representative concentration pathway
SAM	southern Annular Mode
SH	Southern Hemisphere
SIA	sea-ice area
SLCF	short-lived climate forcers
SMD	soil moisture deficit
SOA	secondary organic aerosol
SRM	solar radiation management (a form of geoengineering)
SROCC	Special Report on the Ocean and Cryosphere in a Changing Climate
SSP	shared socioeconomic pathway
SST	sea-surface temperature
TOA	top of atmosphere
TSI	total solar irradiance
TWVC	total water vapour column
WMGHG	well-mixed greenhouse gases

Chemical Elements

Ar	argon
Al	aluminium
Be	beryllium
C	carbon
Ca	calcium
Cl	chlorine
F	fluorine
Fe	iron
H	hydrogen
K	potassium
Mg	magnesium

N	nitrogen
Na	sodium
O	oxygen
Rn	radon
S	sulphur
Si	silicon
Th	thorium
U	uranium

Important Chemical Species

CH_4	methane
CO	carbon monoxide
CO_2	carbon dioxide
CO_3^{2-}	carbonate ion
$CaCO_3$	calcium carbonate
$CaSiO_3$	calcium silicate
CFC-11	$CFCl_3$
CFC-12	CF_2Cl_2
H_2CO_3	carbonic acid
HCO_3^-	bicarbonate ion
HCl	hydrogen chloride (hydrochloric acid)
HNO_3	nitric acid
H_2O	water
H_2S	hydrogen sulphide
H_2SO_4	sulphuric acid (battery acid)
NH_3	ammonia
N_2O	nitrous oxide
NO	nitric oxide
NO_2	nitrogen dioxide
NaCl	sodium chloride (common salt)
OH	hydroxyl radical
O_3	ozone
SiO_2	silicon dioxide (silica/quartz)
SO_2	sulphur dioxide
SO_4^{2-}	sulphate ion

1   Planet in Peril

Human activities, principally through emissions of greenhouse gases, have unequivocally caused global warming, with global surface temperature reaching 1.1°C above 1850–1900 in 2011–2020. Global greenhouse gas emissions have continued to increase, with unequal historical and ongoing contributions arising from unsustainable energy use, land use and land-use change, lifestyles and patterns of consumption and production across regions, between and within countries, and among individuals (*high confidence*).

Widespread and rapid changes in the atmosphere, ocean, cryosphere and biosphere have occurred. Human-caused climate change is already affecting many weather and climate extremes in every region across the globe. This has led to widespread adverse impacts and related losses and damages to nature and people (*high confidence*). Vulnerable communities who have historically contributed least to current climate change are disproportionately affected (*high confidence*).

Continued greenhouse gas emissions will lead to increasing global warming ... Every increment of global warming will intensify multiple and concurrent hazards (*high confidence*). Deep, rapid, and sustained reductions in greenhouse gas emissions would lead to a discernible slowdown in global warming within around two decades ... (*high confidence*).

1.1   Our Dynamic Environment

The words you have just read are three of the key conclusions from the Summary for Policy Makers of the Synthesis Report of the IPCC's 6th Assessment Report (AR6), and while they may be updated in subsequent Reports, they are unlikely to be watered down. It is the aim of this book to help you understand exactly how, and why, the collective body of scientists has come to these, and many other, conclusions.

Addressing this aim requires that you understand two facets of the nature of our planet. Firstly, you need to understand how our planet 'behaves' under what we might choose to call 'normal conditions'. Secondly, you need to understand how human activity has moved us away from that normal. The chapters that follow will take you through both of these. To help prepare you, these first four sections will set the scene.

1.1.1   Cycles and Balances

Although we don't often stop to think about it, we live on a dynamic planet: many facets of its reality are in a constant state of flux. Many of these are cyclic, like day and night, or the seasons. Others are far more subtle. Some very important aspects of our environment are actually the result of dynamic balances: equal but opposite tendencies pushing us 'left and right'. Let us start by taking you through some of these.

The first is the diurnal cycle, which we now understand as simply the rotation of the Earth on its axis. Yet, for centuries this logical idea was rejected, in favour of a cosmology which placed the Earth – and more specifically, us – at the centre of the Universe, with the cosmos revolving about our heads. When this was challenged, in the late sixteenth and early seventeenth centuries, by Copernicus, Kepler, Galileo, and others, they were bitterly resisted by the Church. Galileo was condemned by the Inquisition, and only rehabilitated 350 years later. Associated with the heliocentric model of the cosmos, or at least the solar system, is the fact that the Earth revolves around the Sun in an annual cycle. The additional fact that the Earth's rotational axis is inclined away from the vertical (i.e. is not at right angles to the Earth's orbital plane) then explains the seasonal cycle.

It took many centuries before these ideas were finally accepted. After all, they seem to contradict what our senses tell us: what we might refer to as our common sense. The key step in this re-evaluation came with Newton's laws of motion, which showed that the heliocentric model could be explained by natural laws, plus the necessary mathematics (which Newton also invented). A large part of this book is all about some other physical processes that, at first sight, seem either confusing, or contradictory. It is our task to take you through these, and give you a good grounding in the basic science that underpins them.

What other dynamic cycles, and balances, should we examine? The Earth's climate is the result of a number of such cycles and balances, such as the seasonal cycle. Far more important, however, are the flows of energy, into and out of the Earth and its atmosphere. We certainly understand the inflow of solar radiant energy: we see it, we feel it, and we understand how it is the fuel source to power both plant growth and photovoltaic cells. But this inflow must be balanced by a corresponding outflow, or else the Earth would steadily heat up, and evaporate!

This outflow is also in the form of radiant energy, but it is in the infrared spectral region, which the photoreceptors in our eyes have not evolved to see: probably just as well, or we might not be able to sleep at night. What our senses are also not able to tell us is that most of this outflow is trapped by gases in our atmosphere, with much of it returned to the surface, keeping our planet hospitable for life. This is the greenhouse effect, which we shall examine in some detail in this book, and the relevant gases are known as radiatively active gases, or more colloquially as greenhouse gases.

1.1.2 Chemical Cycles

This brings us to the next cycle/balance. Without giving the matter too much thought, many of us might have been tempted to regard the composition of our atmosphere as fixed, or permanent. So now we come to matters of definition. The major gases in our atmosphere, oxygen and nitrogen, have remained in their current mixing ratios (or mass fractions) for millennia. However, that does not mean that the individual molecules have been in the atmosphere that long. For example, the average lifetime for an oxygen molecule is a few thousand years; tiny compared with the age of the Earth.

Carbon dioxide molecules are constantly being exchanged with the biosphere, with a cycle time of around a decade; short, even by human timescales. Carbon dioxide also dissolves in water, and there is an equilibrium exchange between the atmosphere and the oceans. Carbon and its compounds take part in a range of more complex cycles, which we will examine in Chapter 3, and over millions of years its atmospheric concentration has risen and fallen, for reasons we partly, but not always fully, understand.

One factor has been the slow accumulation of carbon in the form of coal, oil, and methane (natural gas), much of it during the Carboniferous Period, 300–360 million years ago. Over the past couple of centuries, a not-insignificant fraction of this carbon, which took 60 million years to accumulate, has been returned to the atmosphere. So here is a balance that has been disrupted by human activity: one whose impacts we are now coming to appreciate.

There are other trace gases in our atmosphere that are also in various states of flux. Some of these gases manifest themselves as air pollution in large industrial centres and crowded cities. The ozone layer, which protects all terrestrial life, also undergoes dynamic production and destruction processes, some of which may be altered by gases such as chlorofluorocarbons, and the balance shifted. This is one danger we appear to have been alerted to in time, and the Montreal Protocol was put in place to protect us. (All of this will be covered in Chapter 10.)

1.2   Recent Imbalances

The Earth has gone through a series of glaciations and deglaciations over the past two and a half million years. Modern humans appeared during the Last Interglacial, and survived the last glacial, which concluded about 11,000 years ago. Yet that episode is not in our consciousness, as both agriculture, and civilisation, have developed during the Holocene, a comparatively brief period of relatively stable climate, with the available data suggesting that both temperature and atmospheric composition have remained largely stable. However, things now appear to be changing, and not necessarily for the better.

Today we are well aware of the fact that both temperature and composition are changing, and we understand that one is, somehow, contributing to the other. So, let's have a look at

Figure 1.1 Atmospheric concentration of CO_2 from measurements at Mauna Loa, USA. *Source*: data provided by NOAA Global Monitoring Laboratory, Boulder, CO, USA.

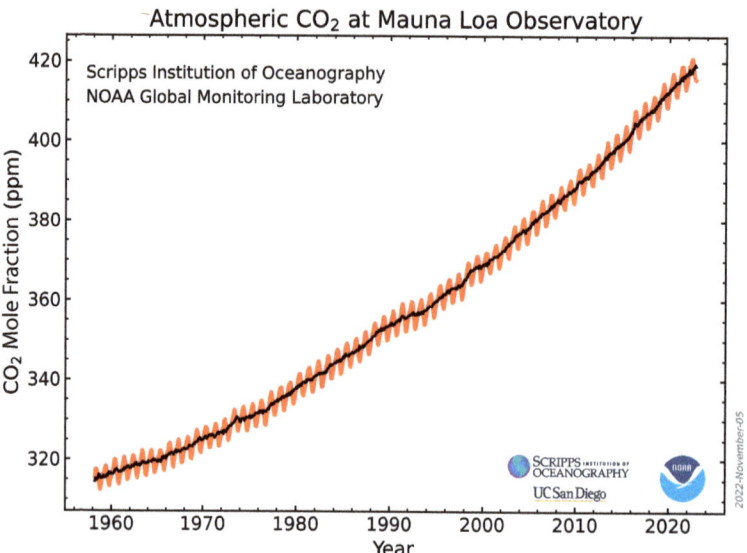

Figure 1.2 Temperature anomalies, 1850–2021. *Source*: data from NOAA/PSL.

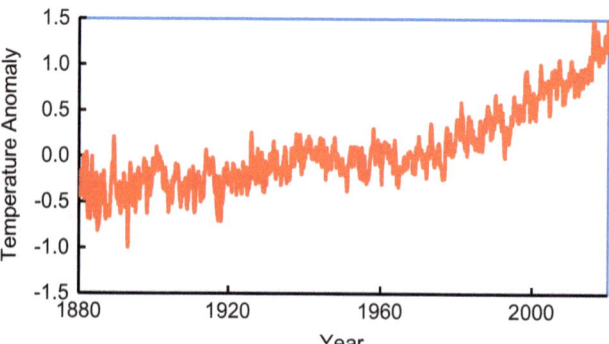

those changes. Figure 1.1 shows the atmospheric concentrations of the major greenhouse gas, carbon dioxide, for the past ~65 years, based on 'direct' measurements at Mauna Loa, USA. In later chapters we'll add the appropriate data for the other two key greenhouse gases, methane and nitrous oxide. (Air bubbles trapped in the ice in both Greenland and Antarctica, which allow us to track back many millennia, show that the CO_2 concentration was 280 ± 5 ppm for the previous several millennia.) So let's examine the data. In just 65 years, the concentration has gone from less than 320 ppm to around 420 ppm; a 30% increase in a geological blink of an eye.

What is relevant, of course, is the more recent rise in temperature, or to give it the full name which that single word usually represents, the **global mean surface temperature**. Figure 1.2 shows the reliable data we now have, going back around 150 years. (These data are actually presented in the form of so-called **anomalies**: deviations either side of a suitably chosen baseline, in this case the average over the period 1960–1990.) We see a mix of behaviours:

fluctuations, plus an overall increase. The fluctuations, with a cycle time of a few years, are examples of **climate variability**, mostly caused by changes in ocean temperature distributions, such as El Niño, about which science is learning more and more. These so-called ocean modes of variability will be covered in Chapter 13.

By far the most significant takeaway from these data, though, is the rapid rise in temperature, especially in the past 50–60 years, to temperatures which, to the best of our knowledge, have not been experienced since human civilisation began. Not so easy to present in graphical form has been the increasing severity of weather-related disasters: heatwaves; wildfires; droughts and floods. While most of us have only recently become aware of how widespread have been such changes, the insurance industry has been watching with a degree of trepidation for at least 30 years.

These two data sets pose a direct question for science: and by that we mean for scientists. Is there a connection – i.e. does one, perhaps, drive the other – or is it just chance? Perhaps there is a third factor that is driving both? While we are sure you are aware of the accepted answer to this question, it is the purpose of this book to give you a deeper insight into the relevant science, and the work of the many scientists involved.

1.2.1 A Little History

Interest in understanding our planet in a scientific way is around two centuries old. The first attempts to measure and map temperatures, for example, go back to the German naturalist Alexander von Humboldt (1769–1859). The fact that the atmosphere retains heat, much like a greenhouse (this analogy, although widely used as you know, is far from perfect) dates back to the French mathematician Joseph Fourier (1768–1830).

The sixth Assessment Report of the IPCC (see Section 1.5) provides an interesting figure {figure 1.6} showing a timeline of some of the key steps, both in observations related to our changing climate, and to our understanding of the (enhanced) greenhouse effect (as it is known). We include it here as Figure 1.3.

Three of the people noted in that figure deserve a further mention. In addition, there are some additional milestones that should be added.

Eunice Newton Foote (1819–1888) was an American scientist, and perhaps the first to realise that rising levels of CO_2 had the potential to alter the Earth's climate. Her publications on this in 1856 and 1857 are regarded as among the first by a woman in the field of physics. She was also active in women's rights, and the abolition of slavery.

John Tyndall (1820–1893) was an English experimental physicist who measured the light-scattering properties of small particles (including the Tyndall effect). He also measured the ability of both CO_2 and water vapour to absorb infrared radiation, an idea that had been suggested by Fourier in 1822. In 1868, Jozef Stefan developed his theory of blackbody radiation: a subject we will develop in some detail in Chapter 8.

Swedish Chemist Svante Arrhenius (1859–1927) was the first to attempt to quantify the effects of the increase in atmospheric CO_2 that would inevitably arise from the use of coal following the Industrial Revolution. He concluded that doubling the level would lead to a

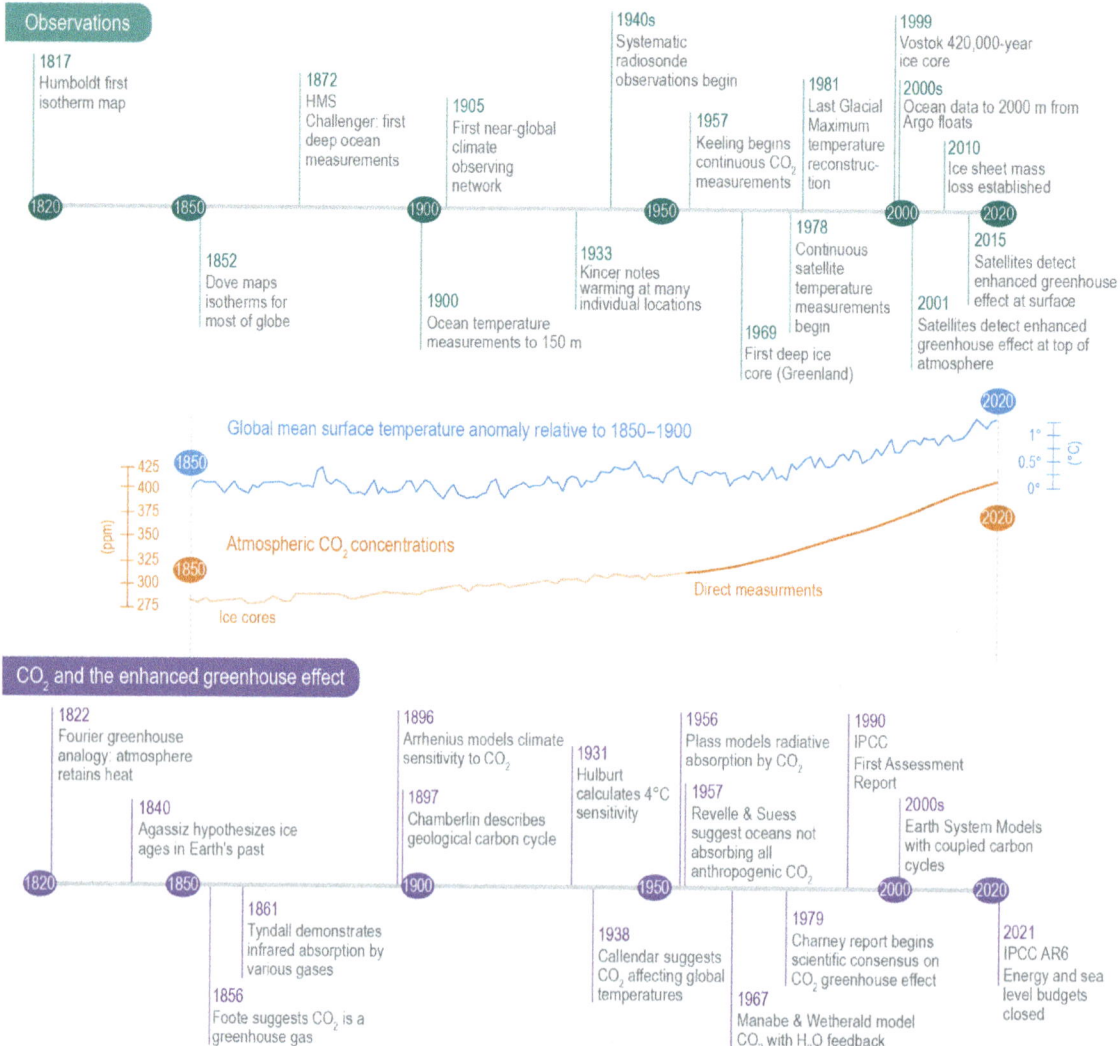

Figure 1.3 Climate science milestones, between 1817 and 2021. Top: milestones in observations. Middle: curves of global surface air temperature, relative to 1850–1900, and atmospheric CO_2 concentrations from Antarctic ice cores; direct air measurements from 1957 onwards. Bottom: milestones in scientific understanding of the CO_2-enhanced greenhouse effect. *Source*: see Acknowledgements at the end of this chapter.

temperature rise of about 5°C. This number, known as the **climate sensitivity**, has been refined over the years that have followed, of course, but his result is considered to be not all that wide of the mark.

In the late 1950s, Roger Revelle began to popularise the idea that global warming might be a problem. In 1958, Keeling started his CO_2 measurements on Mauna Loa. The late 1980s saw seven of the eight warmest years to that point, kick-starting serious interest in the issue.

On 23 June 1988, NASA scientist James Hansen testified before the US Congress, and told the world that 'global warming has started'.

1.3   The Earth System

Our atmosphere is merely the most obvious part of what we may refer to as the climate system, or more generally, the Earth System. Those words are meant to convey to our readers a number of pieces of information. The first is that we are talking about a suite of interconnected components. We will need to understand both how each of these behaves and, more importantly, how they all interact to produce the world we live in. Equally, we hope, these words make it clear that this analysis must be based on the underlying scientific principles – mainly physics and chemistry, but also ecology and economics – which govern them.

We will need to understand how the chemical composition of our atmosphere is changing: not just the data in Figure 1.1, but the underlying causes. And while we have already pointed to our use of fossil fuels as the primary cause in the rising levels of carbon dioxide, that is not the only cause. We also know that CO_2 undergoes rapid exchanges with the biosphere: are these implicated in any way with rising concentrations? What are the sources of methane and nitrous oxide? How might they be changing?

Of course, these are not the only compositional changes our atmosphere is experiencing, with air pollution and acid rain well documented. We also know that certain man-made chemicals are having an impact on the ozone layer. Are there any interconnections between these various environmental insults, or are they totally independent? Perhaps more to the point, could any attempts to address one of these issues actually exacerbate one of the others?

Our atmosphere acts like a machine, which acts to transport heat from low latitudes to high. (The oceans also contribute to this.) In the process, it also transports water vapour, creating the world's jungles and deserts. If we are to construct believable models with which to study a changing climate, we must start with the weather.

The climate system is also a thermodynamic system, meaning that we must carefully study the flows of energy (and chemical substances), on relevant timescales. This invariably means that we start by focusing on short time frames – weather – and so just the atmosphere. But as we look to understand changes on longer timescales, we will need to add the cryosphere (i.e. the ice sheets), the oceans, and the biosphere. Finally, the Earth System incorporates the role of humans, through our economic activity and resulting emissions.

Figure 1.4 ({figure 1.1} from IPCC AR6) shows a visual representation of the climate system's four components. To the right of the image is a simplified representation of the changes that have been taking place in this system starting at 1850 (the base time period for that report): changes in atmospheric composition, temperature, precipitation patterns, glacial melting, sea level, and the interior of the ocean. In Chapter 18 we will take you through these changes in great detail.

The economy of the twentieth century has largely been powered by fossil fuels, mainly coal and oil, and while the twenty-first century has seen a growing move towards renewable energy sources – wind and solar – this largely remains the case today. Coal and oil represent stores of

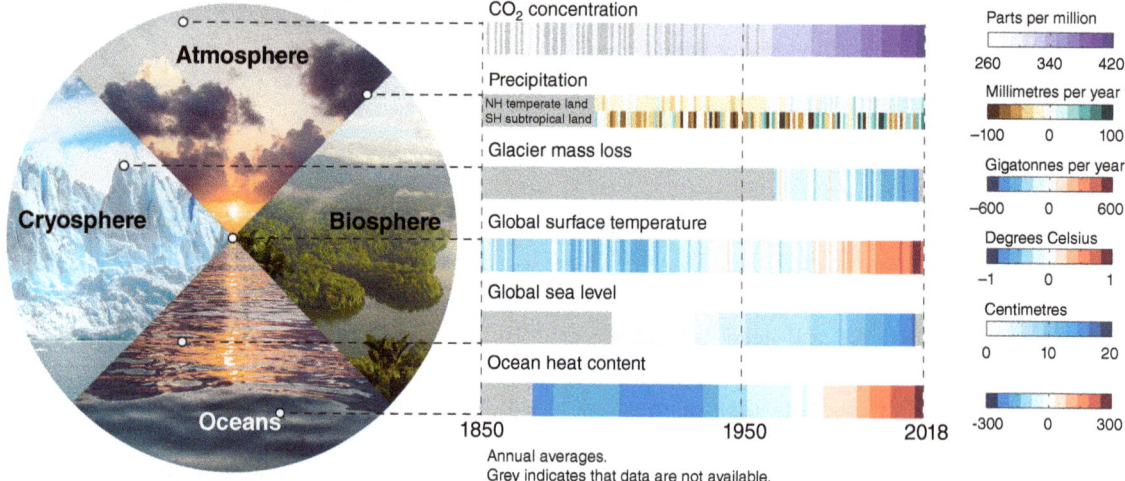

Figure 1.4 Changes are occurring throughout the climate system. Left: main realms of the climate system: atmosphere, biosphere, cryosphere and ocean. Right: six key indicators of ongoing changes since 1850, or the start of the observational or assessed record, through 2018. Each strip (except for precipitation) indicates the global annual mean anomaly for a single year, relative to a multiyear baseline (except for CO_2 and glacier mass loss, which are absolute values). Grey indicates where data are not available. *Source*: see Acknowledgements at the end of this chapter.

carbon which were deposited 300+ million years ago. We burn them to produce heat – i.e. energy – but that also releases carbon dioxide as a by-product; one that most chose to forget for many decades. The rise in CO_2 shown in Figure 1.1 shows the result: much has ended up in the atmosphere. After all, where else was it to go? In fact, some has gone into the ocean, and some has been taken up by the biosphere, but much remains in the atmosphere.

So now let's turn to the temperature data, and especially the temperature rise, because this is having a direct effect on the cryosphere, the biosphere, and on us. We mentioned previously that the Earth's global climate is maintained by the balance between the inflow and outflow of radiant energy, and further that the outflow is, in turn, controlled by the presence of certain constituents of our atmosphere; the greenhouse gases. Because the concentrations of these gases are manifestly on the increase, we must expect some impact on the radiation flows.

This time the science is much more complex than simply charting our usage of fossil fuels, and will occupy a key component of this book. When you have finished reading, we hope you will have a much greater appreciation of both the science, and the efforts of scientists, to go from basic concepts, to scientific principles, to quantitative modelling, allowing us to not only appreciate the connections between the data in Figures 1.1 and 1.2, but also to make statements – and provide warnings if deemed necessary – about what might lie ahead. Of course, the task of climate science is much more than just temperature rise, as was indicated by Figure 1.4. So, we must develop a full understanding of all of the realms of the climate system, so that we can understand these changes.

We have taught university physics for a combined half a century, and appreciate its pedagogy. It is one of the basic sciences, and is invariably developed in a logical, step-by-step approach of going from the basic definitions and laws to the various applications – at least when we are afforded sufficient class hours.

Atmospheric science, which we have also taught, is more intertwined, being somewhat like a jigsaw puzzle, with intermeshing pieces. Chemistry plays a role; thermodynamics plays a role; fluid (i.e. Newtonian) mechanics plays a role; but most importantly, at least for our concerns, the science of electromagnetic radiation plays a key role, and even quantum theory. In the chapters ahead we will try to present the necessary pieces of science in a systematic manner. Note that we have also written the text book *Physics of Radiation and Climate* [PRC] (CRC Press/Taylor and Francis, 2016), where much of this science is developed in greater detail.

1.4 The Road Ahead

This book will introduce you to the key branches of science that underpin all serious efforts to understand our climate, and why it is changing: chemistry; physics; radiant energy. To help guide you, we have divided the book into five parts, and will now provide an outline of what to expect from each, and why it is important. Some of the chapters present material that is both essential and – we hope – easily understandable. Others present subject matter that is either somewhat more technical, or a little more peripheral. However, we believe that every chapter deserves at least one class hour devoted to it, even if only to discuss the concepts.

The order of presentation we have chosen is not set in stone, and instructors may choose alternate paths. For example, covering the physics chapters before the chemistry chapters makes equal sense to the path we have chosen.

1.4.1 Chemistry

As already noted, the chemical composition of our atmosphere has been changing over the past century or more, with the increase in greenhouse gases the most conspicuous. So, this is the logical place to start. Many chemical reactions that take place in the atmosphere are in balance, at least most of the time. However, human activity is complicating the situation in ways that have the potential to shift the balance of some of these processes. We need to understand how and why.

In Chapter 2 we start with the composition of the atmosphere. We then introduce you to some key photochemical reactions – reactions which require solar energy as an input – and also the OH radical, which helps remove unwanted pollutants. Air pollution is more a local/regional problem, but it can still teach us some valuable lessons.

In Chapter 3 we focus on the greenhouse gases; where they come from naturally, and how they may be removed from the atmosphere: their sources and sinks; and their short- and long-term cycles. This is essential in order to place the human contributions in a proper perspective.

Chapter 4 looks at aerosols; small particles and droplets in the atmosphere, with sizes of a few microns or less. These particles are now recognised as key environmental components, with many impacts, both as pollutants causing serious health effects and also having a direct impact on solar radiation and climate. (This was our primary research interest for many years.)

1.4.2 Physics

The atmosphere is a physical system, so we need to cover the key branches of physics that are relevant to climate science: thermodynamics – the intersection of heat and motion – and fluid mechanics – how, and why, a fluid moves.

In Chapter 5 we tackle the first of these subjects: thermodynamics. We firstly focus on the key variables of pressure, temperature, and density: their interconnections and atmospheric vertical profiles. This leads on to questions around vertical stability and convective motions.

In Chapter 6 we add water vapour and latent heat to the mix, with their many consequences, leading on to clouds and their formation: key players in the climate we experience on daily to seasonal timescales.

In Chapter 7 we look at the forces at work on a segment of the atmosphere, and how it responds. The key driver of all atmospheric motion is temperature, or more specifically temperature gradients. This creates what is referred to as a thermal circulation, which can range from a sea breeze to the global circulation of the entire atmosphere. This is the key to regional weather and climate, and the hydrological cycle.

1.4.3 Radiant Energy

Radiant energy, the energy carried by light and other electromagnetic waves, is, of course, a major branch of physics. However, we see it as being so central to climate that it deserves its own part (Part III). It is, after all, the flows of radiant energy into, and especially out of, our atmosphere that are at the core of the greenhouse effect. So, if we are to understand this central fact of our existence, we need to understand the science of radiant energy, and how it interacts with our atmosphere. This part comprises three chapters of basic science, plus two applications chapters.

We start in Chapter 8 with the basic physics of electromagnetic radiation, with a particular emphasis on how substances emit radiation depending on their temperature: known as thermal radiation. The laws of thermal radiation are the key to determining how much radiation the Earth emits to space, in order to balance incoming solar insolation. It turns out that the Earth, as an entity, is distinctly cooler than the average surface temperature, a dilemma which is 'solved' by the atmospheric greenhouse effect. To help you understand this science, we introduce you to simple models of how it works. This chapter is central to the book, and to climate science in general.

In Chapter 9 we look at solar radiation, and its passage through our atmosphere. We start with a quick look at the Sun's output, and the Earth's orbit around it, which determines how much insolation we receive. We then look at processes in the atmosphere that may impact on its inflow, such as absorption by water vapour, and scattering by molecules, particles, and clouds. Finally, we take the opportunity to examine ideas that have been mooted to help reflect some of this energy back to space, to help cool the planet: known as geoengineering.

In Chapter 11 we address the outflow of the Earth's thermal radiation, and its interactions with various gases in the atmosphere; the key to both the natural greenhouse effect and to global warming. Why are gases such as carbon dioxide and water vapour greenhouse gases, while oxygen and nitrogen are not? To answer this vital question, we need to lead you through some basic quantum physics. Equally important is to understand how much the greenhouse effect will change as the concentrations of the relevant gases change.

Chapter 10 looks at the vital ozone layer, which is a product of the interaction of UV radiation with the oxygen molecules in the stratosphere. We start by outlining the reactions that create, and destroy, ozone. Then we look at the threats that have been perceived from the release of certain chemicals into the stratosphere, particularly the CFCs. The Antarctic ozone hole came as a big surprise, involving chemistry that had not been anticipated. Finally, we examine the Montreal Protocol, which was designed to address this threat: fingers crossed.

Now that we understand the physics of electromagnetic radiation, and how it interacts with the constituents of our atmosphere (and the Earth's surface), we will put that knowledge to work. In Chapter 12 we look at remote sensing, and the satellite technology that supplies Numerical Weather Prediction models with so much vital data, one of the keys in the steady improvement in the nightly forecast. Satellites, supported by ground-based instruments, also supply an enormous quantity of 'environmental' data on the changing composition of the atmosphere, and the health of our planet; especially, of course, its climate.

1.4.4   The Climate System

Now that we have all the necessary pieces of science under our belts it is time to turn to their application in understanding the climate, and how/why it is changing. For convenience, we have split the remaining eight chapters into two parts, the first focusing on the climate system as an amalgam of many parts, and the second focusing on understanding recent climate change, and on what the future may hold.

In Chapter 13 we look at circulation within the oceans, both the surface currents and the deep ocean. However, the most important contribution to climate made by the oceans – at least at regional scales – is via their so-called 'modes of variability'. The best known of these is the El Niño/La Niña phenomenon of the central Pacific. Scientists have identified significant modes in all oceans, and we outline those which are the most significant.

Whenever you disturb a complex physical system – and that includes the Earth's climate system – some of the changes that occur may feed back, either enhancing or damping the initial disturbance. The climate system has many important feedbacks, which will be outlined in Chapter 14. We will also introduce a simple Energy-Balance Model, which will illustrate much of what has just been covered. We encourage you to code it up in your favourite computer language.

In Chapter 15 we introduce you to the models used for 'serious' climate studies, starting with the general circulation models used for weather prediction. After that we look at the other pieces that are needed to model climate, as opposed to weather, which include the oceans, cryosphere, land surface, and even the necessary biogeochemistry. We also discuss the suite of models that underpin the actions and conclusions of the IPCC.

In Chapter 16 we look at how climate has changed in the past, before the arrival of modern humans. We firstly take a brief tour into the deep past, and some of the fascinating changes that occurred back then: the co-evolution of the biosphere and the geosphere. After that we focus on the more recent past; the Tertiary and Quaternary periods, and especially the (current) Ice-Age cycle. There are two lessons we need to learn in this chapter. One is the cause of that cycle, and does that still operate? But a more important question for future generations is the connections between CO_2 concentrations, temperature, and sea level.

1.4.5   Climate Change

Finally, we turn to the central challenges of the book: how is our environment changing; how is our climate changing; can we reliably connect these two; and finally, how is our climate likely to change during the present century? That is to say, what is our science telling us?

In Chapter 17 we look at the observed changes in those components of the climate system that may impact on the planetary energy budget, and hence 'force' the climate away from the state it has been in for the past few millennia. This includes both the natural drivers such as volcanic activity and the anthropogenic drivers: greenhouse gases (both major and minor), aerosols and their interactions, and land-use changes. All of these need to be quantified.

In Chapter 18, we ask two questions: How is our climate changing? And, why is our climate changing? To address the first question, we examine the available data on the indicators of change: temperature, precipitation, ice and snow, oceanic heat content and sea level, ocean pH, and changes in the biosphere. To answer the second question, we take the information from Chapter 17, 'run it through' our climate models (to produce 'climate simulations'), and see how well it matches the conclusions from the first question.

In recent decades, many of us have become aware that Mother Nature is not happy, and is making her displeasure known in the increase in 'wild weather', otherwise known as extreme events: droughts and heatwaves; storms and floods. In Chapter 19 we take a quick look at the data to check whether such changes are not just our imagination going into panic mode. We back this up with a number of informative case studies.

Finally, we turn to the question to end all questions: what does the future hold? In order to answer that question, we would need to know how the forcing agents we examined in Chapter 17 – greenhouse gases, aerosols, land use – are going to change in the decades ahead. Needless to say, we don't: that is not science, but rather economics, politics, and technological change. But we can't just give up. Instead, in Chapter 20 we turn to 'scenarios': logical combinations of guesses as to the possible path humanity might decide on (or stumble on). Then we run our climate models, and simulate that possible future world: known as **projections**, rather than predictions. Some of the results are, needless to say, rather scary.

1.5   The Intergovernmental Panel on Climate Change

Before we look in more detail at the physical processes that are at the heart of both climate science and climate change, we will spend a little time examining the role of the IPCC, and the international agreements that many countries have entered into, even if only 'on paper'.

The IPCC does not undertake any research of its own; rather, its role is to examine all the available **peer-reviewed** evidence that has been produced by scientists around the world. This has essentially three aspects:

1. Assessing the current state of knowledge about climate science and climate change, including the magnitudes and confidence levels of changes in atmospheric composition, and changes in the environment.
2. Modelling efforts designed to assess our understanding of the science that connects the changing composition of the atmosphere with the changes in the climate system.
3. Modelling efforts to 'predict' how climate may evolve in the coming decades: this involves both an analysis of the outputs from different climate models, as well as guidance on some of the key model inputs.

1.5.1 Assessment Reports

The IPCC is organised into three Working Groups (their boundaries sometimes shift a little), and has produced six (sets of) major reports, with ever-increasing author teams from all corners of the globe:

- The First Assessment Report (FAR) in 1990;
- The Second Assessment Report (SAR) in 1995;
- The Third Assessment Report (TAR) in 2001;
- The Fourth Assessment Report (AR4) in 2007;
- The Fifth Assessment Report (AR5) in 2013/14; and
- The Sixth Assessment Report (AR6) in 2021/2022 (the Synthesis Report, covering all three Working Groups, was released in March, 2023).

plus, a number of supplementary reports dealing with specific issues, and the Special Report on the Oceans and Cryosphere in a Changing Climate (SROCC) released in September 2019. The vast majority of the material in this book is concerned with the field covered by Working Group I (WGI), the Physical Science Assessment Working Group, and most of the results presented in Parts IV and V have been extracted from its latest Report.

The task of WGI is to present, in the clearest possible terms, our knowledge of the science of climate change, together with the best estimate of the climate changes that are likely to occur over the twenty-first century as a result of human activity. In order to produce Reports that would be taken seriously, it was felt necessary to involve as broad a collection of the world's scientific community in their production as possible.

In addition to the detailed report, each Report also includes a Summary for Policy Makers (SPM), the wording of which is approved in detail by a plenary meeting of the Working Group, the object being to reach agreement on both the science and the best way to convey the science to policy makers with both accuracy and clarity.

1.5.1.1 Evolution

If you take some time and read large sections of the six WGI Reports, you will notice a clear evolution in time. Firstly, the Reports have become larger, usually with more chapters and more lead authors, plus larger support teams. Mostly this is a reflection of the fact that sufficient science has been accumulated on a topic for it to deserve its own chapter, rather than a section or two in a broader chapter. As knowledge accumulates in any area of science, the questions we ask become more nuanced. Some physical processes have been 'subdivided', allowing questions to be asked about each part separately. Occasionally, two items that had been examined separately have been combined, as their mutual interactions are seen to be too complex to disentangle. This has also led to some changes in definitions, which can make it a little difficult to compare results from one Report to the next.

The other way in which climate science in general, and the IPCC in particular, has evolved is in the range of processes and systems that have been studied. This has come about as a result of an improved understanding of these phenomena and also the continual increase in computing power, which has allowed them to be incorporated into the modelling efforts. The earliest models evolved from weather models, and were atmospheric models with the simplest of surfaces: land, ocean, and ice. By the time of the FAR, the available models had components for the atmosphere, coupled to relatively simple models of the ocean, sea ice, and land surface. Aerosols were not introduced until the SAR, despite their known effects on solar radiation fluxes. A carbon cycle and dynamic vegetation were introduced in the TAR, while atmospheric chemistry and land ice were included in AR4 and AR5. Horizontal resolution has also increased significantly over that time frame.

1.5.2   The Sixth Assessment Report

Much of the material in the final chapters of this book is drawn from the latest IPCC Report, AR6, Working Group I. Figure 1.5 {figure 1.1} shows the basic structure of WGI: its 12 chapters, Atlas, Summary for Policy Makers, and its connection to WGII (Impacts, Adaptation and Vulnerability) and WGIII (Mitigation of Climate Change). There are also a number of Annexes, some containing valuable data. In total, WGI is over 2,350 pages, and most chapters have a huge reference list of the peer-reviewed science it was built on. Annex VIII is a list of acronyms, so if we ever fail to define something, try there. Note that any specific references to AR6 (e.g. figures or tables), or indeed any IPCC Report, will be enclosed in curly brackets: {...}.

1.5.2.1   Calibrated Uncertainty Language in AR6

In the latter chapters of this book we will be quoting or paraphrasing some statements and conclusions from AR6. Many of these statements will reflect the degree of confidence the scientific community has in these conclusions {box 1.1, page 169}.

Two metrics are used to communicate the degree of certainty in key findings, namely

1. *Confidence*: a quantitative measure of the validity of a finding, based on the type, amount, quality, and consistency of evidence (e.g. data, mechanistic understanding, theory, models, expert judgement) and the degree of agreement.
2. *Likelihood*: a quantitative measure of uncertainty in a finding, expressed probabilistically (e.g. based on statistical analysis of observations or model results, or both, and expert judgement by the author team or from a formal quantitative survey of expert views, or both).

Following AR5, the IPCC has quantified the words it uses, in terms of probability. The assessed level of confidence is expressed using five qualifiers: *very low, low, medium, high,* and

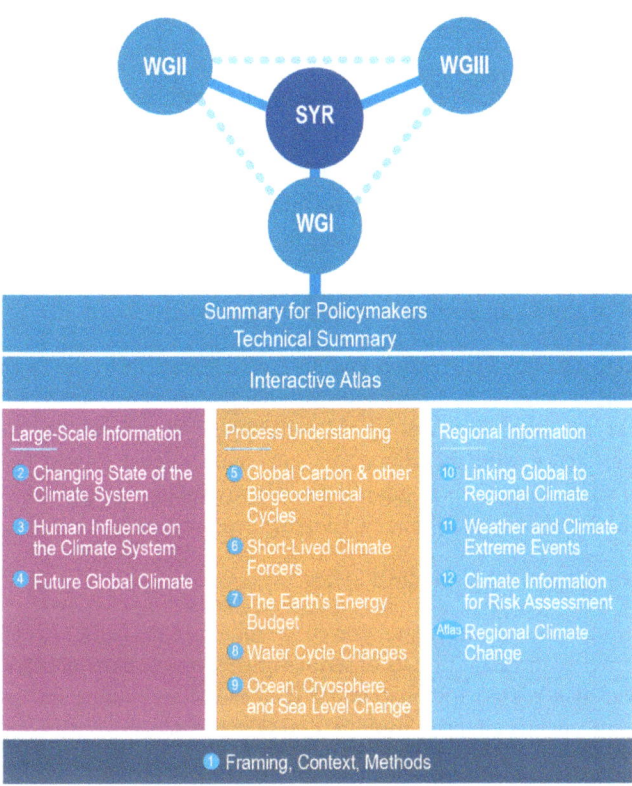

Figure 1.5 The structure of the AR6 WGI Report. Shown are the three pillars of the AR6 WGI, its relation to the WGII and WGIII contributions, and the cross-working group AR6 Synthesis Report (SYR). *Source*: see Acknowledgements at the end of this chapter.

very high. When confidence in a finding is assessed to be *low*, this does not imply that confidence in its opposite is *high*, and vice versa. Similarly, *low confidence* does not imply distrust in the finding; instead, it means that the statement is the best conclusion based on currently available knowledge. Further research may change the level of confidence in any finding.

Terms used to indicate the assessed likelihood of an outcome include: *virtually certain*: 99–100% probability; *very likely*: 90–100%; *likely*: 66–100%; *about as likely as not*: 33–66%; *unlikely*: 0–33%; *very unlikely*: 0–10%; *exceptionally unlikely*: 0–1%. In some instances, multiple combinations of confidence and likelihood are possible to characterise key findings.

Finally, many key results are numerical, such as the warming expected under a given set of circumstances. In most cases, uncertainty is quantified using 90% uncertainty intervals, and reported in square brackets [x to y]. That is, it is 90% certain that the actual value lies within that range. In many situations – but not all – the uncertainty is considered symmetrical, with the most likely value being in the middle. In such cases the range may be expressed in the $\pm$ form. Any time that we quote/paraphrase a conclusion from AR6 that involves one of these measures of either confidence or likelihood, we will use italics, as AR6 does.

For further information on the IPCC, and indeed of Climate Change in general, we strongly recommend *Global Warming, The Complete Briefing* by John Houghton. Sir John chaired or co-chaired Working Group I for the first three Reports – FAR, SAR, and TAR – and has released an updated version of his book after each Report (until his death in 2020). His record of contributions to our science, including Professor of Atmospheric Physics at Oxford University and Director General of the UK Met Office, is second to none.

1.6   The UN Climate Convention

In June 1992, The United Nations Conference on Climate and Development was held in Rio de Janeiro, and over 160 countries signed the United Nations Framework Convention on Climate Change (FCCC). It came into force on 21 March 1994. It is designed to set the agenda for action to slow and stabilise climate change. Signatory nations recognised the reality of global warming, and agreed that action to mitigate the effects needs to be taken, and developed countries should take the lead.

The Objective of the Convention is contained in Article 2:

The ultimate objective of this Convention and any related legal instruments that the Conference of the Parties may adopt is to achieve, in accordance with the relevant provisions of the Convention, stabilization of greenhouse gas concentrations in the atmosphere at a level that would prevent dangerous anthropogenic interference with the climate system. Such a level should be achieved within a time-frame sufficient to allow ecosystems to adapt naturally to climate change, to ensure that food production is not threatened and to enable economic development to proceed in a sustainable manner.

Article 3 includes agreement that the Parties

take precautionary measures to anticipate, prevent or minimize the causes of climate change and mitigate its adverse effects. Where there are threats of serious or irreversible damage, lack of full scientific certainty should not be used as a reason for postponing such measures, taking into account that policies and measures to deal with climate change should be cost-effective so as to ensure global benefits at the lowest possible cost.

In Article 4, each of the signatories agreed

to adopt national policies and take corresponding measures on the mitigation of climate change, by limiting its anthropogenic emissions of greenhouse gases and protecting and enhancing its greenhouse sinks and reservoirs. These policies and measures will demonstrate that developed countries are taking the lead in modifying longer-term trends in anthropogenic emissions.

Each signatory also agreed to communicate detailed information on its policies and measures as referred to above, as well as its projected anthropogenic emissions by source and removals by sinks of greenhouse gases, with the aim of returning to 1990 levels.

1.6.1   The Paris Agreement

To tackle climate change and its negative impacts, world leaders at the UN Climate Change Conference (COP21) in Paris reached a breakthrough on 12 December 2015: the historic Paris Agreement. The Agreement sets long-term goals to guide all nations:

- substantially reduce global greenhouse gas emissions to limit the global temperature increase in this century to 2°C while pursuing efforts to limit the increase even further to 1.5°C;
- review countries' commitments every five years;
- provide financing to developing countries to mitigate climate change, strengthen resilience, and enhance abilities to adapt to climate impacts.

The Agreement is a legally binding international treaty. It entered into force on 4 November 2016. Today, 192 countries plus the European Union have joined the Paris Agreement.

The Agreement includes commitments from all countries to reduce their emissions and work together to adapt to the impacts of climate change, and calls on countries to strengthen their commitments over time. The Agreement provides a pathway for developed nations to assist developing nations in their climate mitigation and adaptation efforts while creating a framework for the transparent monitoring and reporting of countries' climate goals.

The Paris Agreement provides a durable framework guiding the global effort for decades to come. It marks the beginning of a shift towards a net-zero emissions world. Implementation of the Agreement is also essential for the achievement of the Sustainable Development Goals.

It works on a five-year cycle of increasingly ambitious climate action carried out by countries. Every five years, each country is expected to submit an updated national climate action plan – known as a **Nationally Determined Contribution**, or NDC. In their NDCs, countries communicate actions they will take to reduce their greenhouse gas emissions in order to reach the goals of the Paris Agreement. Countries also communicate in the NDCs actions they will take to build resilience to adapt to the impacts of rising temperatures.

To better frame the efforts towards the long-term goal, the Paris Agreement invites countries to formulate and submit **long-term strategies**. Unlike NDCs, they are not mandatory. The operational details for the practical implementation of the Paris Agreement were agreed on at the UN Climate Change Conference (COP24) in Katowice, Poland, in December 2018, in what is colloquially called the Paris Rulebook, and finalised at COP26 in Glasgow, Scotland, in November 2021.

Addressing the challenges of the Paris Agreement fits within the remit of the IPCC, mainly within WGIII Mitigation. Again we quote from the Summary for Policy Makers of the Synthesis Report:

Policies and laws addressing mitigation have consistently expanded since AR5. Global GHG emissions in 2030 implied by nationally determined contributions (NDCs) announced by October 2021 make it *likely* that warming will exceed 1.5°C during the 21st century and make it harder to limit warming below 2.0°C. There are gaps between projected emissions from implemented policies and those from NDCs and finance flows fall short of the levels needed to meet climate goals across all sectors and regions (*high confidence*).

Acknowledgements

Figures 1.3 {1.6}, 1.4 {1.4}, 1.5 {1.1}:

IPCC, 2021: *Climate Change 2021: The Physical Science Basis. Contribution of Working Group I to the Sixth Assessment Report of the Intergovernmental Panel on Climate Change* [Masson-Delmotte, V., P. Zhai, A. Pirani, S. L. Connors, C. Péan, S. Berger, N. Caud, Y. Chen, L. Goldfarb, M. I. Gomis, M. Huang, K. Leitzell, E. Lonnoy, J. B. R. Matthews, T. K. Maycock, T. Waterfield, O. Yelekçi, R. Yu, and B. Zhou (eds.)]. Cambridge University Press, Cambridge, United Kingdom and New York, NY, USA, 2,391 pp. doi:10.1017/9781009157896.

Part I

Chemistry

For centuries, humans have been using the atmosphere as a dumping ground, starting with our first use of fire. After the Earth came out of the Last Glacial Maximum ~11,000 years ago, things began to ramp up, as people congregated in settled communities. The arrival of the Bronze Age, about 5,000 years ago, added the waste products (and their odours) from the smelting of copper and other ores to the (local) atmospheric burden. The first voices of complaint were probably heard in the ancient societies of the Nile and Mesopotamia. The Greeks and Romans expected their leaders to keep the air in their towns clean.

In London in the Middle Ages, sea coal replaced wood as an energy source in lime kilns and forges, adding to the airborne stew. Various laws were passed, with little real success. The invention of the steam engine around 1700 by Savery and Newcomen, and its subsequent improvement by Watt, changed everything. Now we needed fuel to drive machines, not just as a source of heat, and the fuel at hand was coal. The 'dark, satanic mills' were mostly built in the English Midlands, so that Londoners could breathe easy. But what no one knew at the time was that along with the visible pollutants, there was also an invisible one: carbon dioxide.

Part I of this book is about changes in the chemistry of our atmosphere, whether that be due to soot and other pollutants, CO_2 and other greenhouse gases, or more recent chemicals such as the CFCs. How much are we emitting? Does it simply accumulate? If not, what is its fate? So, we start in Chapter 2 by looking at the composition of our atmosphere, and some of the key chemical reactions/processes that happen naturally. Many of these are driven by the energy of the Sun, and hence show diurnal cycles.

There are many chemical substances that cycle through our atmosphere naturally, and have done for many millions of years, the most important being carbon dioxide. This cycle was in balance until the Industrial Revolution: how much has it been perturbed? Our atmosphere also contains solid particulates, known as aerosols. Some enter the atmosphere directly, but most are produced by chemical reactions between gaseous species already in the atmosphere. They can affect visibility, health, but also the climate, by reflecting some sunlight back to space.

2   Atmospheric Composition and Chemistry

The first of the three fundamental areas of science that we identified as being central to an understanding of how, and why, our climate is changing is chemistry; and specifically, the chemistry of our atmosphere. In this chapter we will start our journey by examining the current composition of the atmosphere, and then turn our attention to some of the most important chemical reactions that take place in the unpolluted atmosphere. In particular, we will introduce you to the hydroxyl radical, nature's garbage collector.

As well as the three well-known greenhouse gases (the focus of Chapter 3), the IPCC refers to a wide range of other substances as short-lived climate forcers, including chemically reactive gases such as methane, ozone, nitrogen oxides, carbon monoxide, etc., and aerosols (the focus of Chapter 4). The atmospheric fate of all these species needs to be understood.

After that, we will examine the polluted atmosphere, particularly smog and acid rain. While this topic might not seem directly related to climate change, there are some useful lessons to be learned. We also include a short discussion on how we use isotope data to help narrow in on some of the more important processes in our environment.

2.1   Composition of the Atmosphere

'What is our atmosphere composed of?' While an apparently straightforward question, the answer is not so simple. There are many subtleties that need to be resolved, such as 'where do you choose measure it', and 'over what time frame'. Some of the gases in our atmosphere are classified as 'permanent', as their concentration varies very little from place to place and from day to day. Others are more variable, either in place, or time; often both. Examples of these are pollutants with short residence times that are mainly found near their point of emission, and stratospheric ozone, which is in a state of photochemical balance. Other gases are also involved in photochemical reactions, and thus display diurnal cycles.

We also need to understand that the molecules of the major gases in the Earth's atmosphere are not 'permanent': rather, they are in a constant state of flux, with molecules entering and leaving at rates that are in balance on timescales very much longer than a typical molecular residence time. For example, carbon dioxide is constantly being exchanged with the biosphere, which simply wouldn't exist without it. Other exchanges are more subtle, but no less important.

Table 2.1 **Atmospheric permanent gases.**

Gas	Chemical formula	Volume fraction	Residence time	Major sources
Nitrogen	N_2	78.084%	2×10^7 years	Biological
Oxygen	O_2	20.946%	3–4,000 years	Biological
Argon	Ar	0.934%		Radiogenic
Carbon dioxide	CO_2	400 ppmv	~5 years (see Section 3.5.1)	Biological, combustion
Neon	Ne	18.2 ppmv		Primordial
Helium	He	5.24 ppmv	3×10^6 years	Radiogenic
Krypton	Kr	1 ppmv		Primordial
Xenon	Xe	0.1 ppmv		Primordial
Methane	CH_4	1.7 ppmv	9 years	Biological, agriculture
Hydrogen	H_2	0.56 ppmv	~2 years	Biological, anthropogenic
Nitrous oxide	N_2O	0.31 ppmv	150 years	Biological, anthropogenic
Carbonyl sulphide	OCS	0.5 ppbv	~5 years	Biological

Source: Box and Box, *Physics of Radiation and Climate*, CRC Press, 2016, reproduced by permission of Taylor & Francis Group.

In addition, there are many important chemical reactions taking place in our atmosphere that contribute to this complex balance.

The most variable atmospheric component is water vapour, which can comprise up to 4% of air in some locations. Its variability, of course, has nothing to do with human activity, but is governed by saturation vapour pressure, and how this varies with temperature: this will be covered in Chapter 6. Because of this variability, we treat it separately from 'dry air'.

Table 2.1 lists the permanent gases in our atmosphere: those gases with essentially stable and uniform concentrations. As well as the chemical formula and volume fraction (mixing ratio), this table also includes the mean residence time and the major source(s) of these gases. Volume fractions are either per cent for the major gases, or parts per million/billion/trillion by volume for the trace gases. 'Parts per ... by volume' may also be interpreted as a molecular fraction, but not a mass fraction.

The gases in Table 2.1 are all sufficiently long-lived to attain globally uniform mixing ratios, horizontally and vertically (at least throughout the troposphere and stratosphere – the lowest layers of the atmosphere: Chapter 5). They remain in these fixed proportions, despite their differing molecular weights, due to turbulent mixing. However, as we move upwards, the air becomes less dense and the molecular mean free path increases, until a point is reached where such processes cease to operate efficiently.

One might expect that the **noble gases** would be present in the atmosphere in the proportions they had when the Earth was formed, proportions largely dictated by stellar nucleosynthesis. While this is true of neon, krypton, and xenon, helium and argon have interesting stories to tell. Helium is sufficiently light that it is able to diffuse away to space in a relatively short time on geological timescales, and is only maintained as an atmospheric component by the α-decays of members of the uranium and thorium decay chains. (Hydrogen also diffuses to space, and is

Table 2.2 **Variable atmospheric gases.**

Gas	Formula	Volume fraction	Residence time	Major sources
Ozone (strat.)	O_3	Up to 10 ppmv		Photochemical
Ozone (trop.)	O_3	10–100 ppbv	Days–weeks	Photochemical
Carbon monoxide	CO	40–200 ppbv	~60 days	Anthropogenic, photochemical
Non-methane hydrocarbons		5–20 ppbv	Variable	Biological, anthropogenic
Halocarbons		~4 ppbv	Variable	Anthropogenic
Hydrogen peroxide	H_2O_2	0.1–10 ppbv	1 day	Photochemical
Reactive nitrogen species (see text)	NO_y	10 pptv–1 ppmv	Variable	Lightning, soil, anthropogenic
Formaldehyde	CH_2O	0.1–1 ppbv	~1.5 hours	Photochemical
Ammonia	NH_3	10 pptv–1 ppbv	2–10 days	Biological
Sulphur dioxide	SO_2	10 pptv–1 ppbv	Days	Photochemical, anthropogenic
Dimethyl sulphide	$(CH_3)_2S$	10–100 pptv	0.7 days	Biological: oceanic
Hydrogen sulphide	H_2S	5–500 pptv	1–5 days	Biological: swamps, etc.
Carbon disulphide	CS_2	1–300 pptv	~5 days	Biological, anthropogenic
Hydroxyl radical	OH	0–0.4 pptv	~1 s	Photochemical
Hydroperoxyl radical	HO_2	0–5 pptv		Photochemical

Source: Box and Box, *Physics of Radiation and Climate*, CRC Press, 2016, reproduced by permission of Taylor & Francis Group.

mainly resupplied by biological processes.) Argon's high atmospheric concentration is the result of the radioactive decay of potassium. (These processes will be discussed in Section 2.5.)

Table 2.2 presents important data on a suite of other gases that have variable mixing ratios in our atmosphere. One of the key reasons for this variability is that almost all are highly reactive, with residence times of a week or less. Thus, we would expect their concentrations to vary considerably, especially geographically, depending on the strength and variability of the relevant sources. Many of these sources are intimately connected with human activity, especially fossil fuel burning and other industrial and agricultural practices.

2.2 Background Tropospheric Chemistry

There are a number of very important chemical reactions that take place in the background (non-polluted) troposphere (the lowest atmospheric layer), although they may be enhanced (or occasionally suppressed) by anthropogenic inputs.

2.2.1 The NO_x–O_3 System

Table 2.2 lists a species with the title **reactive nitrogen species**, NO_y, which is a mixture of reactive gases. We firstly define

$$NO_x \equiv NO + NO_2 \tag{2.1a}$$

which are the primary species, and then define

$$NO_y \equiv NO_x + NO_3 + N_2O_5 + HNO_3 + PAN \qquad (2.1b)$$

where PAN is peroxyacetyl nitrate, which will be discussed later.

It should be noted that some of these species have an odd number of electrons (a consequence of nitrogen being atomic number 7, while oxygen is atomic number 8), and thus are technically **radicals.** All radicals have an unpaired electron, which makes them highly reactive.

Nitric oxide, NO, is produced by the reaction

$$O_2 + N_2 + energy/heat \rightarrow 2NO \qquad (2.2)$$

as well as in soils, and is the primary input of reactive nitrogen to the atmosphere. (The heat/energy needed for this reaction might be supplied from a lightning stroke, or via the exhaust system of a motor vehicle, for example.)

One of its main reaction pathways is oxidation with ozone:

$$NO + O_3 \rightarrow NO_2 + O_2 \qquad (2.3)$$

The nitrogen dioxide, NO_2, produced in this reaction may then be photolysed by solar radiation with wavelengths less than ~420 nm (blue and UV light):

$$NO_2 + hf \rightarrow NO + O \qquad (\lambda < 420\ nm) \qquad (2.4)$$

where hf denotes a photon of light.

The photon concept will be introduced in Chapter 8. The key idea to note here is that, the shorter the wavelength of light, the higher the energy of the corresponding photon. That is why, in the above reaction, we indicate the need for wavelengths *shorter* than 420 nm (although in this case the cut-off is not sharp).

Finally, the oxygen atom can join with an oxygen molecule to re-produce ozone:

$$O + O_2 + M \rightarrow O_3 + M \qquad (2.5)$$

Here M is a 'spectator molecule', which is required to ensure the conservation of both energy and momentum in reactions such as this. Reactions (2.3), (2.4), and (2.5) may be said to constitute a photochemical dance, or balance, with oxygen atoms constantly changing partners.

Clearly, photolysis reactions can only take place during daytime, so there are important diurnal variations. During the night, reaction (2.4) switches off, removing the source of O needed for the formation of ozone. The following reaction system may then become important:

$$NO_2 + O_3 \rightarrow NO_3 + O_2 \qquad (2.6)$$

$$NO_2 + NO_3 + M \leftrightarrow N_2O_5 + M \qquad (2.7)$$

$$N_2O_5 + H_2O(aq) \rightarrow 2HNO_3(aq) \qquad (2.8)$$

The last reaction occurs on aerosol or droplet (i.e. aqueous) surfaces, indicated by (aq) as a suffix, and is thus a sink of reactive nitrogen. After sunrise, NO_3 is rapidly photolysed. Reaction (2.7) is quite reversible, as implied by the double-headed arrow, so N_2O_5 is also quickly removed when its source is switched off.

2.2.2 The Hydroxyl Radical

The hydroxyl radical – OH – is one of the most important of the trace species in the troposphere, even though its globally averaged concentration is only a few tenths of a pptv (typically ~10^{12} OH m^{-3}; or 3 OH per 10^{13} air molecules). Reaction with OH provides a major sink for many reactive atmospheric trace species, leading to it being dubbed 'nature's garbage collector'. Because of this reactivity, the average lifetime of an OH radical is ~1 second, or even less.

Production of the OH radical starts with the high-energy photodissociation of ozone:

$$O_3 + hf \rightarrow O_2 + O(^1D) \qquad (\lambda < 320 \text{ nm}) \tag{2.9}$$

Here, $O(^1D)$ – referred to as 'O singlet D' – is an excited (think 'ultra-energetic', or even 'ultra-aggressive') state of the oxygen atom. (Note that the stratospheric ozone layer absorbs almost all high-energy photons with wavelengths less than ~300 nm, and steadily decreasing amounts of radiation up to ~330 nm, so not many of the necessary photons make it to ground level.)

Most $O(^1D)$ atoms lose their excess energy in collisions, heating the air, and recycle via reaction (2.5) to form ozone again. However, up to ~10% react with water vapour (or occasionally other molecules), depending on availability, to form a pair of hydroxyl radicals:

$$O(^1D) + H_2O \rightarrow 2OH \tag{2.10}$$

Note that this reaction is only possible because of the extra internal energy of $O(^1D)$.

Among the many reactions involving the hydroxyl radical, the following should be noted:

$$NO_2 + OH + M \rightarrow HNO_3 + M \tag{2.11}$$

$$CO + OH \rightarrow CO_2 + H \tag{2.12}$$

The hydrogen atom then joins an oxygen molecule to form the hydroperoxy radical:

$$H + O_2 + M \rightarrow HO_2 + M \tag{2.13}$$

Carbon monoxide (CO) is a relatively high concentration variable (trace) gas, with sources including the oxidation of hydrocarbons, biomass burning, and fossil fuel burning (although not from well-regulated power plants). The recent increase in atmospheric CO can be largely linked to human activity. Concentrations are much lower in the Southern Hemisphere.

Its dominant sink is reaction (2.12). This reaction is one of the major sinks of OH (at least in non-urban, non-forested areas). Outside the tropics, CO accumulates during winter, when OH concentrations are lower, but is rapidly depleted in spring. Concerns have been expressed that increasing levels of CO could reduce those of OH, along with its vital cleansing power.

2.2.3 Organic Compounds

Organic compounds are built around the carbon atom, which typically forms four bonds that may be tetrahedrally directed (unless it is in the form of double bonds; e.g. CO_2: O=C=O). It may bond with hydrogen, oxygen, nitrogen, sulphur, halogens and – most importantly – other carbon atoms. Hydrocarbons are a subclass containing only carbon and hydrogen. There are large sources of hydrocarbons, both natural and anthropogenic, and some play interesting roles in the atmosphere.

The most important hydrocarbon is methane (a greenhouse gas), which is listed in Table 2.1 as a permanent gas, with a residence time of ~9 years. The others tend to be more reactive, and are collected in Table 2.2 as 'non-methane hydrocarbons'. Sources of methane include wetlands, landfills, domestic animals, termites, biomass burning, and leakages from natural gas pipelines and coal mines.

Oxidation of methane also starts with attack by OH to form the methyl radical:

$$CH_4 + OH \rightarrow CH_3 + H_2O \tag{2.14}$$

Further reactions convert CH_3 to CH_3OO, then to CH_3O, and then to HCHO (formaldehyde).

There are many other organic compounds found in trace concentrations in both urban and rural atmospheres. Many tree species release terpenes, with the generic formula $C_{10}H_{16}$. One of the most important organic compounds, especially over the oceans, is dimethyl sulphide (DMS): CH_3SCH_3. This is released by phytoplankton and associated bacteria in the ocean, and is oxidised via a complex reaction chain to SO_2.

2.2.4 Sulphur Gases

The biosphere is a major contributor to the atmosphere of reduced sulphur gases, including DMS, OCS, CS_2, and H_2S (hydrogen sulphide). These are then oxidised to SO_2: for example, the oxidation of H_2S to SO_2 proceeds via the following reactions:

$$H_2S + OH \rightarrow H_2O + HS \tag{2.15}$$

$$HS + O_3 \rightarrow HSO + O_2 \tag{2.16}$$

$$HS + NO_2 \rightarrow HSO + NO \tag{2.17}$$

$$HSO + O_3 \rightarrow HSO_2 + O_2 \tag{2.18}$$

$$HSO_2 + O_2 \rightarrow HO_2 + SO_2 \tag{2.19}$$

Volcanoes and biomass burning are direct sources of SO_2. However, by far the largest present-day source is fossil fuel combustion. The smelting of metal ores such as Cu_2S (chalcocite), but also ores of lead (galena, PbS), silver (argentite, Ag_2S), mercury (cinnabar, HgS), and zinc (sphalerite ZnS), among others, is a significant regional contributor. Sulphur dioxide is then oxidised by

$$SO_2 + OH + M \rightarrow HOSO_2 + M \tag{2.20}$$

$$HOSO_2 + O_2 \rightarrow HO_2 + SO_3 \tag{2.21}$$

$$SO_3 + H_2O \rightarrow H_2SO_4 \tag{2.22}$$

Note that this last reaction mostly takes place in the aqueous phase.

2.2.5 IPCC Conclusions

Any chemical substance that may affect the planetary energy balances, either positively or negatively (and even if only indirectly, via its interactions), needs to be investigated; and especially any recent trends. Carbon monoxide is a significant pollutant, and is converted into CO_2 in a couple of months. Ozone is a greenhouse gas found in both the troposphere and stratosphere. Methane has a lifetime of 'only' about a decade before it is converted into CO_2, again a greenhouse gas. While the radiative effects of the other substances mentioned at the start of the chapter are all minor, any long-term trends may be important. And the role of the OH radical is clearly central to the fate of many trace species.

AR6 concludes that, over the decade 2010–2019, strong shifts in the geographical distribution of emissions have led to changes in atmospheric abundances of highly variable, short-lived climate forcers (*high confidence*). Evidence from both satellite and surface observations shows strong regional variations in trends of ozone, aerosols, and their precursors. In particular, tropospheric column amounts of NO_2 and SO_2 continued to decline over North America and Europe, and increase over Southern Asia, but have recently declined over Eastern Asia. Global CO abundance has continued to decline.

There is no significant trend in the global mean tropospheric concentration of the OH radical, from 1850 to around 1980 (*low confidence*), but OH has remained stable or shown a positive trend since the 1980s (*medium confidence*). Although global OH cannot be measured directly, it is inferred from a range of models constrained by emissions, and from inversion models.

2.3 Atmospheric Pollution

As well as the chemical species listed in Tables 2.1 and 2.2, our atmosphere contains an ever-growing list of variable trace species, almost all of which originate directly or indirectly from humanity's industrial or agricultural activities. This list keeps growing because new compounds are constantly being synthesised, but also because the tools available to detect them in low concentrations are constantly improving.

The subject of atmospheric pollution, just one aspect of the much broader subject of pollution in general, is a major interest for many scientists and engineers, covering the emission of a wide range of substances, their interactions in the atmosphere, control technologies that might be adopted to reduce emissions, and the regulatory protocols that government agencies have imposed.

2.3.1 Historical Perspectives

The term smog was coined in 1905 by Des Voeux, a member of London's Coal Smoke Abatement Society, to describe the mix of smoke and fog that was evident in a number of cities across Great Britain. He later described events in the autumn of 1909 in Glasgow and Edinburgh that killed more than 1,000 people. The smoke component was due to emissions from coal burning in a number of industries. Episodes of this type of pollution, now known as **London smog**, were the key to some important regulatory developments.

A number of such events have been recorded in London: in December 1873 (270–700 deaths above the average); January 1880 (700–1,100 excess deaths); December 1892 (4,000 excess deaths); December 1952 (4,000 excess deaths), among others. (Note that it is difficult to attribute a death directly to air pollution; rather, we have to rely on death statistics to reach steadily more convincing conclusions. The same is true of heatwave deaths, and even epidemic deaths.) There were also a number of notable events in other countries: the Meuse Valley in Belgium in 1930; and Dora, Pennsylvania in 1948.

However, it is the London event of December 1952 that is considered something of a turning point. Visibility was so low that London streets were dark at noon, and it was necessary to have buses guided by a lantern.

2.3.2 Photochemical Smog

Photochemical smog, also known as Los Angeles smog (although this terminology is now considered somewhat misleading), is a complex mix of gases and small aerosol particles, some emitted directly, and others produced by photochemical reactions. Until the late 1940s, no one even knew what photochemical smog was – except that it was 'unique' to Los Angeles. This started to change in 1948 when Caltech biochemist Arie Haagen-Smit started working on the problem with a new range of measurement techniques (so often the key to scientific progress). The respected journal *Atmospheric Environment* now offers the Haagen-Smit prize in his honour.

A key characteristic of photochemical smog is (relatively) large concentrations of high-molecular-weight organic compounds, particularly aromatics (e.g. benzene). In addition, the concentration of reactive nitrogen compounds is higher than background air, leading to (much) higher concentrations of ozone. We may refer to *reactive organic gases* (ROGs), or *non-methane hydrocarbons*, although the two are not fully synonymous. (Both classes explicitly omit methane, which is comparatively unreactive on the timescales involved here.)

Primary pollutants are broken down into peroxy radicals (which we denote by RO_2, R being rest of the molecule) via reactions with OH and/or O_3. The main steps can be summarised in the reaction sequence (RH is the primary hydrocarbon):

$$RH + OH \rightarrow R + H_2O \tag{2.23}$$

$$R + O_2 + M \rightarrow RO_2 + M \tag{2.24}$$

These radicals can now enter the ozone–NO_x cycle via such reactions as

$$RO_2 + O_2 \rightarrow RO + O_3 \tag{2.25}$$

$$RO_2 + NO \rightarrow RO + NO_2 \tag{2.26}$$

The NO_2 formed in this reaction can be photolysed to provide an oxygen atom as we saw in Section 2.2.1, creating an ozone molecule.

Formaldehyde (HCHO) is also a key species in the reaction chain for some smaller molecules, and is a major eye irritant (or lachrymator; tear-producer). For example, from the methoxy radical produced following reaction (2.14), we have (in less than a microsecond)

$$CH_3O + O_2 \rightarrow HCHO + HO_2 \tag{2.27}$$

Formaldehyde now produces ozone precursors by such reactions as

$$HCHO + hf \rightarrow HCO + H \quad (\lambda < 334 \text{ nm}) \tag{2.28a}$$

$$HCHO + hf \rightarrow CO + H_2 \quad (\lambda < 370 \text{ nm}) \tag{2.28b}$$

$$HCHO + OH \rightarrow HCO + H_2O \tag{2.29}$$

$$HCO + O_2 \rightarrow CO + HO_2 \tag{2.30}$$

$$H + O_2 + M \rightarrow HO_2 + M \tag{2.31}$$

CO and HO_2 form ozone through the reactions

$$CO + OH \rightarrow CO_2 + H \tag{2.32}$$

$$H + O_2 + M \rightarrow HO_2 + M \tag{2.33}$$

$$NO + HO_2 \rightarrow NO_2 + OH \tag{2.34}$$

$$NO_2 + hf \rightarrow NO + O \tag{2.35}$$

$$O + O_2 + M \rightarrow O_3 + M \tag{2.36}$$

Note that reaction (2.34) competes with

$$HO_2 + O_3 \rightarrow OH + 2O_2 \tag{2.37}$$

Thus, in clean environments ozone is removed, but in more polluted environments ozone is created!

One important component of photochemical smog is peroxyacetyl nitrate, known as PAN. Its major production pathway is via acetaldehyde:

$$CH_3CH(=O) + OH \rightarrow CH_3C(=O) + H_2O \tag{2.38}$$

$$CH_3C(=O) + O_2 + M \rightarrow CH_3C(=O)O_2 + M \tag{2.39}$$

$$CH_3C(=O)O_2 + NO_2 + M \leftrightarrow CH_3C(=O)O_2NO_2 + M \tag{2.40}$$

In these formulae, $C(=O)$ means an oxygen atom double-bonded to a carbon atom.

In Figure 2.1 we show the structure of acetaldehyde, $CH_3CH(=O)$, and peroxyacetyl nitrate, $CH_3C(=O)O_2NO_2$. (Note that the nitrogen atom shares a single bond with one of the oxygen atoms, and a double bond with the other, which oscillate back and forth; known as a resonance hybrid.)

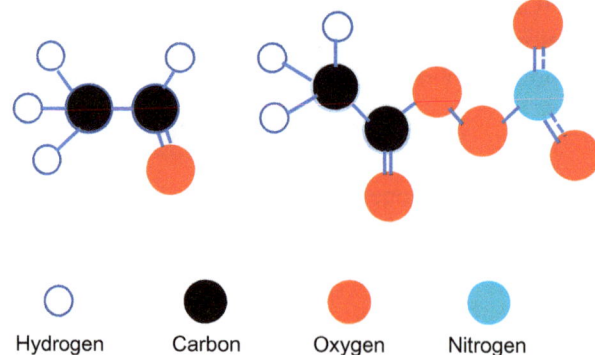

Figure 2.1 Molecular structure of acetaldehyde (left) and peroxyacetyl nitrate (right).

Mixing ratios of PAN in clean air are typically 2–100 pptv. In rural air downwind of urban sites they may rise to 1 ppbv. In polluted air, mixing ratios are typically 10–20 ppbv, and may rise to 35 ppbv. While PAN does not cause severe health effects (its main impact is as an eye irritant) it does damage plants by discolouring their leaves.

2.3.3 Mortality

On 25 March 2014, the World Health Organization (WHO) released a new assessment of the death toll from air pollution: the WHO now reports that ~7 million premature deaths occur annually that may be linked to air pollution, or 1 in 8 global deaths: double the previous estimate. The reasons for this significant increase are based on a better understanding of both human exposure and of the links between air pollution and a number of diseases.

Of these deaths, 3.3 million were due to indoor pollution, and 2.6 million to outdoor pollution, with low- and middle-income countries of South-East Asia and western Pacific regions bearing the greatest burden. Households cooking over coal, wood or biomass stoves were at particular risk. Deaths were broken down into five categories, and the following percentages attributed:

	Outdoor	Indoor
Stroke	40%	40%
Ischaemic heart disease	40%	26%
Chronic obstructive pulmonary disease	11%	22%
Lung cancer	6%	6%
Acute lower respiratory infections in children	3%	12%

More research is constantly being examined by the WHO, and released online.

2.3.4 Smog Chamber Studies

Smog chambers are used to study the processes whereby an initial composition of reactive gases, sometimes including aerosol particles, undergoes (photo)chemical transformations to

Figure 2.2 The CSIRO smog chamber. *Source*: image courtesy of CSIRO.

produce new species, especially pollutant gases and secondary aerosols (Chapter 4). There are a number of such chambers around the world, varying in size and other specifics. Some are permanent and some are not; the latter may be used outdoors. To illustrate the use of smog chambers, we will examine the CSIRO chamber in Australia: Figure 2.2.

The chamber could be filled with appropriate concentrations of various (purified) gases. An external UV light module of 40 blacklight tubes is located at each end of the chamber, all fitted with a polished reflector. The tubes emit over the range 350–390 nm, with a peak at 366 nm. A range of instrumentation is available to measure the products created. One research project performed in this chamber was an investigation of the photo-oxidation of fuel vapour from both unleaded petrol (gasoline), and also blends containing 5% and 10% ethanol.

2.3.5 Acid Precipitation

While photochemical smog is largely a local issue, often confined by the interrelation between topography and meteorology, acid rain (or 'acid deposition') is its regional cousin. Its chief source is the SO_2 emitted by the burning of fossil fuels (NO emissions may also be important), and then oxidised and transported over somewhat longer distances.

Rainwater is, in fact, naturally acidic. Atmospheric CO_2 may dissolve in cloud droplets and form carbonic acid:

$$CO_2 + H_2O(aq) \rightarrow CO_2(aq) + H_2O(aq) \leftrightarrow H_2CO_3(aq)$$
$$\leftrightarrow H^+ + HCO_3^- \leftrightarrow 2H^+ + CO_3^{2-} \tag{2.41}$$

Carbonic acid is a relatively weak acid, as indicated by the reversibility of most of these steps. Natural rainwater has a pH of 5.0–5.6, compared to distilled water with a pH of 7.0.

In polluted environments, gaseous HNO_3 is produced, as well as HCl, via some of the reaction mechanisms discussed above. (HCl may be produced by the reaction of other acids on sea salt particles; NaCl: Section 4.2.1.) These gases may then dissolve in cloud droplets, and dissociate. Aqueous-phase oxidation of SO_2 with dissolved hydrogen peroxide (H_2O_2) can also take place via the following sequence:

$$SO_2(g) \rightarrow SO_2(aq) \tag{2.42}$$

$$SO_2(aq) + H_2O(aq) \leftrightarrow H_2SO_3 \leftrightarrow H^+ + HSO_3^- \tag{2.43}$$

$$HSO_3^- + H_2O_2(aq) \rightarrow H^+ + SO_4^{2-} + H_2O \tag{2.44}$$

Acid rain (or fog) has pH values anywhere between ~2.0 and 5.5. One of the main areas affected by acid precipitation is the north-eastern United States, where pH values of 4.2–4.5 have been regularly measured across Ohio, western Pennsylvania, and neighbouring states. Figure 2.3 shows the scale of pH, from battery acid (H_2SO_4) to caustic soda (NaOH), showing where natural rainwater and acid rain sit.

While concerns have been raised over a number of the impacts of acid deposition, we will briefly mention three. The simplest of these is the impact on buildings and related structures. Acids erode sandstone, limestone, marble, copper, bronze, and brass. Perhaps the biggest concern is with damage to buildings and sculptures of historical or cultural significance, such as the Parthenon in Athens.

Acids also damage plant and tree leaves and roots. A deposit of sulphuric acid forms a liquid film of low pH which erodes the cuticle wax, leading to the desiccation of the leaf. Gas-phase acids or raindrops may enter forest groundwater, damaging plant roots in two ways. Firstly, they may dissolve, and carry away, important mineral nutrients such as Ca, Mg, K, and Na. Secondly they may react with hydroxide minerals such as $Al(OH)_3$ and $Fe(OH)_3$, releasing Al^{3+} and Fe^{3+}, respectively. At high concentrations these ions are toxic to root systems. Such effects

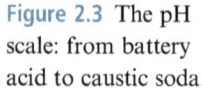

Figure 2.3 The pH scale: from battery acid to caustic soda.

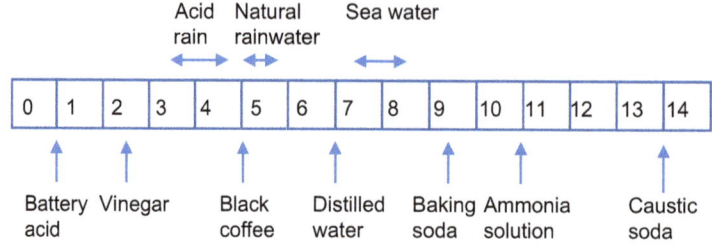

Figure 2.4 Impacts of acid rain on forest. *Source*: United States Geological Survey.

are widespread in Poland, the Czech Republic, Germany, and parts of the eastern USA, for example (Figure 2.4).

When acid rain accumulates in lakes and rivers, their pH can be significantly lowered. Most aquatic insects, algae, and plankton cannot survive at a pH below 5, with consequent effects up the food chain. pH levels of less than 5.5 have been associated with reproductive failures and mutations in fish and amphibians. Lake acidification has been a particular problem in Scandinavia, where the pH fell by 1 unit during the 1950s and 1960s. By the late 1970s about one-quarter of Sweden's lakes had been badly affected, although this situation is now improving.

2.4 Responses

Controlling, and preferably reducing, atmospheric pollution requires the use of a range of emission control technologies, depending on the pollutant and the source. As all such measures cost the emitter money (at least in the short term), regulations are almost always essential if society is to achieve the desirable goal of a healthy environment.

Most readers will be familiar with the frog-in-a-pot story. If the frog had been placed in a pot of very hot water it would have jumped straight out. However, when placed in a pot of cold water which is then steadily heated, the frog eventually dies. In many ways, humans are not much different. When things get worse slowly, we don't seem to notice, or simply treat it as inevitable: 'the price of progress' (or, at least, of a good economy/jobs). However, when hit by a dramatic event, we sit up and take notice.

Although there had been regulations in the UK and the USA for more than a century, it was the London episodes in 1952 which led to concerted efforts to regulate, and to enforce. In 1956, the UK Clean Air Act was passed, which controlled both household and industrial emissions. It dealt only with smoke and did not cover SO_2, although measures to reduce one would have reduced the other. This Act was amended in 1968 to require industries burning fossil fuels to build tall chimneys to aid dispersion.

In 1963, the US Congress passed the Clean Air Act: 'to improve, strengthen and accelerate programs for the prevention and abatement of air pollution'. In particular, it gave the Federal Government the authority to reduce interstate air pollution. This could be achieved by specifying emissions standards for stationary sources such as power plants and steel mills, plus the use of technologies to remove sulphur from coal and oil. This act was followed by the Motor Vehicle Air Pollution Control Act of 1965. In the United States, it has often been California (and more specifically Los Angeles) that has led the way – not surprising, given both its own very obvious problems and the pioneering work done at Caltech.

2.4.1   Air Quality Standards

The WHO has recently released its updated Global Air Quality Guidelines, with the specific objectives that include:

- To provide evidence-informed recommendations in the form of AQG levels, including an indication of the shape of the concentration–response function in relation to critical health outcomes, for $PM_{2.5}$, PM_{10}, ozone, nitrogen dioxide, sulphur dioxide, and carbon monoxide, for relevant averaging times.
- To provide interim targets to guide reduction efforts towards the ultimate and timely achievement of the AQG levels for countries that substantially exceed these levels.

Table 2.3 displays these standards. The table covers the pollutant gases CO, NO_2, O_3, and SO_2, as well as particulates: the notation PM_x means 'particulate matter with diameters less than x μm'. Particulates will be covered in Chapter 4, including their health effects.

Table 2.3 **WHO air quality goals and interim targets.**

Pollutant	Averaging period	Target 1	Target 2	Target 3	Target 4	Goal
$PM_{2.5}$ μg/m^3	Annual	35	25	15	10	5
	24 hour	75	50	37.5	25	15
PM_{10} μg/m^3	Annual	70	50	30	20	15
	24 hour	150	100	75	50	45
O_3 μg/m^3	Peak	100	70			60
	8 hour	160	120			100
NO_2 μg/m^3	Annual	40	30	20		10
	24 hour	120	50			25
	1 hour					200
SO_2 μg/m^3	24 hour	125	50			40
CO μg/m^3	24 hour	7				4
	8 hour					10
	1 hour					35

Table 2.4 **Air Quality Standards for the UK, EU, USA, and Australia.**

Pollutant	Averaging period	WHO goal	UK	EU	USA	Australia
$PM_{2.5}$ µg/m^3	Annual	5	25	20	12	8
	24 hour	15			35	25
PM_{10} µg/m^3	Annual	15	40	50		25
	24 hour	45	50	50	150	50
O_3 µg/m^3	Peak	60				
	8 hour	100	100		138	
	4 hour					157
	1 hour					197
NO_2 µg/m^3	Annual	10	40	40	100	56
	24 hour	25			188	226
	1 hour	200	200	20		
SO_2 µg/m^3	Annual					52
	24 hour	40	125	125		209
	1 hour		350	350	196	524
CO µg/m^3	24 hour	4				
	8 hour	10	10	10	10	10
	1 hour	35			40	

National (and State) Governments set Air Quality Standards, which are essentially targets that they are hoping to achieve. Table 2.4 provides these for the UK, European Union, USA, and Australia, along with the WHO goals as a benchmark: you will see that even these advanced countries have a long way to go! If some of the numbers seem rather odd, it is because some jurisdictions use parts per million, and some use imperial units: we have converted all to micrograms per cubic metre.

A search of the web will quickly lead you to similar standards for other countries, mostly expressed in different ways, to make the exercise more 'interesting'. (Note that many jurisdictions include other pollutants, such as lead: we have chosen those that are common to all.) The WHO keeps data on air quality from many cities across the globe: they can also be found on their website. You may find some of the data 'sobering'.

Primary Standards provide public health protection, including the health of sensitive sections of the population such as asthmatics, children, and the elderly. **Secondary Standards** provide public welfare protection, including protection against decreased visibility, and damage to animals, crops, vegetation, and buildings. The requirements are, not surprisingly, rather complex. Note that it is essential that measurements of all substances be made with standardised equipment, using standardised measurement protocols, by trained technicians who are usually not PhD scientists.

It needs to be appreciated that these are 'standards' – i.e. goals – which governments hope to see achieved: there can be a number of explanations for 'exceedances'. For example, a heavy dust storm, or nearby forest fire, may lead to elevated particulate levels – rarely can anyone be held to account for this. If desired Air Quality Standards are not being met, regulators might call for the relocation of certain industries, the mandatory use of more highly refined fuels, the use of more sophisticated catalytic converters, or other technologies.

2.4.2 Regulation

Regulation can take many forms, depending on the 'target(s)', including bans on specific sources/pollutants, setting pollutant concentration limits/goals, and tradable permits to pollute. **Tradable permits** proved a valuable mechanism to reduce the SO_2 emissions which contribute to acid precipitation in the north-east of the United States. Some power stations found it profitable to upgrade, or even rebuild, significantly reducing their emissions in the process, leaving them with permits that they could then sell (helping to cover their costs of upgrading) to other power stations, which were not in such a position.

All requirements to reduce the emission of pollutant gases (and particulates) to the atmosphere have come at a cost: economic, and often political. Invariably those who have to bear at least the initial costs – usually industry – have objected. Of course, these costs must then be passed on to consumers, and some businesses may not be in the position to easily raise the capital necessary to meet the requirements. An important issue for policy makers should always be to not place undue burdens on one sector of society, and allow others to escape. If one jurisdiction (e.g. one nation, or state) imposes weaker/cheaper requirements on its industries, that may constitute an unfair trade advantage. Note that it is now considered legitimate to impose measures such as tariffs on imports from countries with weak environmental requirements.

Some ozone precursors have natural sources, such as vegetation. Some industry groups, and their political allies, opposed to further regulation have occasionally suggested cutting down trees as the solution to poor air quality! This is an area where careful modelling is needed, to be able to predict the impacts of any and all suggested changes. To give just one example, the push to make greater use of biofuels, both to reduce dependence on oil imports and to reduce greenhouse gas emissions, is likely to result in a new suite of by-products: how will they react with the current mix of (trace) gases in their local environment?

Controlling the emission of CO_2 and other greenhouse gases poses similar challenges. Those who may have to pay the initial costs have worked hard to persuade their political friends to delay and weaken any regulation. Thus, it is always valuable to learn any lessons that are available from other 'battles' fought and won (or lost). In Chapter 10 we will look at the Montreal Protocol, which has been highly successful in greatly reducing the threat to the ozone layer.

2.5   Isotope Studies

As well as looking at the chemical nature of substances in our environment, there are many valuable lessons to be learned by looking at the presence, within some of these substances, of a range of isotopes, both unstable and stable.

Radioactive isotopes decay at well-defined rates, which can be determined either experimentally or, at least in principle, theoretically. If we know how much of an isotope was present in certain locations at some time in the past, and measure how much is present today, we can use this information for dating purposes.

2.5.1   Natural Radioactivity

When the Earth was created, it contained a wide range of unstable nuclides (isotopes), which were produced in the same stellar nuclear furnace as all the stable species. By now most of these have decayed away to negligible levels. However, a number of key radioactive elements and isotopes are still present due to their long half-lives, and their effects are felt in subtle ways. Table 2.5 lists some of the key isotopes and their properties of relevance in science today.

2.5.1.1   The Uranium–Thorium–Lead System

Lead, element 82 in the Periodic Table, is almost the last stable element that Nature is able to provide. The key reason for its stability is that the number of protons, 82, is a 'magic number', representing a closed shell: the nuclear equivalent of a chemically stable inert gas. As a consequence, it has four stable isotopes, with 122, 124, 125, and 126 neutrons. (One table actually lists a half-life of $\sim 10^{17}$ years for the lightest of these, ^{204}Pb, which qualifies as stable by most reckonings!) One hundred and twenty-six neutrons is also a magic number, so that ^{208}Pb is 'doubly magic', and thus extra stable. As a result, we ought to expect one more stable nuclide, which is ^{208}Pb plus either a proton or a neutron. It turns out to be the extra proton, with the result that bismuth-209 is actually the heaviest stable isotope in the universe.

Nature also provides three heavier non-stable (i.e. radioactive) isotopes with half-lives comparable to the age of the Earth: thorium-232, uranium-235, and uranium-238. As indicated in Table 2.5, each of these undergoes a series of α- and β-decays until they reach one of the stable isotsopes of lead. Figure 2.5 shows the decay chain for ^{238}U. Others may be found on the web.

All of the daughter intermediates have half-lives that are many orders of magnitude shorter than the initial parent's half-life. It is for this reason that Table 2.5 lists only the parent and final (i.e. stable) daughter. Samples of ancient rock may be dated from the ratios of the original radioactive isotopes remaining compared to the relevant lead isotope end points.

Table 2.5 Properties of selected radioactive isotopes.

Parent	% of the natural abundance	Daughter	Process	Half-life	Useful dating range
^{238}U	99.27	^{206}Pb	8α, 6β	4.5×10^9 years	10^7 years → A_E
^{235}U	0.72	^{207}Pb	7α, 4β	0.7×10^9 years	10^7 years → A_E
^{232}Th	100	^{208}Pb	6α, 4β	14×10^9 years	10^7 years → A_E
^{222}Rn		^{210}Pb	3α, 2β	3.8 days	0–40 days
^{147}Sm	15	^{143}Nd	α	1.1×10^{11} years	10^8 years → A_E
^{87}Rb	28	^{87}Sr	β	49×10^9 years	10^7 years → A_E
^{40}K	0.01	^{40}Ar	EC	1.3×10^9 years	5,000 years → A_E
^{14}C	$\sim 10^{-10}$	^{14}N	β	5,730 years	0–70,000 years
^{7}Be		^{7}Li	EC	53.4 days	0–1 years
^{10}Be		^{10}B	β	1.4×10^6 years	

Notes: A_E indicates the age of the Earth; EC indicates electron capture.

Figure 2.5 The uranium-238 decay chain.

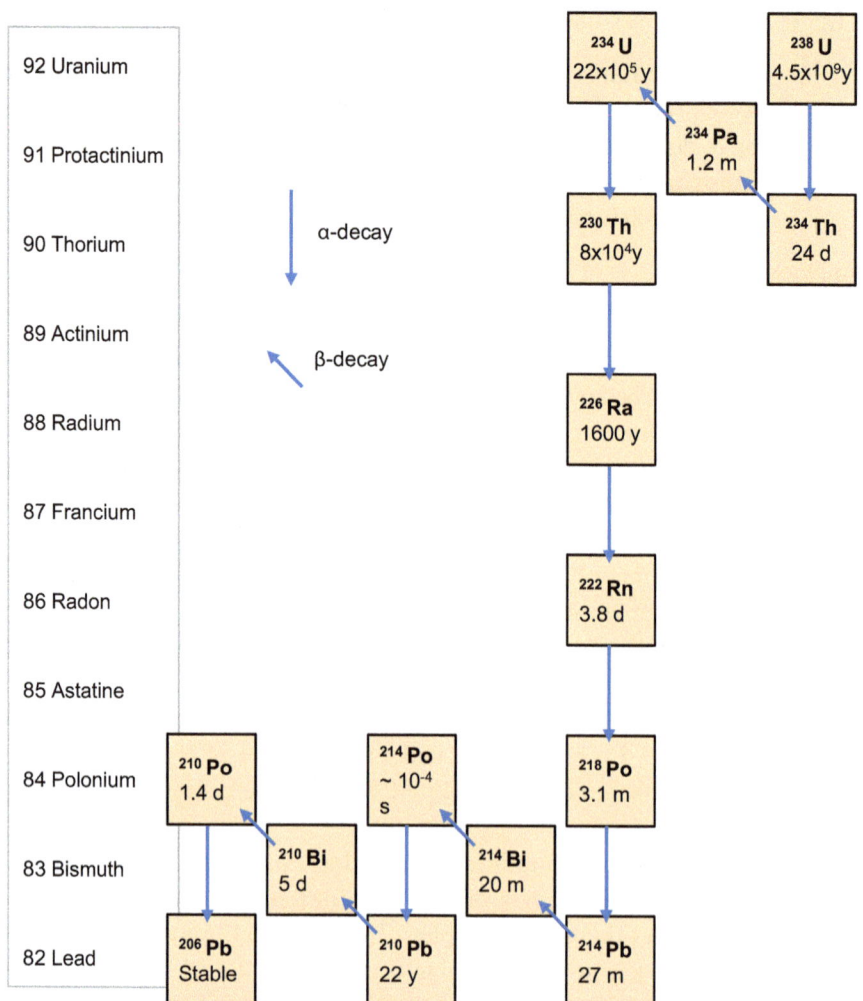

Table 2.5 also includes one of the intermediates in the ^{238}U decay chain, radon-222, and lists its daughter as ^{210}Pb, rather than the end point of the chain which is ^{206}Pb. The reason for listing ^{210}Pb as the daughter is that its half-life of 22.5 years is much longer than ^{222}Rn's half-life of 3.8 days. (Note that the intermediate nuclides between ^{222}Rn and ^{210}Pb have half-lives of minutes, or less.)

The reason for including radon in that table is that it is a gas. Uranium is present (in small and variable amounts) in soils around the world, so that radon is constantly being emitted by the land surface (unless it is ice-covered). Thus, measurements of this isotope in an air sample can provide insight into whether it has passed over land or ocean in its recent past. This concept is used by Australia's Cape Grim Baseline Air Pollution Station to indicate which air samples have come from the Southern Ocean, and may therefore be considered 'background air': i.e. unpolluted.

2.5.1.2 Potassium–Argon

All elements have unstable isotopes, although most are only found/produced in nuclear physics laboratories. An important exception is potassium, of which one part in 10,000 is potassium-40, with a half-life of 1.25 billion years. $^{40}K_{19}$ is an odd–odd nuclide (19 protons, 21 neutrons), which can actually decay via two paths,

$$^{40}K \rightarrow {}^{40}Ca + e + \bar{\nu}$$

$$^{40}K + e \rightarrow {}^{40}Ar + \nu$$

with a fixed branching ratio (10.72% to ^{40}Ar). The second of these processes is known as electron capture, a form of β-decay.

When lava erupts from a volcano, we might expect any argon that had accumulated in the material before the eruption to be able to escape, as it is a gas. Thus, a measurement of the $^{40}K/Ar$ ratio in old lava beds can give a good indication of the time of the eruption. In this way it has also been used to date some hominid fossils.

The emitted argon, having no reactive sinks either in the air or at the surface, has been able to accumulate over the eons, now making up 1% of our atmosphere. Argon has three stable isotopes; however, 99.6% of the Earth's argon is ^{40}Ar, strongly suggesting that a major source has been ^{40}K decay. (The Sun's argon is mostly ^{36}Ar, with less than 15% ^{40}Ar.)

2.5.2 'Induced' Radioactivity

2.5.2.1 Carbon-14

Cosmic rays are constantly striking the top of the Earth's atmosphere, and some of them induce nuclear reactions. The best known of these is the creation of carbon-14:

$$^{14}N_7 + n \rightarrow {}^{14}C_6 + p$$

About 10 kg of ^{14}C is produced in the atmosphere each year.

This ^{14}C atom may then combine with oxygen to form a molecule of $^{14}CO_2$. The chemistry of this molecule is the same as any other CO_2 molecule, and it may find itself incorporated into a plant, and later eaten by a herbivore, which may in turn be eaten by a carnivore. Because the half-life of ^{14}C, 5,730 years, is two (or more) orders of magnitude longer than that of most life forms, all living things on Earth, including humans, have essentially the same fraction of their carbon as ^{14}C.

The equilibrium atmospheric abundance of ^{14}C (i.e. fraction of all carbon in the atmosphere) was ~1.5×10^{-12} a century ago, but this decreased until 1954 due to the burning of fossil fuels which are, of course, fully depleted of ^{14}C. After that date the abundance increased as a consequence of nuclear testing. However, this is an example of a (mushroom) cloud with a silver lining. Because the ^{14}C pulse from nuclear testing was much shorter than the ^{14}C lifetime, it has been possible to use it as a 'passive tracer' to study how quickly CO_2 is exchanged out of the atmosphere.

Carbon-14 is an excellent dating agent for (dead) biological samples as old as 50,000 years, and this can be extended by some of the more sophisticated accelerator-based technologies recently developed. However, the technique needs calibration, as it is known that the cosmic ray flux is not constant, but varies for a number of reasons. Perhaps the best approach is to use tree ring data (dendrochronology), where available. Because of the 'contamination' of atmospheric ^{14}C by nuclear testing, radiocarbon dates are usually quoted as BP – Before Present – where 'Present' is the year 1950.

This is a good place to point out that all human (and other) bodies are naturally radioactive. A typical 70 kg body experiences 8,000 radioactive disintegrations per second, about half each from ^{40}K and ^{14}C. (Potassium is essential in all living organisms to facilitate signalling across cell membranes.) In terms of dosage, the higher-energy electrons from ^{40}K produce a dose more than an order of magnitude larger than ^{14}C. In 10% of decays, ^{40}K also emits a gamma ray, which escapes the body (unlike the electrons). This exposure does make a minor contribution to cancer deaths, shortening our lives by an average of one hour.

2.5.2.2 Beryllium

Cosmic radiation actually creates a range of other radionuclides, of which two will be mentioned here. Very-high-energy cosmic rays may shatter an oxygen or nitrogen nucleus in the upper atmosphere; among the wreckage are beryllium-7 and beryllium-10. Beryllium-7 has a half-life of 53 days, making it a useful tracer of air motions, especially above the troposphere. It has been used to show that certain instances of high ozone in the troposphere were primarily the result of air exchange from the stratosphere, not photochemical smog.

Beryllium-10 has a much longer half-life: 1.39 million years. Because beryllium tends to exist in solutions below about pH 5.5 it will dissolve and be transported to the Earth's surface via rainwater. As the precipitation becomes more alkaline, beryllium drops out of solution. Cosmogenic ^{10}Be thereby accumulates at the soil surface, where its relatively long half-life permits a long residence time before decaying. ^{10}Be and its daughter have been used to examine soil erosion, soil formation from regolith, and the ages of ice cores.

The rate of ^{10}Be production depends on the strength of high-energy galactic cosmic rays. Its influence is lower if the solar wind that is associated with solar activity is higher. The variation of ^{10}Be concentration with time in Antarctic ice samples may be an indicator of solar activity in past centuries. It is also formed in nuclear explosions by a reaction of fast neutrons with ^{13}C in the carbon dioxide in air, and is one of the historical indicators of past activity at nuclear test sites.

2.6   Stable Isotopes

Different stable and unstable isotopes of a particular element have fundamentally the same chemical properties, either in isolation, or as part of a molecule. However, their different masses can lead to subtle physical effects, such as a difference in latent heat. This can produce temperature shifts/effects in processes such as evaporation and condensation. The combined effect of studying selected stable and unstable isotopes is to enable science to gain an understanding of conditions in the past, as well as some useful information on the sources and cycles of greenhouse (and other) gases.

Table 2.6 **Isotopes of hydrogen, carbon, nitrogen, and oxygen.**

Isotope	Abundance (%)
^1H	99.9851
^2D	0.0149
^{12}C	98.892
^{13}C	1.108
^{14}N	99.63
^{15}N	0.37
^{16}O	99.76
^{18}O	0.204

Four of the most important elements in our atmospheric environment, H, C, N, and O, possess more than one stable isotope: see Table 2.6. While the chemical properties of the different isotopologues of a given compound are *almost* identical, there are some subtle physical differences that can be used to provide insight into some of the biogeochemical processes of interest to climate studies.

We should point out from the start that many of the effects are subtle, and require very careful interpretation, preferably by making use of multiple lines of evidence. This is a highly specialist area, and the examples we give below should not be taken as definitive.

2.6.1 Oxygen-18

Oxygen consists of three stable isotopes, with atomic masses 16, 17, and 18, with ^{16}O dominant (99.76%). Oxygen-17 is of little interest as its concentration is much lower than ^{18}O. When water evaporates, a molecule of $H_2^{18}O$ requires a little more energy than a molecule of $H_2^{16}O$. As a result, water vapour in the air contains slightly less ^{18}O than sea water. A similar separation occurs during condensation in clouds. The strength of these separations depends on the temperatures at which evaporation and condensation occur. Measurements of snowfall at different locations may be used to calibrate the method, and show that the concentration of ^{18}O varies by about 0.7 parts per 1,000 for each degree change in the average surface temperature. This result can be used to provide temperature estimates – **proxy data** – from ice cores, of the temperature when the ice layer was deposited.

During major glaciations, water is lost from the oceans and stored in the ice caps, and the remaining water becomes enriched in ^{18}O. Microscopic marine creatures take up oxygen, and carbon, to form calcite (or aragonite) shells or coccoliths – basically $CaCO_3$. The fractionation of isotopes during such processes in sea water has been shown to be temperature-dependent. Again, specialist analysis is needed.

2.6.2 Carbon-13

Carbon-13 is a complex multi-proxy, reflecting many aspects of the climate system, and its changes. Plants are observed to preferentially take up ^{12}C in CO_2, enriching the atmosphere in ^{13}C. Fossil fuels were originally organic matter, and so they contain less ^{13}C than the current atmospheric abundance, by about 18 parts per 1,000: which we may express as $\delta^{13}C = 18‰$. Carbon-13 is used to help understand some of the changes that took place during the glacial–interglacial cycle, as the land available for photosynthesis varied. However, photosynthesis also takes place in the oceans, which needs to be included in any analysis.

2.6.3 Deuterium

Deuterium, or ^{2}H, is sufficiently important that it is given its own symbol, D. While it differs in mass from the more common isotope, ^{1}H, by only one unit, it is, of course, twice as heavy: the largest (stable) isotopic mass ratio. Deuterium may be incorporated into a water molecule, usually in the form HDO. Just as with $H_2{}^{18}O$, HDO tends to be left behind when water evaporates, which occurs mainly in the warmer latitudes, after which much of the vapour moves to higher, and colder, latitudes. As this vapour condenses, the HDO is preferentially, and progressively, removed from the storm system, so that storms near the poles drop snow that is considerably depleted in deuterium. This situation is enhanced during colder periods, because the equator-to-pole temperature gradient is enhanced. This result has been used to study the temperature record incorporated in ice sheets during the last ~1 million years.

2.6.4 Nitrogen-15

Nitrogen is one of the most important elements in all of biochemistry, as it is essential for amino acids. Its incorporation into living matter starts with nitrogen fixation by plants, and their associated microbes, via the enzyme nitrogenase, which converts atmospheric N_2 to $NH_4{}^+$. Nitrogenase discriminates between the nitrogen isotopes. Differences in the nitrogen isotopic ratio indicate which plant species are involved in nitrogen fixation, as they have reduced ^{15}N. By contrast, those species that depend on nitrogen uptake from the soil show enhanced values. During plant decomposition, soil microbes mineralise ^{14}N in favour of ^{15}N, which increases in the undecomposed residue. Denitrifying bacteria further fractionate among the nitrogen isotopes in $NO_3{}^-$.

Summary

For centuries humans have used the atmosphere as a dumping ground, both for toxic pollutants like soot, CO and SO_2, as well as for CO_2 and other greenhouse gases: out of sight, out of mind (although the soot was hardly out of sight). However, over the past century, we have come to realise the folly of such practices.

This chapter has six sections, which can be thought of as three pairs of two. In the first (and most important) we focused on the atmosphere as a chemical stew-pot, and how it processes much of the material that enters it. We have introduced you to the vital role of photochemistry: chemical reactions powered by solar/photon energy. We also introduced you to the OH radical, which does more to keep our environment clean than any other substance.

In Sections 2.3 and 2.4 we have looked at some of the key chemistry of air pollution. While this may not seem all that relevant to the challenge of global-scale climate change, there are many valuable lessons to be learnt. After all, both are political and technological challenges.

Finally, in Sections 2.5 and 2.6 we have presented some basic information on chemical isotopes; invaluable tools in many environmental investigations. In later chapters you will be referred back to this material: for example, to help extend the data presented in Figures 1.1 and 1.2 back in time.

FURTHER READING

There are a number of specialist texts on atmospheric chemistry, such as *Chemistry of Atmospheres* by R. P. Wayne (Oxford University Press, 2000).

At a level somewhat closer to this book, we would suggest:

- *Atmospheric Science, An Introductory Survey* by J. M. Wallace and P. V. Hobbs (Academic Press, 2006).
- *Atmospheric Change, An Earth System Perspective* by T. E. Graedel and P. J. Crutzen (W. H. Freeman, 1993).
- *Atmospheric Pollution; History, Science, and Regulation* by M. Z. Jacobson (Cambridge University Press, 2002).
- *Atmospheric Science for Environmental Scientists* by C. N. Hewitt (Wiley-Blackwell, 2020).

There is one other book we would like to introduce you to: an excellent resource on how to think about environmental problems; and also, a useful source of exercises:

- *Consider a Spherical Cow* by J. Harte (University Science Books, 1988).

REVIEW QUESTIONS

1. List the major gases in the Earth's atmosphere.
2. What fraction of the atmosphere is nitrogen? Oxygen?
3. Which photons have greater energy, red or blue?
4. Why do we regularly talk of NO_x, rather than NO and NO_2?
5. How does the mixture within NO_x vary between day and night?
6. What is the average lifetime of an OH radical?
7. Give two examples of how OH removes unwanted species.
8. List some examples of the impacts of acid rain.
9. Why are radioactive isotopes useful for dating purposes?
10. Why might $H_2^{18}O$ be 'harder' to evaporate then $H_2^{16}O$?

EXERCISES

1. There are approximately 10^{44} molecules in our atmosphere. Use this information, plus the data in Table 2.1, to calculate how many CO_2 molecules enter (and leave) the atmosphere per year/day/second.

2. An oil-fired power station consumes 1 million litres of oil per day. Assume that the oil has a density of 0.8 g cm^{-3}, a composition of $C_{15}H_{32}$, and the combustion reaction involved is

$$C_{15}H_{32} + 23\,O_2 \rightarrow 15\,CO_2 + 16\,H_2O$$

Determine the total amount of gas emitted per mole of oil [CO_2 + H_2O] from combustion plus the nitrogen from the unused fraction of air. Determine the number of moles of oil consumed per day, and hence the total mass of gas emitted per day.

3. In Table 2.4 we gave you the Air Quality Standards for several jurisdictions around the world. If yours was not included, search the web to find it, and compare it with those listed.

4. You should also be able to find daily data for a station near where you live or study. Check this regularly, and keep a daily log of the important data, along with your own observations of the air quality you experience. Towards the end of your course, you should write a report on your data set, and any observations of 'pollution events' which took place in that period.

5. It has been said that 'the solution to pollution is dilution'. How true is this?

3   Sources and Sinks of Greenhouse Gases

This book was written primarily because of concerns that our climate is changing, and many of those concerns centre on changes in the composition of our atmosphere. While the rapid increase in the carbon dioxide concentration is the best known of these changes, we have also alluded to increases in other greenhouse gases such as methane and nitrous oxide. Concerns have also been voiced about possible changes to ocean chemistry, primarily as a result of their uptake of a share of the carbon dioxide produced (primarily) by fossil fuel burning.

In this chapter we will examine the major exchange processes that cycle key elements through our environment, from the atmosphere and oceans, to the biosphere, and even the 'solid' Earth, and back again. Our primary focus will, of course, be on the key element carbon, as this encompasses both carbon dioxide and methane. We will also look at the fascinating element nitrogen, and the greenhouse gas nitrous oxide.

While our primary focus will be on the natural biogeochemistry of these gases, we will present some of the latest data on how these gases have been increasing in our atmosphere in recent decades. Much of this has been extracted from IPCC AR6 Chapter 5, Global Carbon and Other Biogeochemical Cycles and Feedbacks, which contains a wealth of information beyond what we have space to cover.

3.1   The Carbon Cycle

We know that the chemical composition of the atmosphere is not static, but varies on a variety of scales. One of the reasons is that many key elements are constantly being cycled between a number of reservoirs, in response to both biological and even geological processes. Biological processes are reasonably fast, ranging from the almost instantaneous, to the daily cycles of photosynthesis and respiration, to the growth and harvesting of crops, and the long-term growth and decay of forests. By contrast, geological processes, such as rock weathering and plate tectonics, are 'glacial'.

If we are to understand the changes that human activity is making to the composition of our atmosphere, we need to firstly understand these natural cycles, starting with the biological processes. However, the role of geology cannot be ignored, especially if we want to understand climate change in the distant past.

3.1.1 Box Models

Biogeochemical cycles can be very complex, if examined in all their detail; however, we may usefully simplify much of this detail by making use of box models. In such models, the major reservoirs of a species are the boxes, with a quantifiable content (**stocks**). These 'species' may consist of a chemical element in all its chemical forms, or just a single compound, or a group of compounds, such as NO_x. (We may usually assume that all reactions between the species within a reservoir are sufficiently rapid that we may assume an equilibrium distribution, or ratio, of these species.) These boxes are then interconnected by reactions and other processes, with quantifiable **flux rates**.

The most important of these cycles is that of carbon. Carbon, in all its forms, is central to so much of what makes Earth the planet it is: our home. It is, of course, the central element in all life forms, due to its almost unique ability to build complex molecules, most of which are based on the C–C bond. Carbon dioxide is doubly important, both in photosynthesis and in the (natural) greenhouse effect. Methane and carbon monoxide are produced in many natural processes. Finally, carbonate rocks are one of the major rock types (the white cliffs of Dover).

Table 3.1 provides information on the major **carbon reservoirs** in the Earth System. Note that it is simplest to express all reservoirs, and fluxes, as quantities of carbon: usually tonnes or grams, and per year where appropriate. We can collect some of these together, and display one of the simplest box models of the **carbon system** in Figure 3.1, which shows four major reservoirs, and the processes, both biological and geological, responsible for moving carbon from one reservoir to another. This is the big picture view. We will find it easiest to start at this scale, and examine the geological processes, before zooming in on some of the 'boxes-within-boxes' of the biochemistry.

Table 3.1 **Carbon reservoirs in the Earth System.**

Reservoir	Content ($kg\ m^{-2}$)	Residence time
Atmospheric CO_2	1.6	5 years
Atmospheric CH_4	0.02	9 years
Green part of the biosphere	0.2	Days to seasons
Tree trunks and roots	1.2	~Centuries
Soils and sediments	3	Decades to millennia
Fossil fuels	10	–
Organic C in sedimentary rocks	20,000	2×10^8 years
Inorganic C in sedimentary rocks	80,000	10^8 years
Oceans: dissolved CO_2	1.5	12 years
Oceans: CO_3^{2-}	2.5	6,500 years
Oceans: HCO^-	70	200,000 years

Source: reprinted from *Atmospheric Science* by J. M. Wallace and P. V. Hobbs, table 2, with permission from Elsevier.

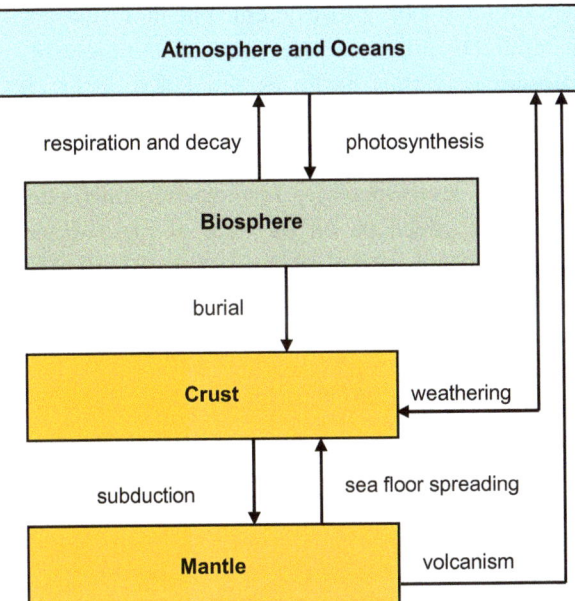

Figure 3.1 Simple box model of the carbon system. *Source*: *Physics of Radiation and Climate*, Box and Box, CRC Press, 2016, reproduced by permission of Taylor & Francis Group.

3.1.2 Geological Processes

The stores of carbon in the Earth's crust are enormous, and the exchange rates are very slow: except, of course, for fossil fuel extraction – but that is not a geological process. Carbon enters both the organic and inorganic reservoirs from the biosphere. Most deposits of organic carbon in coal, oil, gas, and shale were formed in anoxic (oxygen-deficient) conditions in ocean basins. The much larger $CaCO_3$ reservoir is almost exclusively of marine origin.

Organic carbon in sedimentary rocks may be exposed to atmospheric oxidation as a result of weathering. This returns it to the atmosphere as CO_2, completing the so-called **long-term organic carbon cycle**. Current rates of fossil fuel burning are returning as much carbon to the atmosphere in a single year as weathering would achieve in hundreds of thousands of years. While fossil fuels represent only a small fraction of crustal organic carbon, this is still an order of magnitude more carbon than is currently in the atmosphere.

On the longest timescales, plate tectonics and volcanism play important roles in the global carbon picture: the **inorganic carbon cycle**. Limestone deposits on the sea floor are subducted into the Earth's mantle along plate boundaries, where the denser oceanic plates dive below the lighter continental plates (the continents may be said to be floating on the mantle, a bit like an iceberg). At the high temperatures within the mantle, limestone is **metamorphosed**:

$$CaCO_3 + SiO_2 \rightarrow CaSiO_3 + CO_2 \tag{3.1}$$

The CO_2 released is eventually returned to the atmosphere through volcanism, which is relatively common above such subduction zones (e.g. the Pacific rim).

The metamorphic rocks containing $CaSiO_3$ are recycled as newly formed crust which emerges at the mid-ocean ridges. Reaction (3.1), plus weathering, plus the reverse of this reaction, form a closed loop which cycles silicon and calcium, but more importantly cycles carbon, between the atmospheric and the inorganic reservoirs on a timescale of about a hundred million years. (Plate tectonics may thus be an essential feature of our planet in permitting life to remain, and evolve, over the eons.)

Whenever the rate of volcanic ejection exceeds the rate at which calcium ions are made available by weathering, atmospheric CO_2 levels increase. The injection rate is determined by the rate of metamorphism, which in turn depends on the rate of plate movement along convergent subduction zones. The rate of weathering is proportional to the rate of water cycling within the atmospheric branch of the hydrological cycle, which increases with the increased evaporation caused by higher temperatures. Changes in atmospheric CO_2 content in response to variations in these processes, on timescales of tens of millions of years, are believed to have been quite substantial in the geological past.

3.1.3 Biospheric CO_2 Exchanges

The largest fluxes in the global carbon cycle are those involving the exchanges of CO_2 between the atmosphere, and the terrestrial vegetation and the oceans. The terrestrial flux is roughly 100×10^{15} gC/year, which can also be expressed as 100 petagrams (Pg) C/year, or 100 gigatonnes (Gt) C/year, while the oceanic flux is about half that. (We'll provide some more exact numbers in Section 3.3.) The sum of these fluxes is sufficient to provide for the turnover of every CO_2 molecule in the atmosphere in just a few years. We discuss CO_2 in the oceans in the next section.

Atmospheric CO_2 observations, especially in the Northern Hemisphere, show a distinct seasonal cycle. Globally, about two-thirds of terrestrial vegetation occurs in regions with seasonal growth, mostly in the Northern Hemisphere, with the remainder in the wet tropics. Of the ~100 Gt C/year taken up by the terrestrial biosphere, about half is returned, more or less directly, by overnight respiration (like animals, plant cells contain mitochondria). Thus, **net primary productivity** is ~ 50 Gt C/year. The other half is returned, on longer timescales, by detritus (leaf litter, dead trees, etc.) decomposition.

3.1.4 Atmospheric CH_4

Methane is an important greenhouse gas, a fuel source, and an interesting component of the global carbon cycle. It has an atmospheric lifetime of ~9 years before it is converted to carbon dioxide via a series of reactions starting with attack by the OH radical: details in Section 2.2.2. (Some methane is transported to the stratosphere where its oxidation provides the main source of stratospheric water vapour.) Because the majority of emissions occur in the Northern Hemisphere, the Southern Hemisphere data show an understandable lag.

The atmospheric sources and sinks of methane vary widely in space and time, and flux rates are mostly quite small, making the construction of a methane budget quite a challenge. Two approaches are being employed, in the hope of arriving at some form of consistency. The bottom-up approach tries to take measured data that are available, and extrapolate them globally. The top-down approach looks at the global atmospheric data from various sites, including isotope data where available, and runs an inverse model, tracing back to the presumed sources. These are complementary approaches. Table 3.2, extracted from table 5.2

Table 3.2 **Atmospheric methane budget.**

	Top-down	Bottom-up
Sources		
Natural sources	215 (183–248)	371 (245–488)
Wetlands	180 (159–199)	149 (102–182)
Other sources	36 (21–49)	222 (143–306)
Freshwater: lakes and rivers		159 (117–212)
Wild animals, termites		11 (4–18)
Geological (land and ocean)	23 (0–71)	45 (18–65)
Other oceanic		6 (4–10)
Permafrost (excluding lakes and wetlands)		1 (0–1)
Anthropogenic sources	357 (336–375)	356 (335–383)
Agriculture and waste	221 (209–238)	208 (192–230)
Enteric fermentation and manure		109 (106–115)
Landfills and waste		64 (55–77)
Rice		31 (25–37)
Fossil fuels	106 (81–131)	115 (114–116)
Coal		38 (36–39)
Oil and gas		70 (68–73)
Transport		5 (1–11)
Industry		3 (1–5)
Biomass burning and biofuels	30 (22–36)	30 (22–39)
Sinks		
Total chemical loss	514 (474–529)	602 (496–754)
Tropospheric OH		560 (483–682)
Stratospheric loss		31 (12–37)
Tropospheric Cl		11 (1–35)
Soil uptake	37 (27–43)	30 (11–49)
Sum of sources	576 (550–589)	727 (581–872)
Sum of sinks	551 (501–572)	632 (507–803)
Imbalance	21 (18–26)	95

Source: based on data from Table 5.2 of IPCC AR6: see Acknowledgements at the end of the chapter.

of AR6, shows the results of the two approaches for the decade 2008–2017. The units are Tg CH_4 year^{-1}. Isotope studies are invaluable here, as fossil fuel sources (coal mines, natural gas) are fully depleted of ^{14}C, while biogenic sources are not.

3.2 Carbon in the Ocean

3.2.1 Inorganic Carbon

Carbon in the oceanic reservoir is present in three forms: dissolved CO_2 or H_2CO_3 (carbonic acid); carbonate ions paired with metallic ions such as Ca^{2+} and Mg^{2+}; and bicarbonate ions, HCO_3^-. The bicarbonate form is by far the largest of these. The dissolved CO_2 concentration reaches equilibrium with the atmospheric concentration through the reaction

$$CO_2 + H_2O \leftrightarrow H_2CO_3 \qquad (3.2)$$

An increase in the atmospheric concentration raises the equilibration concentration of dissolved CO_2. As a consequence, carbonic acid dissociates:

$$H_2CO_3 \leftrightarrow H^+ + HCO_3^- \qquad (3.3)$$

causing the water to become more acidic. This increase in hydrogen ions causes the equilibrium between carbonate and bicarbonate,

$$HCO_3^- \leftrightarrow H^+ + CO_3^{2-} \qquad (3.4)$$

to shift to the left. The net effect of reactions (3.2), (3.3), and (3.4) in reverse is

$$CO_2 + CO_3^{2-} + H_2O \leftrightarrow 2HCO_3^- \qquad (3.5)$$

This incorporates the added carbon into the bicarbonate reservoir, buffering the increase in ocean acidity. However, the ability of the oceans to take up, and buffer, CO_2 in this way is limited by the supply of ions in the carbonate reservoir.

3.2.2   Primary Production in the Ocean

Net primary production (NPP) refers to the conversion of inorganic carbon, in the form of CO_2, into organic matter, via photosynthesis. It has been estimated that up to half of the world's NPP occurs in the oceans. Comparing the world's oceans with its massive forests, this may seem surprising. However, oceanic NPP is primarily the domain of **phytoplankton**, which are both tiny and ephemeral. This activity takes place in the surface mixed layer, where sunlight, nutrients, and dissolved oxygen are available.

There is some uncertainty as to the global total NPP in the oceans, as it is not easy to measure. Estimates range between 20 and 50×10^{15} gC/year. The highest rates are found in coastal waters, where nutrient-rich rivers enter, and regions of upwelling, where nutrient-rich water is brought to the surface. However, because of their much larger total volume, the open oceans still probably account for the majority of NPP.

Much of our current data on oceanic NPP come from satellite observation. When ocean waters contain little phytoplankton there is little absorption of incident sunlight, and the reflected radiation is blue. However, when chlorophyll is abundant, the reflected light contains a greater proportion of green wavelengths. When suitably calibrated using in-situ data, such satellite data can be used to determine the concentration of chlorophyll, and hence primary production. NASA's Coastal Zone Color Scanner, on board the Nimbus-7 satellite (Chapter 12), was the first space-based instrument to make such measurements, which are now 'routine'.

Most of this marine primary production is consumed by **zooplankton**, and bacteria, in the surface waters. While the zooplankton represent a key step in the food chain that leads to fish – and sometimes the dinner table – the bacteria are consumed by a large population of bacterivores, which mineralise the nutrients and release CO_2 into the surface waters. It appears that the clear majority of the NPP is degraded to inorganic species such as CO_2, NO_3, PO_4, etc. in the surface waters.

Many marine organisms precipitate carbonate in skeletons and protective tissues. The process by which marine organisms incorporate this substance in their shells is

$$Ca^{2+} + 2HCO_3^- \rightarrow CaCO_3 + H_2CO_3 \tag{3.6}$$

Commercial shellfish might be the first thing that comes to mind, but vast quantities of $CaCO_3$ are contained in foraminifera, pteropods, and other small zooplankton found in the oceans. Coccolithophores, a group of marine algae, are responsible for large quantities of $CaCO_3$ reaching the sea floor in the open ocean.

Carbon dioxide is produced in deep water by the degradation of organic materials that have sunk from the surface waters. Because of their long isolation from the surface, the deep ocean is saturated with CO_2, which is more soluble in the lower temperatures and higher pressures that are found there. As a result, these deep waters are subsaturated with respect to $CaCO_3$

$$H_2O + CO_2 \leftrightarrow H^+ + HCO_3^- \leftrightarrow H_2CO_3 \tag{3.7}$$

When skeletal remains sink to the ocean floor, they dissolve:

$$CaCO_3 + H_2CO_3 \leftrightarrow Ca^{2+} + 2HCO_3^- \tag{3.8}$$

The Ca^{2+} ions that are a key part of many marine organisms enter the seas by way of the weathering of rocks by rainwater, and are then carried by rivers to the ocean.

3.2.3 Ocean Acidification and Deoxygenation

As atmospheric CO_2 levels have increased over the past 250 years, so has its oceanic concentration, leading to the potential problem of ocean acidification. It is estimated that in 1750 the pH of the oceans was 8.25, compared with the current value of 8.14. Figure 3.2 shows how the relative concentrations of CO_2, HCO_3^- and CO_3^{2-} vary with pH. For pH between ~8.3 and ~6.3, HCO_3^- dominates. However, as pH decreases, the concentration of CO_3^{2-} drops dramatically. Should the atmospheric CO_2 concentration reach 900 ppm (an unlikely, but not impossible, scenario by 2100) this could have profound effects on marine organisms.

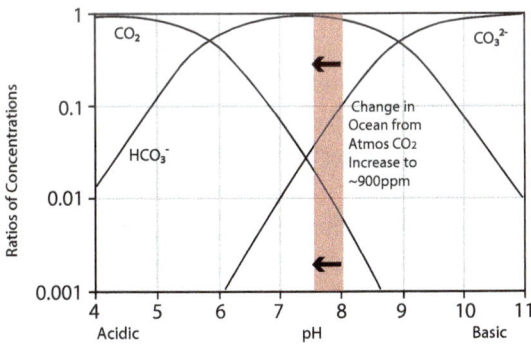

Figure 3.2 Relative concentrations of CO_2, HCO_3^-, and CO_3^{2-} as a function of ocean pH. *Source*: courtesy of Ben McNeil.

The latest conclusions from IPCC AR6 Chapter 5 state that 'ocean acidification is strengthening as a result of the ocean continuing to take up CO_2 from human-caused emissions'. This is driving changes in sea-water chemistry that result in a decrease in pH and associated reductions in the saturation state of calcium carbonate. These trends are becoming clear globally, with a '*very likely*' rate of decrease in pH in the surface layer of 0.016 to 0.020 per decade in the subtropics since the 1980s. Ocean acidification has spread deeper into the ocean, surpassing 2,000 m depth in the northern North Atlantic and in the Southern Ocean.

The IPCC has noted a second change in ocean chemistry as a result of our greenhouse gas emissions: deoxygenation. As water warms, its solubility of gases such as oxygen decreases. To date this has been minor, but is projected to worsen under some of the higher emission scenarios discussed in Chapter 20.

3.2.4 Ocean Fertilisation

Geoengineering, also known as climate engineering, refers to a suite of initiatives that are designed to modify our environment in ways which might offset some of the effects of global warming. According to the Royal Society it is 'the deliberate large-scale manipulation of the planetary environment to counteract anthropogenic climate change'. These initiatives are generally divided into two strategy directions, one of which is known as **carbon dioxide removal** (CDR). We will look at the other group in later chapters.

Removing CO_2 from the atmosphere always requires some input of energy, as this is the thermodynamically lowest energy state of carbon: we burn carbon-based fuels for the simple reason that heat/energy is released in the process. It is also a low-concentration gas, so that just collecting it is costly. Nevertheless, there have been a number of imaginative ideas put forward to do just this, based on the use of cheap renewable energy, at times of an oversupply. The free energy supplied by the Sun, via photosynthesis, is also an option. Growing and harvesting special crops to use as biofuels, followed by carbon sequestration, might be an option.

Closely related to this latter idea is ocean fertilisation. Phytoplankton require sunlight, of course, but also nutrients, which are far more plentiful near continental margins, especially river outfalls, than the mid ocean. If such nutrients could be increased (iron being one of the more important), then more phytoplankton should be the result, shifting carbon to the organic pool. Assuming that some of this will die, and take its carbon to the sea floor, that would shift the balance of inorganic carbon between its atmospheric and oceanic pools, causing more to enter the ocean: potentially a good thing.

How might this be brought about? Some experiments have been carried out supplying iron to a small section of the ocean, with positive results. Nature often performs this experiment for us, for example when a dust storm dumps iron-rich mineral dust, leading to a phytoplankton bloom that may be observed by satellite.

Is this the magic bullet we've been looking for? Modelling suggests that it is unlikely to remove CO_2 at close to the rate we are currently sending into the atmosphere, so it's not the solution some might have hoped for. However, it may have a longer-term use. Once we stabilise our greenhouse gas emissions, and global mean temperatures settle down somewhere between 1.5°C and 2.0°C above pre-industrial levels (we hope!), it would still be a good idea to remove some of that CO_2 and bring temperatures slowly down, to avoid long-term threats to the polar ice caps.

3.3   Anthropogenic Perturbation

Figure 3.3 {FAQ 2.1, figure 2.1 from AR4} shows the concentrations of the three key greenhouse gases for the past 2,000 years: most of these data come from air bubbles trapped in ice. We see that all three were quite stable until roughly 1800, implying that sources and sinks were in balance. Clearly that is no longer the case!

3.3.1   Recent CO_2 Budget

Carbon dioxide is the key greenhouse gas, and its rapid increase in the past few centuries is the primary driver of global warming/climate change. Reducing, and ultimately eliminating, this increase is thus an essential focus of climate mitigation. The primary anthropogenic sources of CO_2 are, of course, fossil fuel usage, land-use changes, and cement production, which involves the following chemical reaction:

$$CaCO_3 + heat \rightarrow CaO + CO_2 \qquad (3.9)$$

IPCC studies the relevant data on sources and sinks, and they are discussed and presented in {figure 5.12 and table 5.1}, both of which we have chosen to condense. Figure 3.4 is a simplified version of {figure 5.12}, showing both the (reconstructed) natural/pre-industrial fluxes of CO_2 from both land and ocean, plus the anthropogenic contributions for the decade 2010–2019.

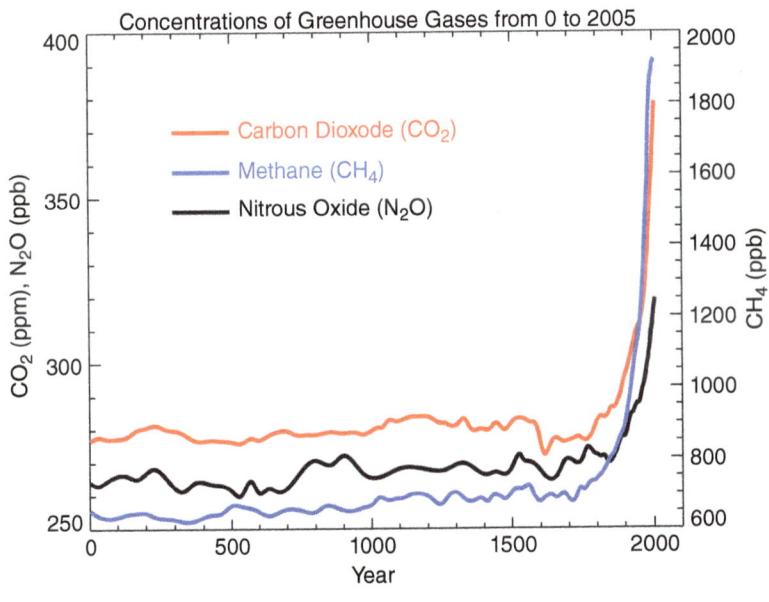

Figure 3.3 Concentrations of CO_2, CH_4, and N_2O over the past 2,000 years. *Source*: see Acknowledgements at the end of the chapter.

Figure 3.4 Global carbon (CO_2) budget for 2010–2019 (based on figure 5.12 of AR6: *Source* see Acknowledgements at the end of the chapter).

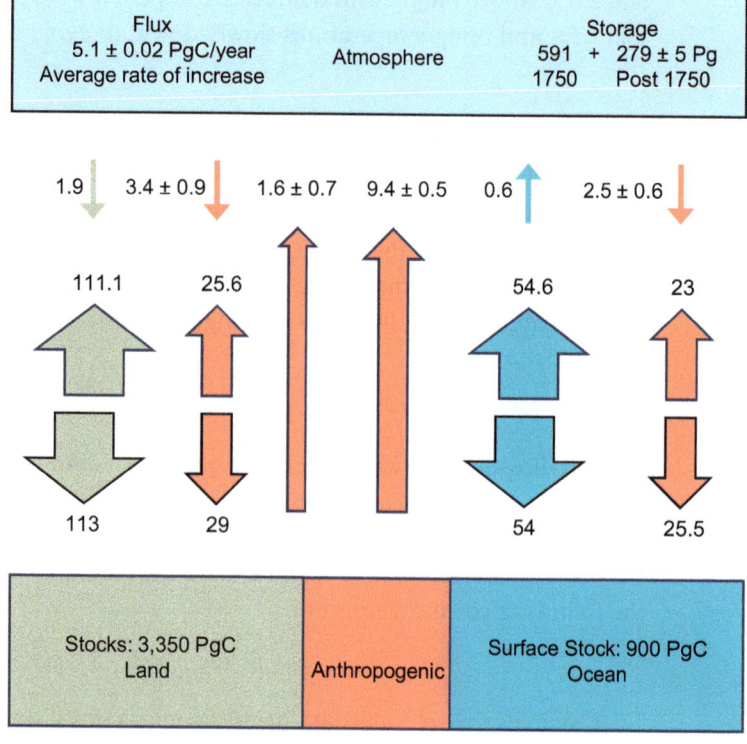

We'll start with the natural fluxes (sage green, and blue). The left-hand section shows gross photosynthesis and total respiration plus fire, with a net flux to the ground of 1.9 PgC/year. Also shown are stocks of 3,350 PgC: made up of 450 PgC in vegetation, 1,700 PgC in soils, and 1,200 PgC buried in permafrost, a subject we will return to in Chapter 14. The right-hand section shows the ocean–atmosphere exchange, with a net flux to the atmosphere of 0.6 PgC/year. Also shown is the surface ocean stock of 900 PgC. (AR6 provides some extra detail on other ocean stocks, and internal fluxes, which we have chosen to omit.) Not included in our version are fluxes to the atmosphere of 1.5 PgC/year from fresh water, and 0.1 PgC/year from volcanism; plus a flux to the ground of 0.3 PgC/year due to rock weathering. This implies an overall flux balance to the atmosphere. (There is also a flux of 0.8 PgC/year carried by rivers to the ocean.)

In the middle are the direct anthropogenic flux rates (red) for 2010–2019: 9.4 ± 0.5 PgC/year from fossil fuels and cement production, and 1.6 ± 0.7 PgC/year from land-use changes. However, we also know that some of this is being taken up by the land surface/biosphere (3.4 ± 0.9 PgC/year) and some by the oceans (2.5 ± 0.6 PgC/year).

Table 3.3, extracted from table 5.1 of AR6, provides the key data on cumulative sources and sinks from 1750 to today, plus the growth rates for the decades 1980–1989 and 2010–2019. The numbers in the final column are, of course, also presented in Figure 3.4.

AR6 offers these conclusions. Over the past six decades, the average fraction of anthropogenic CO_2 emissions that has accumulated in the atmosphere – the airborne fraction – has

Table 3.3 **Carbon dioxide sources and sinks.**

	1750–2019 Cumulative (PgC)	1980–1989 Mean annual growth rate (PgC year^{-1})	2010–2019 Mean annual growth rate (PgC year^{-1})
Emissions			
Fossil fuel combustion and cement production	445 ± 20	5.4 ± 0.3	9.4 ± 0.5
Net land-use change	240 ± 70	1.3 ± 0.7	1.6 ± 0.7
Total emissions	685 ± 75	6.7 ± 0.8	10.9 ± 0.9
Partition			
Atmospheric increase	285 ± 5	3.4 ± 0.02	5.1 ± 0.02
Ocean sink	170 ± 20	1.7 ± 0.4	2.5 ± 0.6
Terrestrial sink	230 ± 60	2.0 ± 0.7	3.4 ± 0.9
Budget imbalance	0	-0.4	-0.1

Source: based on data from table 5.1 of IPCC AR6: see Acknowledgements at the end of the chapter.

remained nearly constant at approximately 44%. The ocean and land sinks of CO_2 have continued to grow in response to the increase in emissions. Variability on interannual to decadal timescales of the regional land and ocean sinks indicates that these sinks are sensitive to climatic conditions, and therefore to climate change in the future.

3.3.2 Sectoral Sources

While the climate does not care where a CO_2 molecule comes from – what country, what activity, etc. – policy makers need such information, as it is an essential input as we formulate policies to reduce our emissions. Much of these data are readily available, as they are required for reporting to various agencies, and as part of the Paris Agreement process. One excellent source is ourworldindata.org; another is climatewatchdata.org.

There is an enormous amount of data, and many ways to present them. One obvious way is by country; or per capita; or per unit of GDP. Such data primarily reflect a country's level of economic and technological development. They might also indicate the use of nuclear energy, or the growth of renewables, replacing coal and oil, although both only affect emissions from electricity generation. For example, one-quarter of all greenhouse gas emissions in the United States come from electricity generation, whereas that figure in Australia is one-third. The USA uses nuclear energy, Australia does not.

We believe that one valuable approach is to show a breakdown of emissions from different sections of the economy, as that might help us understand the challenges we face in getting to net zero. Table 3.4, based on data from the ourworldindata website, provides one useful breakdown.

While we do not have space in this book for a full analysis of these data and their implications, a few comments are in order. Our use of energy accounts for nearly three-quarters of all emissions. This, of course, extends well beyond the generation (and consumption) of electricity, a sector in the process of decarbonisation.

Table 3.4 **Sector-by-sector sources of greenhouse gases.**

Sector	%	%
Energy use in industry		24.2
Iron and steel	7.2	
Chemical and petrochemical	3.6	
Other	13.4	
Transport		16.2
Road transport	11.9	
Aviation	1.9	
Shipping	1.7	
Other	0.7	
Energy use in buildings		17.5
Residential buildings	10.9	
Commercial buildings	6.6	
Unallocated fuel combustion		7.8
Fugitive emissions		5.8%
From oil and gas	3.9	
From coal	1.9	
Energy use in agriculture		1.7
Direct industrial processes		5.2
Cement	3.0	
Chemicals and petrochemicals	2.2	
Waste		3.2
Wastewater	1.3	
Landfills	1.9	
Agriculture, forestry, land use		18.4
Deforestation	2.2	
Cropland	1.5	
Crop burning	3.5	
Rise cultivation	1.3	
Agricultural soils	4.1	
Livestock and manure	5.5	

Source: based on public data from the ourworldindata website.

Road transport accounts for 12% of all emissions. The uptake of electric vehicles (EVs) – slow in some countries, faster in others – will reduce emissions, *provided* that the input electrical energy is carbon-neutral. (Many purchasers of EVs charge them from their own roof-top solar panels and batteries; 'for free'.) These two transitions are, at present, progressing hand-in-hand. Fuel efficiency standards for motor cars and light commercial vehicles may be an important policy driver. For heavy trucks, (green) hydrogen might be the best option.

Some industrial processes generate greenhouse gases directly; the prime example is cement production. The only work-around for this that we know of is carbon sequestration.

Agriculture, forestry, and land use account for 18% of all emissions. Historically it has been deforestation in the so-called New World – the Americas, Australia, and southern

Africa – which has been a driver of atmospheric CO_2 increases. However, this has now slowed, and only accounts for 2.2% of emissions. Emissions from livestock (from both ends of the alimentary canal) account for 5.8%, while soils contribute 4.1%, in the form of nitrous oxide. Improved land-use management is being looked upon as a possible sink for CO_2, and is included in carbon accounting, although some of the assumptions involved do appear to be questionable.

3.4 The Nitrogen Cycle

Nitrogen is unique in the atmosphere in that it exists in four quite separate reservoirs: unreactive N_2; relatively inert N_2O; the reactive acidic species NO_y (a group of species in equilibrium, as we saw in the previous chapter); and the reactive alkaline species NH_3 (ammonia), which is virtually the only such species in the atmosphere. Apart from the obvious reaction between NH_3 and HNO_3 (a minor component of NO_y) there is little interaction between these reservoirs within the atmosphere, and so each needs separate consideration.

More than 99.99% of atmospheric nitrogen is in the form of N_2, and N_2O makes up more than 99% of the remainder. While the other species are present in trace amounts, their presence is crucial to atmospheric chemistry. Ammonia, for example, is the only gas capable of neutralising H_2SO_4 and HNO_3.

Nitrogen fixation is the reduction and incorporation of atmospheric nitrogen in living plant material, which is accomplished by certain bacteria. (Fixed nitrogen refers to nitrogen in chemical forms which may be used by plants and bacteria.) Other bacteria perform the opposite process – **denitrification** – at later stages in a plant's life cycle, returning much of the fixed nitrogen to the atmosphere as N_2 (and some as N_2O). These flows are close to equilibrium, with N_2 fluxes of 100 TgN/year to/from the marine biosphere, and 140 TgN/year on land.

All of these processes are complex, and vary with local conditions. As a consequence, the magnitude of terrestrial gaseous N emissions is very uncertain, with global-scale errors of ±50%. One recent approach to reducing some of these uncertainties makes use of potential $^{15}N/^{14}N$ isotopic effects in some of the biological pathways.

Table 3.5, based on data from IPCC AR5 {table 6.10}, presents a budget for the two reactive N species NO_x and NH_3. The primary sources of NH_3 are decomposition of proteins and urea $[CO(NH_2)_2]$ from animals, plus contributions from soils, the ocean, and biomass burning. As the only alkaline gas in the atmosphere, it readily reacts with acidic species, forming compounds (salts) such as NH_4NO_3, NH_4HSO_4, and $(NH_4)_2SO_4$. These mostly condense, forming aerosol particles, which are then removed by either wet or dry deposition. (*Wet deposition* refers to removal in raindrops, etc., while *dry deposition* refers to removal by direct contact with the surface.) Small amounts are oxidised to NO_x.

3.4.1 Nitrous Oxide

Nitrous oxide, N_2O, is the third most potent greenhouse gas, and so its atmospheric budget merits more careful attention. It also plays a role in stratospheric ozone loss, as will be covered

Table 3.5 Direct sources and sinks of NO$_x$ and NH$_3$.

	NO$_x$ TgN/year	NH$_3$ TgN/year
Emissions to atmosphere		
Anthropogenic sources	37.5	40.1
Fossil fuel combustion and industrial processes	28.3	0.5
Agriculture	3.7	30.4
Biomass/biofuel burning	5.5	9.2
Natural sources	11.3	10.6
Soils (natural vegetation)	7.3	2.4
Oceans	–	8.2
Lightning	4.0	–
Deposition from atmosphere	46.9	53.1
Continents	27.1	36.1
Oceans	19.8	17.0

Source: Based on data from {table 6.10} of IPCC AR5: see Acknowledgements at the end of the chapter.

in Chapter 10. The AR6 headline conclusion {page 676} states that 'Atmospheric concentration of N$_2$O grew at an average rate of 0.85 ± 0.03 ppb year^{-1} between 1995 and 2019, with a further increase of 0.95 ± 0.04 ppb year^{-1} in the most recent decade (2010–2019).' This increase is dominated by anthropogenic emissions, which have increased by 30% between the 1980s and the most recent decade. Increased use of nitrogen fertiliser and manure contributed to about two-thirds of the increase, with fossil fuels/industry, biomass burning, and wastewater accounting for much of the rest.

Since AR5, our understanding of sources has improved, as a result of extended atmospheric observations, improved atmospheric inversions, and updated and expanded source inventories, as well as improved 'bottom-up' estimates of freshwater, ocean, and terrestrial sources. (N$_2$O is also used in anaesthesiology, mostly in maternity wards, but a review of the literature suggested that emissions from this source are well below 0.1 Tg/year N$_2$O.)

Table 3.6, based on tabulated data from IPCC AR6 {table 5.3}, presents an N$_2$O budget for the troposphere, for the decades 1980–1989 and 2007–2016. (Note that the units used are Tg of N per year.) The human perturbation of the natural nitrogen cycle from the use of fertilisers, as well as nitrogen deposition from land-based agriculture and fossil fuel burning, has been the largest driver of the increase of 31.0 ppb between 1980 and 2019.

3.5   The Sulphur Cycle

Sulphur is a major biochemical element, with its own important cycle. Over the past century that cycle has been severely perturbed, mainly as a result of the release of SO$_2$ from the burning of fossil fuels. Table 3.7 provides estimates of the natural fluxes of the major sulphur compounds to the atmosphere. Reduced species (mainly H$_2$S) are oxidised by the OH radical to SO$_2$. Of this, the majority is then oxidised to SO$_4^{2-}$, with the remainder removed by dry deposition.

Table 3.6 **Sources and sinks of tropospheric N_2O (TgN per year).**

	1980–1989	2007–2016
Anthropogenic sources		
Fossil fuel combustion and industry	0.9 (0.8–1.1)	1.0 (0.8–1.1)
Agriculture (including aquaculture)	2.6 (1.8–4.1)	3.8 (2.5–5.8)
Biomass and biofuel burning	0.7 (0.7–0.7)	0.6 (0.5–0.8)
Wastewater	0.2 (0.1–0.3)	0.4 (0.2–0.5)
Inland water, estuaries, coastal zones	0.4 (0.2–0.5)	0.5 (0.2–0.7)
Atmospheric nitrogen deposition on ocean	0.1 (0.1–0.2)	0.1 (0.1–0.2)
Atmospheric nitrogen deposition on land	0.6 (0.3–1.2)	0.8 (0.4–1.4)
Other indirect effects from CO_2, climate and land-use change	0.1 (–0.4–0.7)	0.2 (–0.6–1.1)
Total anthropogenic	5.6 (3.6–8.7)	7.3 (4.2–11.4)
Natural sources and sinks		
Rivers, estuaries, and coastal zones	0.3 (0.3–0.4)	0.3 (0.3–0.4)
Open oceans	3.6 (3.0–4.4)	3.4 (2.5–4.3)
Soils under natural vegetation	5.6 (4.9–6.6)	5.6 (4.9–6.5)
Atmospheric chemistry	0.4 (0.2–1.2)	0.4 (0.2–1.2)
Surface sink	–0.01 (–0.3–0)	–0.01 (–0.3–0)
Total natural	9.9 (8.5–12.2)	9.7 (8.0–12.0)
Total bottom-up source	15.5 (12.1–20.9)	17.0 (12.2–23.5)
Observed growth rate		4.5 (4.3–4.6)
Inferred stratospheric sink		13.1 (12.4–13.6)
Atmospheric inversion		
Atmospheric loss		12.4 (11.7–13.3)
Total source		16.9 (15.9–17.7)
Imbalance		4.2 (2.4–6.4)

Source: Based on data from {table 5.3} of IPCC AR6: *Source*: see Acknowledgements at the end of the chapter.

Table 3.7 **Estimated rates of emission of sulphur to the atmosphere from natural sources (10^{12} g S/year).**

Source	SO_2	H_2S	DMS	CS_2	OCS	Total
Oceanic		0–15	38–40	0.3	0.4	38.7–56.7
Salt marsh		0.8–0.9	0.7	0.07	0.12	1.7–1.8
Inland swamps		11.7	1.0	2.8	1.85	17.4
Soil and plants		3–41	1–5	0.6–1.5	0.2–1.0	5.0–48.5
Biomass burning	7	0–1	0–1		0.1	7.1–9.1
Volcanoes, etc.	8	1				9.0
Totals	15	16.5–70	40–45	4–5	~2.5	80–140

Source: from *Biogeochemistry An Analysis of Global Change*, Schlesinger, table 13.2, with permission from Elsevier.

DMS is the dominant source over the oceans, and is, in fact, the major natural source of sulphur to the atmosphere. As we saw in Table 2.2, most of the sulphur gases have lifetimes of a few days. The exception is carbonyl sulphide – OCS – with a residence time of a few years, and correspondingly high (and uniform) mixing ratio. (It is for this reason that we have placed it in Table 2.1 as a 'permanent gas', rather than Table 2.2 as a 'variable gas', which is the case in some books.) Some makes its way to the stratosphere, where, under the influence of UV radiation, it becomes a major source of sulphate aerosols.

The anthropogenic flux is reasonably well known, and is comparable to the natural fluxes. Since the mid-1970s there have been serious efforts to reduce emissions in Europe and North America, which have mainly resulted in a shift in emissions to the emerging economies of Asia. Ninety per cent of the emissions are in the Northern Hemisphere, implying that the two hemispheres have qualitatively different sulphur cycles.

Summary

Our Earth has had an atmosphere for over 4 billion years, and its composition has changed significantly over that time, a subject we will return to in Chapter 16. Those changes have involved interactions with both the geosphere (i.e. the 'solid Earth') and the biosphere. Such interactions continue today, but at their own pace, of course (see Table 2.1).

Climate science is concerned both with the nature of the climate at various times in the past, the present, and (potentially) in the future, and the processes which may, or may not, bring about change. One very important consideration for the latter is the *rate* of change. If we are to put recent climate change (e.g. Figure 1.2) in a proper context, it is vital to compare the rates of such changes with the rates of changes in the past. We cannot do this unless we can put the changes displayed in Figures 1.1 and 3.3 in proper perspective. This chapter represents a key step in this analysis.

Acknowledgements

Tables 3.2, 3.3, 3.6; Figure 3.4:

Climate Change 2021: The Physical Science Basis. Contribution of Working Group I to the Sixth Assessment Report of the Intergovernmental Panel on Climate Change [Masson-Delmotte, V., P. Zhai, A. Pirani, S. L. Connors, C. Péan, S. Berger, N. Caud, Y. Chen, L. Goldfarb, M. I. Gomis, M. Huang, K. Leitzell, E. Lonnoy, J. B. R. Matthews, T. K. Maycock, T. Waterfield, O. Yelekçi, R. Yu, and B. Zhou (eds.)]. Cambridge University Press, Cambridge, UK and New York, NY, USA, 2391 pp. doi:10.1017/9781009157896.

Table 3.5:

Ciais, P., C. Sabine, G. Bala, L. Bopp, V. Brovkin, J. Canadell, A. Chhabra, R. DeFries, J. Galloway, M. Heimann, C. Jones, C. Le Quéré, R. B. Myneni, S. Piao, and P. Thornton, 2013: Carbon and Other Biogeochemical Cycles. In: *Climate Change 2013: The Physical*

Science Basis. Contribution of Working Group I to the Fifth Assessment Report of the Intergovernmental Panel on Climate Change [Stocker, T.F., D. Qin, G.-K. Plattner, M. Tignor, S. K. Allen, J. Boschung, A.Nauels, Y. Xia, V. Bex and P. M. Midgley (eds.)]. Cambridge University Press, Cambridge, UK and New York, NY, USA, pp. 465–570, doi:10.1017/CBO9781107415324.015.

Figure 3.3:

FAQ 2.1 figure 1 (chapter 2) from Forster, P., V. Ramaswamy, P. Artaxo, T. Berntsen, R. Betts, D. W. Fahey, J. Haywood, J. Lean, D. C. Lowe, G. Myhre, J. Nganga, R. Prinn, G. Raga, M. Schulz and R. Van Dorland, 2007: Changes in Atmospheric Constituents and in Radiative Forcing. In: *Climate Change 2007: The Physical Science Basis. Contribution of Working Group I to the Fourth Assessment Report of the Intergovernmental Panel on Climate Change* [Solomon, S., D. Qin, M. Manning, Z. Chen, M. Marquis, K.B. Averyt, M.Tignor and H.L. Miller (eds.)]. Cambridge University Press, Cambridge, UK and New York, NY, USA.

FURTHER READING

Biogeochemistry – an alternative title for this chapter – is a relatively recent branch of science, with its own literature. Four useful books in this field are:

- *Biogeochemistry, An Analysis of Global Change* by W. H. Schlesinger (Academic Press, 1991)
- *Atmospheric Change, An Earth System Perspective* by T. E. Graedel and P. J. Crutzen (W. H. Freeman, 1993).
- *Earth System Science, From Biogeochemical Cycles to Global Change* by M. C. Jacobson, R. J. Charlson, H. Rodhe, and G. H. Orians (Academic Press, 2000).
- *Atmospheric Science for Environmental Scientists* by C. N. Hewitt (Wiley-Blackwell, 2020).

Once again, a valuable reference is:

- *Consider a Spherical Cow* by J. Harte (University Science Books, 1988).

REVIEW QUESTIONS

1. What are the key chemical processes that appear to help regulate the atmospheric concentration of CO_2 on geological timescales?
2. What fraction of the carbon taken up by plants during the daytime is returned overnight?
3. Our use of fossil fuels is the main source of CO_2 emissions to the atmosphere: what is second?
4. What is the major form of inorganic carbon in the ocean?
5. What are the four nitrogen reservoirs in the atmosphere?

EXERCISES

In Section 3.3.2 we presented one breakdown of our global sources of greenhouse gas (GHG) emissions. There are many other ways to think through this question, both at the global level and by country/state/region. There is a vast array of data readily available online (for example, from the International Energy Agency), and numerous research exercises which can be produced from this. Here are just a few.

1. Find information on GHG emissions per country; per capita; and per unit of GDP.
2. How much of each country's electricity is nuclear generated: does this have an impact on the results of exercise 1?
3. CO_2 has an *effective* atmospheric lifetime of around a century, despite the number shown in Table 2.1 (which reflects the equilibrium exchange with the biosphere). So, if we are to measure a nation's overall contribution to climate change, we need to know that nation's total emissions over, say, the past 75 years or even more. If you search more deeply you will find such data.
4. Find the emissions breakdown for your country/state/region. As we noted above, the breakdown in Table 3.4 is one of many. Another common breakdown is into Electricity; Stationary energy; Transport; Fugitive emissions; Industrial processes and Product use; Agriculture; Waste; Land use, Land-use change, and Forestry.
5. Can you find your country's emissions history? Have emissions fallen recently?
6. What is your country's Nationally Determined Contribution (NDC) under the Paris Agreement? These NDCs are supposed to be updated and 'improved' every five years. Can you find your country's historic NDCs? Have they been improving as hoped?
7. What specific policies does your country have to reduce its emissions? Do they include some form of carbon price?
8. Canada has recently introduced a carbon pricing scheme, which includes payments to households: search out the details. Do you think it has merit?

4   Atmospheric Aerosols

Aerosols, and rising levels of certain aerosol types, are now recognised as important players in our environment, including climate change. Aerosol, or 'particulate matter' (PM), is the collective name for small particles and droplet solutions, with sizes ranging from ~1 nm to ~20 μm, suspended in the air. Individually they are often referred to as aerosols, aerosol particles, or particulates.

Aerosols are usually considered a branch of atmospheric chemistry – its 'condensed phase' – and treated in chapters devoted to that subject. Rather than include aerosols as a section in Chapter 2, we believe that they deserve a chapter of their own. For a start, aerosols vary enormously, in many ways, as we will see, and in this respect, they differ radically from gases. To quote Stephen Schwartz's plenary address to the First Australian Aerosol Workshop in 2005: 'If you've seen one carbon dioxide molecule, you've seen them all: If you've seen one aerosol particle, you've seen one aerosol particle.'

In Section 4.7 we will look at their optical properties (i.e. their ability to scatter and absorb radiation), and the effects these may have on the energy flows that are central to climate. Skim the earlier sections if you wish, but read Section 4.7 carefully. Aerosol particles are also the seeds of all cloud droplets, as will be discussed in Chapter 6, and so are a key component of the hydrological cycle. Finally, aerosols may be a pollution issue (Section 4.3), and are a central component of Air Quality standards, as we have seen. For those readers interested, we close with a section on aerosol research.

4.1   Aerosols: Characterisation

Aerosols may be classified in several different ways. They may be the result of **natural** process, or **anthropogenic** activities, and classified as such. Similarly, aerosols may be injected into the atmosphere directly – **primary aerosols** – or formed in situ by gas-to-particle conversion processes from precursor gases – **secondary aerosols**. While this two-by-two 'classification matrix' may appear straightforward, reality is far more complex than that. For example, many volatile gases, themselves the result of either natural or anthropogenic processes, condense onto the surfaces of existing particles, which could originally have been regarded as primary or secondary, as well as being regarded as natural or anthropogenic. Also, chemical reactions may take place on the surface, and even on the inside, of particles, altering their composition. Thus, the story of aerosols is one of the most complex in the whole of atmospheric science.

Aerosols are also distinguished by their sizes, or rather their **size distributions**. When plotted on a log–log graph they are often seen to break up into two or three modes. The **nucleus mode** covers the range up to ~100 nm; the **accumulation mode** covers the range 100 nm to 1 μm; and the **coarse mode** covers the range above 1 μm. We may also refer to Aitken nuclei for particles smaller than 1 μm, in honour of John Aitken (1839–1919), considered the founder of atmospheric aerosol science, and aerosol measurement techniques. We also use the notation PM_x to denote 'particulate matter with diameters less than x μm'.

Gas-to-particle conversion creates nucleus mode particles, which may then combine to enter the accumulation mode (hence the name). Alternatively, gases condensing on nucleus mode particles can cause them to grow into the accumulation mode. With few exceptions, primary aerosols are in the accumulation and coarse modes.

We will start with a survey of the major aerosol types. The most useful of the possible divisions is into primary and secondary, as this reflects both their origins and, to a large extent, their chemistry. Because the stratosphere is such a different environment from the troposphere, we treat stratospheric aerosols as a category on their own.

4.2   Primary Aerosols

4.2.1   Mineral Dust

About one-third of the Earth's land surface is arid – desert or semi-desert – and concerns have been raised that poor agricultural practices, coupled with droughts, could be causing this to increase. It has been estimated that up to half of all emitted soil dust may be due to anthropogenic activities. Bare ground, especially dry desert soils, are a major source of large dust particles, in response to wind and turbulence. Globally, such soils are the main source of particles from the Earth's surface, providing ~2,000 Tg per year of mineral particles. The major source regions are the Earth's deserts – Sahara, Gobi, Arabian, Central Australian, and many smaller deserts – but more specifically dry lake areas, such as the Bodélé depression in Chad.

Mineral dust particles are mostly in the coarse mode, with sizes ranging from ~1 μm up to 10 or 20 μm, and occasionally even more, depending on wind strength. They are quite irregular in shape, and are often an agglomerate of several mineral grains. While their compositional details vary considerably, they are usually dominated by quartz [SiO_2] and kaolinite [$Al_2Si_2O_5(OH)_4$] (the major clay mineral), and are invariably metal rich.

4.2.2   Sea Salt

The oceans are a major source of aerosols: between 1,000 and 5,000 Tg per year. The average rate of particle production over the oceans is ~100 cm^{-2} s^{-1}. The major mechanism for ejecting oceanic material into the atmosphere is the bursting of small bubbles. Many small droplets are formed when the upper film of the air bubble bursts. Bubbles ~2 mm or larger in diameter eject 100 or more such droplets into the air. These then evaporate, leaving behind a sea salt particle with a

Table 4.1 **Ion composition of sea water and river waters.**

Constituent	Concentration (mg/kg)	Mean residence time (Myr)	Concentration in river water (mg/kg)
Sodium	10,760	75	5.15
Magnesium	1,294	14	3.35
Calcium	412	1.1	13.4
Potassium	399	11	1.3
Strontium	7.9	12	0.03
Chloride	19,350	120	5.75
Sulphate	2,712	12	8.25
Bicarbonate	145	0.1	52.0
Bromide	67	100	0.02
Silicate	2.9	0.02	10.4
Boron	4.6	10	0.01
Fluoride	1.3	0.5	0.10

Source: from *Biogeochemistry, An Analysis of Global Change*, Schlesinger, table 9.1, with permission from Elsevier.

diameter of ~0.3 μm, or less. This is one of the main examples of an accumulation mode primary aerosol. When a bubble bursts, a few larger drops break away in a jet, and are thrown about 15 cm into the air. When these evaporate, they leave behind sea salt particles with diameters greater than 2 μm. These constitute a significant fraction of the sea salt aerosol mass, but because of their size they do not travel very far before they are quickly returned to the sea.

On average, sea water in the open ocean contains ~35 g of dissolved salts per kg of freshwater, making sea water ~2.4% denser than freshwater at the same temperature. Actual values typically vary between 34 and 36 g/kg (34–36‰), for a variety of reasons. However, the most important is the interconnection between evaporation – mostly from tropical and sub-tropical waters – subsequent atmospheric transport, and precipitation, including runoff, at higher latitudes. The highest values are often found at places where water is freezing, as ice is very low in salt. By contrast, some of the lowest values are found in regions of glacial melt.

Table 4.1 provides a list of the major ions found in sea water, and river water, as well as their mean residence times. Note that the Na/Cl ratio of 0.556 is distinctly lower than that of pure NaCl.

These species are not static, as can be seen from the column headed 'Mean residence time'. Some material, especially Ca, is incorporated into marine skeletons and other structures, and may eventually settle on the ocean floor. Other processes also carry minerals to the ocean floor, forming clays, which tend to show relatively high concentrations of Na, K, and Mg. Hydrothermal vents act in the opposite way, enriching the oceans with elements such as Fe and S. Erosion on land leads to rivers carrying various amounts of salts to the oceans, maintaining the concentration mix.

4.2.3 Biological

Solid and liquid particles are released into the atmosphere from animals and plants. These emissions include seeds, pollen, spores, and plant and animal fragments, usually in the

1–200 μm diameter range. By contrast, bacteria, algae, protozoa, fungi, and viruses are mostly <1 μm. Some biogenic particles, such as bacteria from vegetation, are known to be able to nucleate ice in clouds.

4.2.4 Smoke

Smoke from forest and scrub fires is a major source of aerosols. Small smoke particles (organic compounds and elemental carbon) and fly ash are injected directly into the atmosphere. The burning of 1 hectare (10^4 m^2) can release several tonnes of particulates. It is estimated that biomass burning releases ~ 50 Tg of particles into the atmosphere each year, of which ~6 Tg is elemental carbon.

'Black carbon' actually enters the atmosphere in fractal form, with sizes around 15–40 nm, before coagulating to form particles with a size distribution that peaks at around 100 nm diameter. The products of biomass burning are quite sensitive to the conditions: 'flaming' involves high temperatures and oxygen availability, with (near) complete combustion, while 'smouldering' involves lower temperatures and oxygen deficit, leading to incomplete combustion (although this mainly affects the gases emitted, especially the ratio of CO to CO_2).

4.2.5 Anthropogenic

The main anthropogenic sources of primary aerosols are dust and rubber particles from roads, wind erosion of tilled soils, smoke from slash-and-burn agriculture and domestic wood burning, industrial processes such as machining and grinding, and emissions from energy production sources, including soot and fly ash. A very rough estimate of anthropogenic direct emissions of particles <10 μm diameter is ~400 Tg. This is still quite small compared with the major natural sources of primary aerosols.

4.3   Secondary Aerosols

Secondary aerosols form by in-situ condensation of gases with low (saturation) vapour pressures. (The word 'saturation' is often taken as read, which can mislead.) In most cases, these gases are formed by gas-phase atmospheric reactions – in particular, photochemical reactions – as any gas with such a low saturation vapour pressure emitted into the atmosphere would travel only a very short distance before condensing.

This may appear to be a straightforward process, as any supersaturated gas ought to condense without further ado. However, as we will explain in Section 6.3, saturation vapour pressure over a curved surface is higher than over a flat surface. Thus, in many cases gases condense onto the surfaces of existing particles, increasing their mass, but not the total number of particles. This is **heterogeneous nucleation**.

Homogeneous nucleation is the process by which new particles are formed *ab initio*. It may involve just one molecular species, such as sulphuric acid, two gases such as sulphuric acid and water, or three gases such as sulphuric acid, water, and ammonia. Saturation vapour pressure over a particle of radius 1 nm is many times larger than over a flat surface, posing some challenging questions, so that homogeneous nucleation is still a subject of intense research. One interesting concept is that cosmic rays may ionise some molecules, leading to much stronger attractive forces with other molecules.

4.3.1 Sulphates

In Chapter 2 we discussed the major reactions of the various sulphur gases that enter the atmosphere – e.g. DMS, H_2S, OCS, CS_2, SO_2. (See also Table 3.7.) The DMS and H_2S are oxidised to SO_2, and then oxidised to SO_4^{2-} as discussed in Section 2.2. This may then combine with any ammonia in the surrounding atmosphere to form salts such as NH_4HSO_4, $(NH_4)_2SO_4$, and metal sulphates and bisulphates. Condensational growth onto pre-existing particles, possibly involving displacement reactions on the surface, is also possible. Globally averaged, the dominant sink of SO_2 is aqueous-phase reactions in cloud droplets (many of which subsequently re-evaporate, rather than precipitate).

The burning of fossil fuels has led to a very significant increase in SO_2 levels over the past half century, with a corresponding increase in the global and regional levels of sulphate aerosols. As a consequence, they were one of the first aerosol types to be extensively studied, including their sources, properties, and effects. One estimate is that 72% of sulphate aerosol arises from SO_2 emissions from fossil fuel burning, 19% from DMS emissions, 7% from volcanoes, and 2% from biomass burning: but with distinct hemispheric differences, as might be expected. Estimates of global anthropogenic emissions in the 1990s are ~70–90 TgS/year, and total emissions of ~90–125 TgS/year.

4.3.2 Nitrates

Nitrogen exists in the atmosphere in a wide variety of reactive and unreactive species, as we saw in Chapter 2 (and Section 3.4). During daylight hours, NO, NO_2, and O_3 are involved in a tight photochemical dance, although the steps involved can become more complex in heavily polluted environments. However, at night, one of the key reactions switches off, and this can allow more oxidised species to build up, including highly reactive N_2O_5, which reacts with water:

$$N_2O_5 + H_2O \rightarrow 2HNO_3 \tag{4.1}$$

This reaction is particularly efficient in the aqueous phase, in much the same manner as SO_2. NH_4NO_3 forms if H_2SO_4 is fully neutralised and there is excess ammonia. Thus, this species is expected to increase if efforts to reduce SO_2 emissions are successful.

4.3.3 Organics

While both sulphate and nitrate aerosols (secondary inorganic aerosols) are relatively simple chemical mixtures, the opposite is true of **secondary organic aerosols** (SOAs), which are a chemical soup. They are currently one of the most active areas of research in this field. Because of their complexity, much of our knowledge of SOAs has come from smog chamber experiments.

SOAs are formed after the photodegradation of volatile organic compounds (VOCs) in the atmosphere. VOCs can be emitted by vegetation (biogenic emissions such as terpenes), or from anthropogenic sources such as motor vehicles or certain industries: more SOAs are produced from biogenic sources than from anthropogenic sources. In an urban airshed, carbonaceous aerosols can account for ~50% of the average $PM_{2.5}$ mass, and up to 90% of the total organic aerosol mass is estimated to consist of SOA.

4.3.4 Survey

Table 4.2 gives one set of estimates for the emissions of the various aerosol types discussed above. (NMHC = non-methane hydrocarbons.) While these estimates – and they are just estimates – are quite old, there are some valuable lessons that can be gleaned from this data set. Perhaps the most interesting numbers are those at the bottom of three of the columns.

The first column gives emissions in Tg per year, and we see that the anthropogenic fraction is only ~10%. The second column gives the average lifetime in days. For primary particles this is ~4 days, with the exception of sea salt, much of which is exchanged rapidly with the ocean surface. By contrast, for anthropogenic aerosols (which are mostly secondary) the values are almost always longer. By combining data from the first two columns (plus the surface area of the Earth), we arrive at the third column, which is the 'column burden' – the mass of aerosol (in mg) above each square metre of surface: above your head. Because the anthropogenic aerosols have, in general, longer lifetimes than natural aerosols, we find that the anthropogenic fraction of the column burden is now ~20%: double the value for emissions. In short, emissions alone are only half the story! We will discuss the last two columns in Section 4.7.

4.4   Stratospheric Aerosols

The lower stratosphere has an aerosol population, or (stratospheric) aerosol layer, but with some quite different properties from tropospheric aerosols. Because of its large separation from the ground, inputs of primary aerosols are negligible, and inputs of secondary aerosol precursors are also much less. This separation also affects removal processes, and hence residence times, giving a very different character to this layer. Particles with diameters less than 10 µm are largely unaffected by gravity, even in the lower stratosphere, with the result that residence times may be around a year or more.

4.4.1   Background Sulphate Layer

The dominant aerosol type in the stratosphere is sulphate. Large volcanic eruptions are responsible for periodic enhancement of this aerosol, but it also exists as a background aerosol during quiescent periods. The major non-volcanic source of sulphur to the stratosphere is thought to be carbonyl sulphide, OCS. This trace gas is one of the most stable in the

Table 4.2 **Major aerosol types and sources.**

Sources	Emissions, Tg/year	Lifetime, days	Column burden, mg/m^2	Extinction efficiency, m^2/g	Optical depth
Natural					
Primary					
Mineral	1,500	4	32.2	0.7	0.023
Sea salt	1,300	1	7.0	0.4	0.003
Biological	50	4	1.1	2.0	0.002
Volcanic	33	4	0.7	2.0	0.001
Secondary					
Sulphates from biogenic gases	90	5	2.4	5.1	0.013
Sulphates from volcanic SO$_2$	12	5	0.3	5.1	0.001
Organics from biogenic NMHC	55	7	2.1	5.1	0.011
Nitrates from NO$_x$	22	4	0.5	2.0	0.001
Total natural	3,060		46		0.055
Anthropogenic					
Primary					
Industrial dust	100	4	2.1	2.0	0.004
Black carbon (soot, etc.)	20	6	0.6	10.0	0.006
Secondary					
Sulphates from SO$_2$	140	5	3.8	5.1	0.019
Biomass burning (not black carbon)	80	8	3.4	8.0	0.028
Nitrates from NO$_x$	36	4	0.8	2.0	0.002
Organics from anthrop. NMHC	10	7	0.4	5.1	0.002
Total anthropogenic	390		11.1		0.061
Totals	3,450		57		0.115
Anthropogenic fraction	11%		19%		53%

Source: IPCC IS92, table 3.6.

troposphere, with a lifetime of a few years, allowing some to make its way to the stratosphere. There it may be photolysed by UV, or react with free oxygen atoms (a feature of the ozone layer), to release reactive sulphur. This is then oxidised to H$_2$SO$_4$(g), which nucleates to form sulphuric acid–water droplets, with an average diameter of 0.14 µm. This is known as the Junge layer, after Christian Junge (1912–1996), whose 1963 book *Air Chemistry and Radioactivity* was seminal.

4.4.2 Volcanic Sulphate Aerosols

More than 500 volcanoes are currently rated as active across the globe. A large eruption ejects massive amounts of ash and dust, as well as a mixture of gases, including N$_2$, CO$_2$, SO$_2$, H$_2$S, HCl, and OCS. While the ash and dust particles are too large to remain airborne for more than a few days, the gases may rise as far as the stratosphere. The sulphur gases are oxidised to H$_2$SO$_4$(g), which again nucleates, to form a layer that may spread globally.

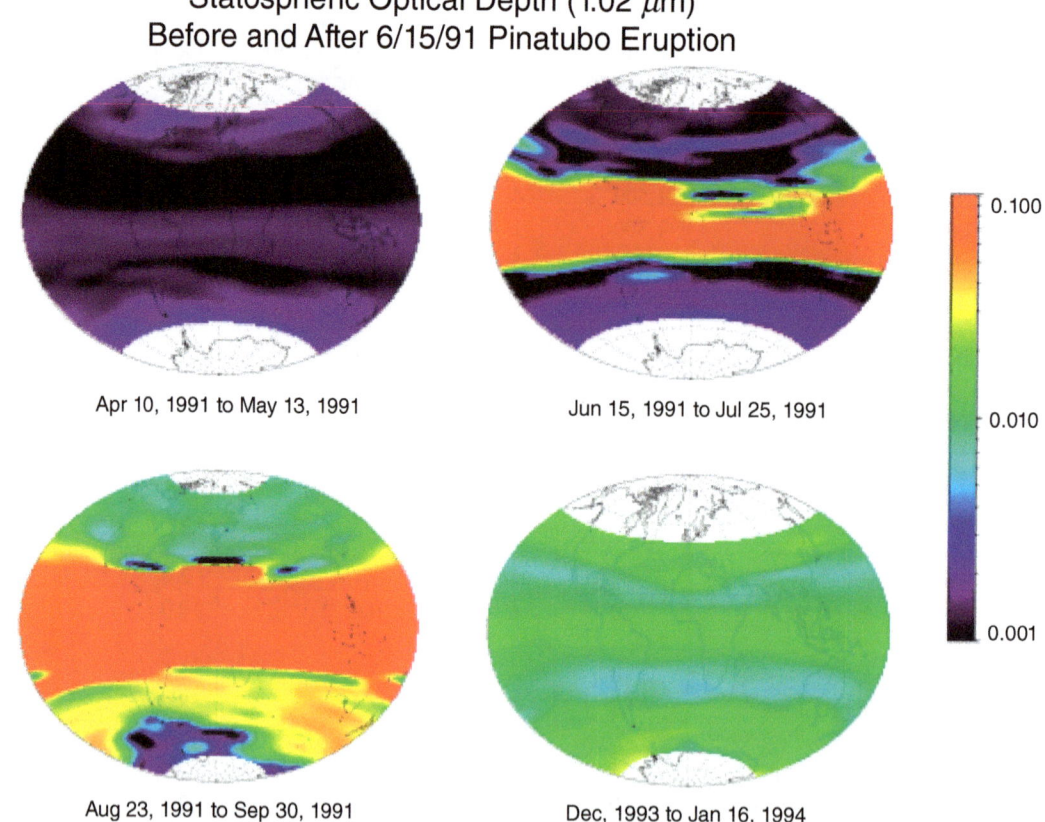

Figure 4.1 Evolution of the stratospheric aerosol loading following the eruption of Mt Pinatubo. *Source*: courtesy of M. Patrick McCormick, Hampton University's Center for Atmospheric Science.

Following the eruption of Mt Agung on the island of Bali in March 1963, there was little major activity, and a relatively 'clean' stratosphere, until El Chichon in March–April 1982, and Mt Pinatubo in June 1991. Pinatubo's effects have been extensively studied using both satellite and ground-based observations. This is beautifully illustrated by Figure 4.1, based on data from the SAM II and SAGE II satellite instruments (Chapter 12). The units involved are optical thickness (measured at a wavelength of 1.02 μm; see below), a measure of total stratospheric (aerosol) loading.

The first panel shows the situation averaged over the period 10 April–13 May 1991; a very 'clean' stratosphere. Mt Pinatubo erupted on 15 June, sending ~20 MT of SO_2 into the stratosphere, which produced an aerosol layer of ~30 MT. Panel 2 shows the situation averaged between 15 June and 25 July: the layer had spread around the tropics, >100 times the pre-eruption values. Panel 3 shows that by 23 August–30 September, the layer had spread globally (lowering global temperatures by around 0.5°C). It was just a few kilometres thick, initially centred at an altitude of about 23 km, and slowly descended over time. Finally, panel 4 shows that, even two and a half years after the eruption, the aerosol loading was still ~10 times the pre-eruption value.

4.4.3 Polar Stratospheric Clouds

The stratosphere over Antarctica in winter is a unique place, as it is isolated from the rest of the atmosphere by the southern Polar Vortex. Temperatures there are much colder than in most of the stratosphere, which is critical in the formation of two types of polar stratospheric clouds, or PSCs. (PSCs were first observed by the SAM II instrument just discussed.) Type I PSCs condense at $-78°C$, and consist of mixed solid and liquid nitric acid trihydrate $[HNO_3(H_2O)_3]$, or NAT, plus some water and sulphuric acid. These particles have typical sizes of ~1 μm, and thus have very low fall speeds (~10 m per day). Type II PSCs form near $-85°C$ and are a mixture of water ice and a little dissolved nitric acid. These particles are ~10 μm, and hence have much higher fall speeds (1–2 km per day). The sinking of PSCs, and especially Type II, removes both water and nitrogen from the stratosphere, referred to as dehydration and denitrification. The role of PSCs in the Antarctic ozone hole will be discussed in Chapter 10.

4.5   Aerosols in the Atmosphere

After aerosols have entered the atmosphere, they are subjected to further physical and chemical processing, with the uptake of water being one very important example. All of these processes affect the aerosols themselves, individually and collectively, creating the aerosol populations observed in situ.

4.5.1 Atmospheric Processing

Particles may combine due to coagulation, and grow due to either the condensation of vapours, or chemical reactions. If they are incorporated into a cloud droplet, they may undergo additional transformations. Coagulation occurs when two particles collide and coalesce. Atmospheric turbulence may enhance the chances of particle collisions. In addition, a heavier particle will fall faster than a lighter particle, and may overtake and capture the lighter one. This is gravitational collection, and is an important process in the formation of raindrops, as we will see in Section 6.3.

When a condensed phase (particle or droplet) occurs in close proximity to the gas phase of a substance, there will be a constant exchange of molecules between the two. In equilibrium, the rates of condensation and evaporation will be equal, and the partial pressure of the gas near the surface will equal the gas's saturation vapour pressure. If that is not the case, gas molecules will either diffuse towards, or away from, the particle/droplet. In the case of a plume source, dilution of the concentrations as the plume disperses may actually allow some volatiles to re-evaporate.

Sulphuric acid gas has a low saturation vapour pressure, and readily condenses onto particles: this is mostly a one-way process. Condensation is mainly onto accumulation mode particles, as they generally have the highest total surface area per volume of air. High-molecular-weight organic species may also condense. Some of these are the result of (photo) chemical reactions that take place during the course of the day, and may take several hours to reach a critical point (a process studied in smog chambers).

The most abundant condensable gas is invariably water vapour, and water is usually a significant component of many aerosols, mostly chemically unbound. At low relative humidity

(RH), inorganic (salt) aerosols are solid. As RH increases, they will remain so until a threshold value is reached – the **Deliquescence Relative Humidity** (DRH) – at which the particle spontaneously absorbs water, producing a saturated aqueous solution. On the other hand, as RH decreases, water evaporates. This process may continue well below the DRH before recrystallisation occurs – a form of hysteresis.

Dissolution is the process whereby a gas in the vicinity of a particle or droplet diffuses to and dissolves in liquid on the surface. Liquid water is the most common solvent. In solution, dissolved molecules may dissociate. The three major acid species – $HCl(g)$, $HNO_3(g)$, and $H_2SO_4(g)$ – are all examples of such gases, which dissolve and dissociate.

The other common gas that may be said to dissociate is ammonia: once dissolved

$$NH_3(aq) + H_2O \leftrightarrow NH_4^+ + OH^- \tag{4.2}$$

The presence of these species in both the gas and aqueous phases may lead to reactions, as the acids look for (and perhaps compete for) opportunities to neutralise. One example of this competition is the chloride displacement reaction

$$H_2SO_4 + 2NaCl \rightarrow Na_2SO_4 + 2HCl \tag{4.3}$$

This is a very common reaction in the atmosphere whenever both species are present. As a result, an aerosol population which might be considered to be marine in origin will often be found to contain 'excess' sulphate: known as **non-sea-salt sulphate**.

4.5.2   Aerosol Modelling

Unlike gas molecules, aerosols come in an enormous variety of shapes, sizes, and compositions, which, as we have just seen, are subject to a variety of processes while they remain in the atmosphere. How do we get a handle on such a heterogeneous mess? The only viable approach is to construct models that, of necessity, simplify many of the (hopefully) less important details. Central to most models is a size distribution, coupled to a suitable set of chemical composition assumptions. Such models provide an economical summary of the aerosol in a region, and may also be used in chemical transport modelling for air pollution and other studies.

Aerosol particles come in a wide range of sizes, potentially covering three or more orders of magnitude in radius, assuming we can consider the particles to be more or less spherical. Any plot of particle numbers versus size would inevitably show a rapid fall-off as a function of radius. (A large dust storm is one exception.) For this reason, size distributions are presented in a number of ways. For example, a plot of particle surface area versus radius is usually more informative than particle number versus radius.

A number of authors have produced models of the different types of aerosols. Such models usually start with a mathematical size distribution, plus some model of the chemical composition from which a refractive index may be determined. This is necessary for the next step, which is to compute the scattering and absorption properties for the aerosols, which are important for their optical/radiative processes (Section 4.8).

The mathematical models used to represent the size distribution range from the very simple, to the relatively sophisticated. One common model is known as the lognormal distribution, which allows users to adjust the magnitude, mode radius, and width of the distribution: sometimes two such models are fitted to measured data, one each for the fine and coarse modes. However, such precision can be hard to justify.

At the other extreme is the power-law fall-off, known as the **Junge distribution**:

$$n(r) = cr^{-\nu} \tag{4.4}$$

For large values of the exponent ν the distribution is dominated by small particles; for small values of ν it is dominated by large particles.

4.5.3 Transport and Removal

Aerosols move under the influence of gravity, but it is air resistance, and turbulent air motions, that are invariably dominant. To give some perspective, a 1-μm diameter particle would require almost 1 year to fall 1 km in still air; a 10-μm particle would require a few days. Thus, we might expect particles up to at least this size to be essentially 'airborne': as we move to progressively larger sizes, gravitation steadily takes over.

Once formed/injected, aerosols are transported by the airflows they experience during the time they remain in the atmosphere. If they are caught in the passage of a front, they are likely to be pushed to somewhat higher altitudes, above the boundary layer (the lowest ~1 km). Saharan dust is regularly detected in the Americas, and Gobi Desert dust is measured on the west coast of the United States. Despite the examples just cited, primary particles are normally transported shorter distances than secondary particles.

Aerosol particles are removed from the atmosphere by sedimentation, dry deposition, and wet deposition. **Sedimentation** – falling under the effects of gravity – is very slow for particles smaller than about 20 μm. However, this does provide some sort of 'upper limit' to the sizes of aerosol particles, and the number of particles above this size is small, and drops off rapidly with increasing size. **Dry deposition** is the process whereby particles are carried by molecular diffusion or turbulent motion to trees, buildings, car windows, grass, or the ocean surface. There they rest on and adhere to (or perhaps react with) these surfaces.

Wet deposition, or rainout, is the process whereby aerosol particles are incorporated into raindrops which precipitate. One aerosol particle will have formed the initial 'seed' for the cloud droplet, and more may be collected by the drop through coagulation processes, or by **scavenging**. Wet deposition probably accounts for over 80% (by mass) of aerosol removal from the atmosphere. Visibility often improves after a rain shower due to the scavenging of much of the aerosol in the area.

4.6 Aerosol Effects

Aerosols are one of the most complex players in the entire Earth System. We may classify the effects and interactions of aerosols in (and on) the environment into four major areas.

1. Atmospheric chemistry. Aerosol particles are complicated mixtures of chemical species, in contact with their surroundings. Chemical reactions take place on their surface, changing

the nature of both condensed and gas-phase chemistry. Aerosols provide a sink for many species when they are removed from the atmosphere. However, some aerosol types, such as sea salt, actually provide a source to the atmosphere (e.g. HCl).

2. Cloud droplet formation. Homogeneous nucleation of water vapour cannot take place under normal conditions, and an aerosol 'seed' is needed. This process will be discussed in some detail in Section 6.3. Wet deposition is the main mechanism for removing aerosols, and with them a range of substances, from the atmosphere.

3. Interaction with electromagnetic radiation. These effects are clearly relevant to the subject of climate change, and will be developed in the next section.

4. **Human health**. This is not really within the realm of physics, but rather pathology. Thus, Air Quality Standards include a component on 'particulate matter'.

4.6.1 Aerosols and Air Quality

Aerosol particles may contain a variety of organic and inorganic substances, including some which may be harmful to health. Some hazardous organic substances include benzene, polychlorinated biphenyls, and polycyclic aromatic hydrocarbons. Hazardous inorganic substances include metals and sulphur compounds. Metals cause lung injury, bronchiorestriction, and increased incidence of infection.

One global concern is the health hazard linked to the indoor burning of biomass, animal waste and coal: a reality for large sections of the world's population. The World Health Organization estimates that, of the 2.7 million people who die each year from air pollution, 1.8 million die in rural areas where the largest source of mortality is the indoor burning of biomass and coal. According to the WHO, more than 3 billion people rely on solid fuels, including coal, biomass, and animal waste, for cooking and heating. As a result, the 'World Health Report 2002' stated that 2.7% of the global disease burden and 4.6 million deaths per year can be attributed to indoor air pollution.

Stepping outdoors, studies in the 1970s found a link between cardiopulmonary disease and high levels of aerosols and sulphur gases. Later studies even found a link between low concentrations of aerosols and ill health, including that short-term increases of 10 μg m^{-3} of particulate matter were associated with ~1% increase in daily mortality, higher hospitalisation and health-care visits for respiratory and cardiovascular disease, and enhanced outbreaks of asthma and coughing.

The developed world still has a strong dependence on fossil fuels, especially for road transportation. The WHO and the World Bank reported that, worldwide, more deaths are attributable to air pollution than to motor vehicle accidents. In June 2012, the Agency for Research on Cancer, part of the WHO, classified diesel exhaust as carcinogenic, based on sufficient evidence that exposure is associated with an increase in the risk of lung cancer.

Small particles can pass through the nasal passages, and lodge in the lung alveoli, the deepest part of the lung where oxygen exchange with the blood supply takes place. Coal miners may develop black lung disease over many years of exposure to coal dust, which builds up in the lungs, making breathing difficult. In the United States, black lung kills an average of 2,000 coal workers per year.

The first public health/Air Quality Standards focused on PM_{10}. Some studies found that there may be no threshold below which such particulates don't contribute to health problems. The problem is that most PM_{10} is not particularly hazardous, and the damage is primarily caused by the smaller particles. The current US national ambient Air Quality Standards (NAAQS; Table 2.4) has targets for both $PM_{2.5}$ and PM_{10}; Australia has recently introduced a $PM_{2.5}$ standard.

Some experts are now suggesting that we should focus even more finely, on PM_1. This may well be desirable, but it needs to be remembered that associated with any regulatory standard is an appropriate measurement protocol. As we move to ever finer sizes, it becomes progressively more difficult, and expensive, to produce instruments that will supply the relevant data, with field-ready reliability and minimal hands-on involvement.

4.7 Aerosol Optics

Aerosols are now recognised as one of the key players in our changing climate, mainly via their ability to scatter, and at times absorb, solar radiation. In this section we will outline some of the key aspects of how these interactions occur, and the key variables in aerosol optics.

Virtually any object, from an atom to a planet, will scatter and absorb electromagnetic radiation, with the details depending strongly on both the wavelength of the radiation, and the size and refractive index of the scatterer. Aerosol particles are an important contributor when trying to understand all of the interactions of solar radiation within our atmosphere.

4.7.1 Optical Parameters

The interaction between a particle and electromagnetic radiation is characterised by a number of parameters, the most important being its **scattering** and **absorption cross-sections**, σ_s and σ_a, a measure of how much radiation they scatter, or absorb. (Both have units of area, and are a measure of how big the target 'appears' to the radiation.) **Extinction** is the sum of scattering plus absorption: $\sigma_e = \sigma_s + \sigma_a$. **Single scattering albedo** is scattering divided by extinction; that is, the fraction of the 'intercepted light' that is scattered into new directions: $\varpi = \sigma_s/\sigma_e$. The **phase function** describes the angular distribution of any scattered radiation.

For a spherical particle of uniform composition – a 'reasonable' assumption for many, but not all, aerosol particles – these parameters may be computed relatively easily using Mie theory. Large mineral dust particles are more of a challenge!

Without going into unnecessary details, we do need to say a little about these optical properties. For many types of aerosol, absorption is quite small, so that the single scattering albedo is close to 1.0, as is often assumed. The exception, of course, is soot – black carbon – where it is absorption that dominates. Mineral dust particles are also reasonably absorbing, depending on their composition. Australia's iron-rich mineral dust absorbs more than Saharan dust, for example.

How do these properties vary with particle size? In fact, it is usually the ratio of particle size to the wavelength of the incident radiation that is the defining parameter. For particles that are large compared with the wavelength, the extinction cross-section is close to twice the geometric cross-section, πr^2. The unexpected factor of 2 is a quirk of how a wave is scattered by, and around, an object (Babinet's principle).

For particles much smaller than the wavelength, the extinction cross-section becomes very small, as the wave can, effectively, bend around the particle. The molecules in our atmosphere are a good example, and our atmosphere is largely transparent to solar radiation (provided we may ignore any absorbing gases). There is one exception to this rule, and that is for highly absorbing materials like soot. Even for particles just a tenth the wavelength, their absorption cross-section is relatively large (and their single scattering albedo close to 0.0).

That leaves an in-between region: particles with sizes more or less the same as the wavelength. These particles have scattering cross-sections that are typically four times their geometric cross-section, making them very efficient scatterers. For many practical purposes the most useful parameter is the **mass extinction efficiency**, which is simply the extinction cross-section divided by the mass. For example, a large mass extinction efficiency would be a good property of a smoke screen material, as it would mean maximum screening for the mass of stuff (bang for the buck, so to speak).

For solar (visible) radiation, with wavelengths in the range ~350–750 nm, accumulation mode aerosols are highly efficient scatterers – which usually means secondary aerosols. Coarse mode aerosols invariably have lower efficiencies. (For terrestrial radiation with wavelengths in the range ~5–50 μm, coarse mode aerosols may have 'reasonable' efficiencies, while the smaller modes are largely invisible.)

4.7.2 Optics of Populations

Aerosols come in a wide range of sizes, so the optical properties – and effects – of an aerosol population will reflect the contributions of all particles in that population. Hence, we must add all their contributions, which effectively means integrating over the size distribution. Focusing just on extinction, and remembering that the cross-section depends on both particle radius, r, and radiation wavelength, λ, we may formally write

$$\tau(\lambda) = \int \sigma_e(r, \lambda) n(r) dr \qquad (4.5)$$

The result of this integration is known as the optical thickness (Equation (4.6)), if we assume that $n(r)$ is the size distribution for the entire atmospheric column. (We may also talk about the optical thickness of a defined layer of atmosphere.) Optical thickness is a measure of the attenuation of a light beam as it traverses a medium, such as an atmospheric column (assuming overhead sun):

$$I = I_0 e^{-\tau} \qquad (4.6)$$

In general, Equation (4.5) can only be evaluated numerically, using Mie theory. However, if we assume the Junge distribution introduced above, the calculation simplifies, at least to some extent, and we may show that

$$\tau(\lambda) = C\lambda^{-\alpha} \qquad (4.7)$$

where $\alpha = \nu - 3$ is known as the **Angstrom coefficient**. (The coefficient C still requires numerical calculation, but that is not the key issue here.)

This simple result is quite valuable. It tells us that if we can make a series of measurements of optical thickness at a range of wavelengths, and then plot them in log–log form, the slope of the graph tells as a great deal about the aerosol size distribution: if α is large, the aerosol population is dominated by small particles, and vice versa. Of course, if the data do not reasonably approximate a straight line on such a graph, it tells us that the Junge distribution is not so useful in that particular situation.

Table 4.2 contains a column giving the mass extinction efficiency (at a mid-visible wavelength of 550 nm) for each of the major aerosol types, assuming a relative humidity of 70–80% (with some minor adjustments). We see that for primary aerosols the values are mostly ~1–2 m^2/g, while for secondary aerosols the values are mostly equal to, or greater than, 5 m^2/g. The stand-out exception to this is black carbon, as we hinted at above.

The final column is the optical thickness (at 550 nm), and is obtained by simply multiplying the mass extinction efficiency by the column burden. At the bottom of this column, we see that the anthropogenic fraction has now jumped up to ~50%: a direct consequence of, firstly, the longer lifetimes of the anthropogenic contribution, and secondly, their higher mass extinction efficiencies.

4.7.3 Implications

So we see that about half of all light scattering by aerosols is due to the anthropogenic contribution: what are the consequences of this? One consequence is for visibility, as high aerosol loadings (whether natural or anthropogenic) can impair visibility. Much of the aerosol layer is below ~1 km, potentially within our 'line of sight'. The US NAAQS secondary standards are designed, among other aims, to 'protect against decreased visibility', particularly in areas of scenic beauty such as the Grand Canyon.

Friends of ours recently toured India, and took the photo of the Taj Mahal in Figure 4.2. Compare it with any photo you can find online: most will show blue skies. If you do a broader search, we're sure you can find even 'worse' examples.

Most of the radiation intercepted by aerosol particles is scattered into new directions, as governed by the phase function. While most of this radiation is scattered in the (near) forward direction, that is, down towards the surface (within, say, 20° of the incident beam direction) – the phase function is said to be strongly 'forward-peaked' – roughly 15% is scattered in backward directions, and hence back to space. (If you are able to look close to the sun on a clear day you will see a whitish halo, known as the solar aureole; the result of aerosol scattering. Measurements of this scattered light are another method for learning about the aerosol size distribution.)

An increase in aerosol loadings would have the effect of increasing the planetary albedo, and hence cooling the Earth: this is known as the **aerosol direct effect**. In a 'steady state' situation this is not an issue, as this is just one component of the albedo. However, as anthropogenic emissions of certain aerosol types (or their precursor gases such as SO_2) have been increasing in

Figure 4.2 Photo of the Taj Mahal on a typical smoggy day. © Sarah Fitzherbert 2017.

the past, this represents a (negative) **radiative forcing**. The IPCC uses the term **aerosol–radiation interaction.**

This may appear to be a 'good thing', as it is (partly) counteracting the effects of the increases in greenhouse gases. However, such a view may be naive. An alternative view is that this aerosol cooling may have been masking some of the greenhouse warming from us – potentially deceiving us. Should anthropogenic aerosol emissions be reduced over the coming decades, and there are many reasons to believe this is likely (including those discussed above), we may suddenly find ourselves exposed to the full force of the greenhouse gas increases. Clearly it is vital to quantify the radiative effects of both aerosols and gases, especially when their concentrations are changing, in order to understand the size of the masking. These vital issues will be addressed in some detail in Chapter 17.

As global warming has steadily worsened in recent years, many remedies have been suggested, with a rapid reduction in greenhouse gas emissions being the most obvious, and clearly most desirable. Unfortunately, that does not seem to be happening, at least not at a rate to give anyone confidence that many harmful effects can be avoided. One 'get out of jail card' which has been suggested is the deliberate injection of aerosols into the stratosphere, where their lifetimes are around a year, in order to increase the planetary albedo. This is an example of solar radiation management, one of the two key thrusts of **geoengineering**, and will be discussed in more detail in Section 9.4.

4.8   Investigating Aerosols

Research into the formation and properties of aerosols, and especially their spatial and temporal variability, is one of the main areas of *experimental* research in the whole of climate science. We therefore believe that a few pages should be devoted to this topic. We will start with a quick look at some of the most important instrumentation that may be deployed in various investigations, and then discuss some of the fundamental laboratory research being done. Further subsections will cover studies of aerosols 'in the field' where they affect our environment; local, regional, and beyond.

4.8.1   Aerosol Instrumentation

A differential mobility analyser (DMA) uses electrical mobility to size/count particles. The particles are electrically charged, then passed between two electrically charged plates of a certain length, which produce an electric field in between. This acts on the particles, producing an acceleration proportional to their charge-to-mass ratio. Those with low inertia (i.e. mass) will hit the walls before escaping the region of the field, while those with higher inertia will escape. If the particles are all singly charged, this means that particles up to a certain mass will be collected, and those with greater mass will pass through.

DMA technology can also be used to select individual particles for further attention. For example, a low-humidity air stream may be passed through a DMA, and particles of a certain size selected. These may then be sent on to a chamber with pre-selected conditions such as relative humidity, or it may contain a vapour such as H_2SO_4 at a certain vapour pressure. In such an environment, many particles will grow, as discussed above, and their new size then determined by a second DMA. Such a set-up is referred to as a TDMA, or tandem DMA: Figure 4.3.

Inertial impactors are widely used for the size-selective collection of aerosol particles. One reason for this is that their principle of operation is simple: a jet of particle-laden air is directed at a flat impaction plate. Large particles are collected on the plate, while smaller particles follow the airflow out of the impaction region and are not collected. The particles in the impactor are classified by their aerodynamic diameter, an important parameter in many fields of study, including health. Finally, it is easy to collect particles in discrete size ranges by passing the air stream through a series of stages (a stage consists of a nozzle and an impaction plate), with each stage successively collecting particles smaller than the one before. Figure 4.4 shows the 12-stage micro orifice uniform deposition impactor (MOUDI) used by one of our graduate students during a field campaign in Central Australia.

4.8.2   Aerosol Scattering

Routine monitoring is also undertaken for air quality purposes, although the primary focus is usually just on concentrations, as required by regulations. A nephelometer is an instrument that draws in a stream of particle-laden air, and passes it in front of a laser beam. The small scattering volume is surrounded with light detectors that measure the scattering coefficient,

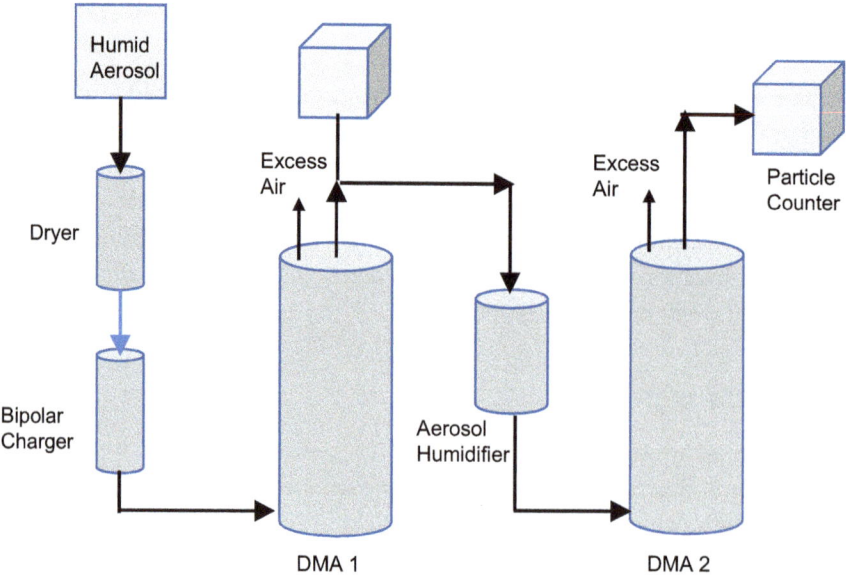

Figure 4.3 Schematic diagram of a tandem differential mobility analyser.

which has contributions from both molecules and aerosols. However, the molecular contribution is known, so the aerosol contribution may be easily obtained. The scattering coefficient is a secondary property of an aerosol population, but it is important for many reasons. One aspect of Air Quality Standards, as we have seen, is to protect visibility, especially in areas of high 'visual amenity'. For this reason, some Air Quality Standards specify aerosol scattering coefficient as a component.

4.8.3   Field Studies

Most routine field observations involve the deployment of sampling equipment at a site, and the return of samples for analysis. This might involve equipment that is permanently housed at a site, or a short collection campaign. Bulk samples, collected on substrates such as filters, may be subjected to a range of analysis techniques designed to provide information on the average chemistry of the samples. It is difficult, of course, to use such bulk data to draw conclusions as to the chemical associations involved, and all but impossible to draw any conclusions about the properties of individual particles. Nevertheless, bulk analysis provides much of the information required for a range of tasks. Some of the analysis techniques employed will now be outlined.

4.8.3.1   Analysis Techniques

Accelerator-based ion beam analysis (IBA) techniques are well suited to the fast, non-destructive, multi-element analysis of filters obtained from aerosol samplers. Such filters are

Figure 4.4 Our micro orifice uniform deposition impactor in the field (with permission from Majed Radhi).

sufficiently thin, as seen by a proton beam, that potential multiple interactions within the target may be ignored. Thus, any measured signal can be considered as proportional to the amount of 'target'. Four complementary measurement techniques are usually applied in IBA: PIXE (proton-induced X-ray emission), which studies the X-rays emitted from inner shell electrons; PIGE (proton-induced gamma-ray emission), which studies the gamma rays emitted following nuclear processes; PESA (proton elastic scattering); and Rutherford backscattering. Each is sensitive to elements in different regions of the Periodic Table (including some overlap).

4.8.3.2 Routine Monitoring

Ion beam techniques have been used in routine monitoring programmes at a number of locations across the world. In the United States, an interagency consortium of Federal Land Managers and the EPA established a national visibility and aerosol monitoring network in 1988, known as IMPROVE – Interagency Monitoring of Protected Visual Environments. While initially based in the west, it now has over 100 sites.

Figure 4.5 An Aerosol Sampling Program (ASP) sampler being checked by a technician. *Source*: see Acknowledgements at end of chapter.

In Australia, the Australian Nuclear Science and Technology Organization (ANSTO) has established the ASP (Aerosol Sampling Program), collecting and analysing biweekly $PM_{2.5}$ samples from a number of sites in New South Wales. Figure 4.5 shows a technician checking the sampler at Liverpool, a suburb of Sydney, which switches on and off to allow collection on Wednesday and Sunday. Figure 4.6 shows the IBA laboratory: the accelerator is at the rear: the sample holder and detectors are at the front.

ANSTO has also deployed the same instrumentation during the Asia–Pacific Aerosol Database project as part of the Regional Co-Operative Agreement of the International Atomic Energy Agency (IAEA). This involved a sampling and analysis campaign from 2000 to 2010 in Australia, Bangladesh, China, India, Indonesia, Korea, Malaysia, Mongolia, New Zealand, Pakistan, Philippines, Sri Lanka, Thailand, and Vietnam.

4.8.4 Fingerprinting

How might we make sense of the large volumes of data collected in programmes such as IMPROVE and ASP? Clearly it makes little sense to treat each measurement set independently. Instead, a statistical approach is called for. The database can be thought of as a matrix – a two-dimensional array – with the different measurement sets being one direction, and the 20+ elements being the other. The concentrations/measurements are then the entries in the array. There is software available that can analyse such a database and look for patterns: in this case combinations of elements, in distinct *concentration ratios*, which between them account for a large fraction of all the data in the measurement sets. These patterns are referred to as fingerprints.

It is then up to scientists to interpret these concentration ratios as representing particular aerosol types. For example, sodium and chlorine in the expected ratio would indicate sea salt. On the other hand, sodium with very little chlorine would suggest sea salt which has undergone the type of chemical processing mentioned above: reaction (4.3). Sulphur without sodium might imply industrial sources. Potassium plus black carbon is an indicator of smoke. The crustal elements silicon,

Figure 4.6 The Ion Beam Laboratory of the Australian Nuclear Science and Technology Organization, Menai, NSW 2234, Australia. *Source*: see Acknowledgements at end of chapter.

aluminium, iron, titanium, etc., are a clear indicator of soil/mineral dust. If all (or most) of the fingerprints can be interpreted in this way, we may refer to them as **sources**.

Once we have let our statistical software work its magic, the analysis of individual samples becomes straightforward: it can be decomposed into so much from this source, plus so much from that source, etc. (There will always be a small residual which cannot be accounted for; nothing is perfect.) This is referred to as source apportionment, and can prove invaluable for regulators, who can check whether polluted days are the result of a dust storm or fire, which are beyond our direct control, or industrial activity, which may require intervention. We may also observe seasonal cycles, such as a rise in the level of smoke in winter, coming from wood-fired heaters, or greater photochemical processing in summer.

ANSTO now has 20 years of data from its Liverpool site, and has been able to extract a total of nine fingerprints; these are shown in Figure 4.7 (BC is black carbon). Figure 4.8 shows their time series. A few comments on each are in order.

1. Soil. Dominated by crustal elements such as Al, Si, Ca, Ti, Mn, and Fe; low in BC. Its time series is clearly episodic, relating to dust storms, etc.
2. Secondary sulphate. Dominated by S, H, and BC. Its time series is partially cyclic, a reflection of stronger photochemistry in summer.

Figure 4.7 Composition of the ASP fingerprints. *Source*: see Acknowledgements at end of chapter.

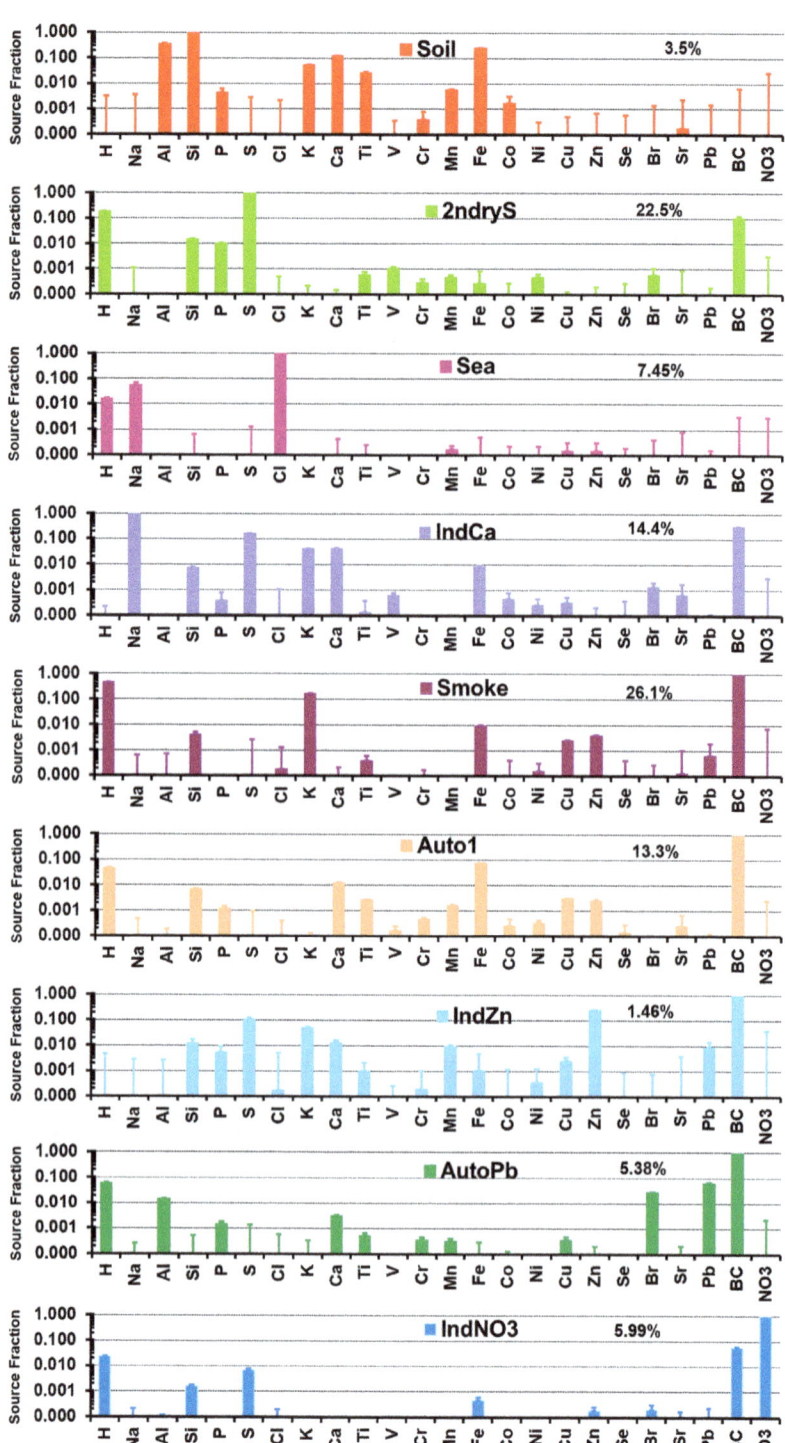

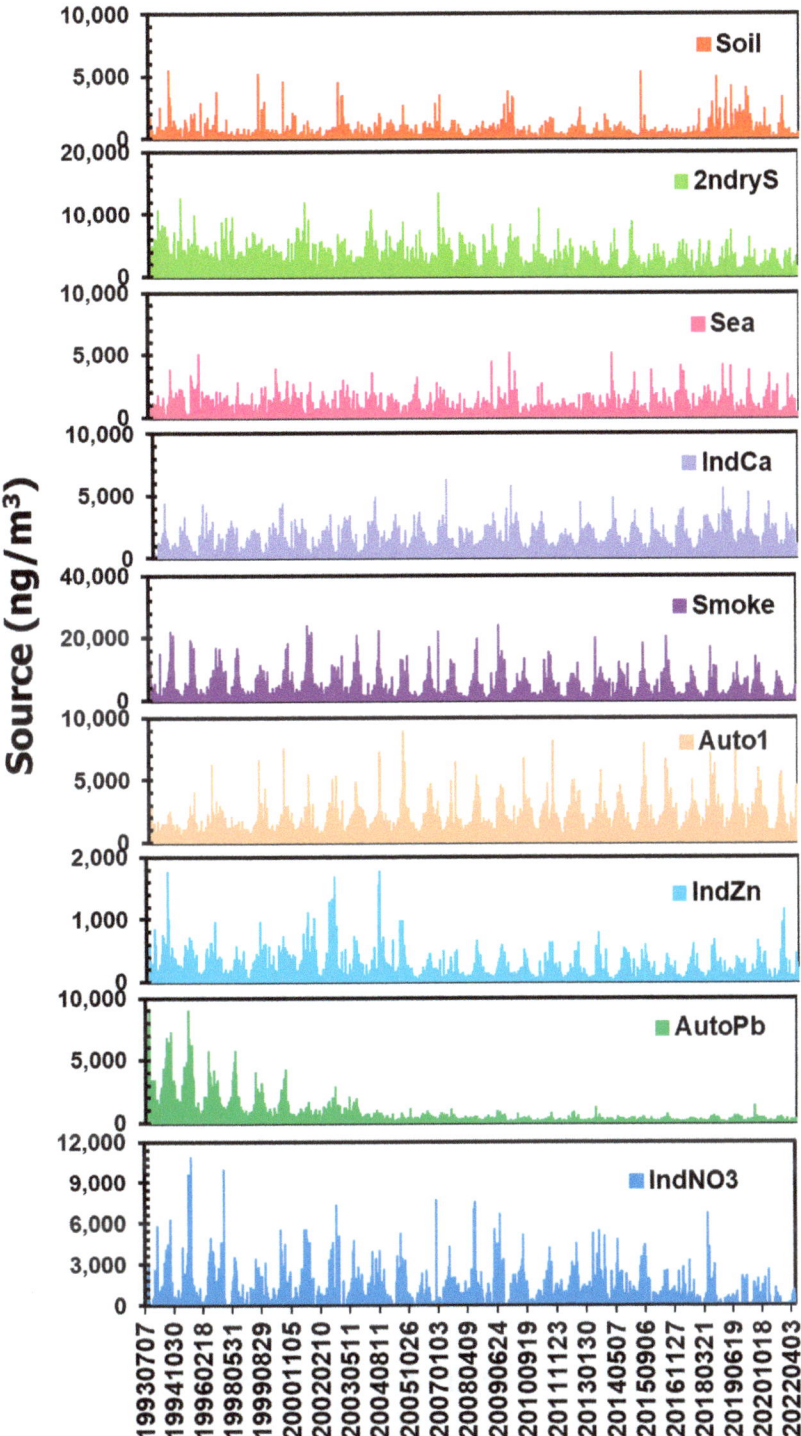

Figure 4.8 Time series of the ASP fingerprints. *Source*: see Acknowledgements at end of chapter.

3. Sea. Dominated by Na and Cl in the expected ratio. Driven by onshore winds, especially in summer.
4. IndCa. Significant levels of Ca from cement production, south of Sydney. Its time series will reflect the seasonal wind regime.
5. Smoke. High in H, K, and BC, plus soil elements, suggesting biomass burning. Time series again seasonal with a winter peak, as expected.
6. Auto1. Dominated by BC, Fe, and H. Time series again cyclical.
7. IndZn. High levels of Zn and other metals from metals manufacture and use.
8. AutoPb. Dominated by BC, Pb, Br, Zn, and H, indicative of older vehicles (and light aircraft) burning leaded petrol. A clear reduction over time.
9. IndNO3. Dominated by NO_3 from NH_4NO_3 used in fertiliser and explosives manufacturing (guard it carefully).

Summary

Aerosols, while clearly a component of atmospheric chemistry, are manifestly different from the gases we looked at in the previous two chapters, and the reasons for this should, by now, be clear.

Like Chapter 2, this chapter can be thought of as three parts. In the first part we looked at the nature – and more to the point, the variety – of aerosols, from their injection into the atmosphere, to their processing within it. The key takeaway for our readers here is the difference between primary and secondary aerosols. You can refer back at any time for more details.

In the second part we looked at the environmental effects of aerosols, with the most important effect being their scattering (and absorption) properties. It is for this reason that aerosols, and rising levels of certain aerosol types, are now recognised as important players in our environment, including climate change. Aerosols are also a pollution issue, and thus are a central component of Air Quality Standards.

In the final part we turned our attention to some of the current areas of research in this field, including inputs from scientists we have had the pleasure to work with (including, of course, our own graduate students). We then examined some of the key approaches to the study of aerosols 'in the field', especially the analysis of bulk samples, and how we make sense of the data.

Acknowledgements

Table 4.2:

From Jonas, P. R., R. J. Charlson and H. Rodhe. 1994: Aerosols. In: *Climate Change 1994 Radiative Forcing of Climate Change and An Evaluation of the IPCC IS92 Emission Scenarios* [J. T. Houghton, L. G. Meira Filho, J. Bruce, Hoesung Lee, B. A. Callander, E. Haites, N. Harris and K. Maskell (eds.)], Cambridge University Press, UK.

Figures 4.5, 4.6, 4.7, and 4.8:

Provided by Professor David Cohen, Distinguished Research Scientist, Centre for Accelerator Science, Australian Nuclear Science and Technology Organisation, Menai, NSW 2234, Australia.

FURTHER READING

Most books on atmospheric chemistry also discuss aerosols, so the three books listed at the end of Chapter 2 would be worth consulting. We can also recommend:

- *Atmospheric Aerosols* by S. Twomey (Elsevier Scientific Publishing Co., 1977).
- *Atmospheric Aerosols. Global Climatology and Radiative Characteristics* by G. A. d'Almeida, P. Koepke and E. P. Shettle (A. Deepak Publishing, 1991).
- *Optics of the Atmosphere* by E. J. McCartney (J. Wiley and Sons, 1976).
- *Atmospheric Aerosols: Characteristics and Radiative Effects* by S. Ramachandran (CRC Press, 2017).

For more information on the IMPROVE network, check out

- https://aqrc.ucdavis.edu/improve

For those who might wish to delve deeper into the peer-reviewed literature, we suggest:

- Synthesis of information on aerosol optical properties. H. Liu et al. *Journal of Geophysical Research*, vol 113, D07206, doi:10.1029/2007JD008735, 2008

REVIEW QUESTIONS

1. What are the two key minerals in mineral dust aerosol?
2. What are the main chemical elements found in sea salt?
3. What is the difference between homogeneous nucleation and heterogeneous nucleation?
4. What are the main sulphur gases that lead to aerosol formation in the atmosphere?
5. How is HCl 'released' into the atmosphere?
6. Which aerosols are most dangerous to health?
7. What sizes of particles are the most 'efficient' at scattering sunlight?
8. What is optically special about black carbon?
9. Explain, briefly, what TDMA means.

EXERCISES

1. Verify that the data in the third column of Table 4.2 can be obtained from the first two columns. Also confirm that the final column can be obtained from the third and fourth columns.
2. Try to find any information from your local air quality monitoring as to what types of aerosols are prevalent in your area. These are likely to vary considerably, as it will depend on where you live/study: urban/rural; coastal/inland; local industry. The material in this chapter should enable you to make sense of the data.
3. ACE-Asia was one of a number of International Field Campaigns. Prepare a report on the aims and outcomes of this project, starting with the reference An overview of ACE-Asia: Strategies for quantifying the relationships between Asian aerosols and their climatic impacts. B. Huebert, *Journal of Geophysical Research*, vol 108(D23). doi:10.1029/2003JD003550, 2003.

Part II

Physics

In Part I we looked at the chemistry of our atmosphere (with a brief excursion below the ocean surface). This is, of course, the air that we breathe, and its health is our health. At the same time, the atmosphere is a physical system, which must obey the relevant branches of physics: which, of course, means all of them, but in Part II we will focus on two.

The key branch of physics that deals with a gaseous body at rest, or at least whenever we can ignore bodily motion, is thermodynamics. The properties of a gas may be defined by its pressure, temperature, and density: for an atmosphere, their vertical profiles are important. Thermodynamics then tells us that these are not independent variables, but are linked by physical laws. Pressure decreases as we rise through the atmosphere, and another piece of physics tells us how rapidly. Finally, whenever energy (heat) may enter or leave an atmospheric segment, another thermodynamic result tells us the consequences. All of this is explored in Chapter 5.

Water, a substance both vital and remarkable, can exist in our atmosphere in all three phases. How much vapour may exist is a direct function of the (local) temperature. The evaporation and condensation of water vapour absorbs/emits large amounts of energy, much of which is the power source of storms. Clouds are collections of liquid droplets, but their formation is another of the remarkably complex stories that we need to explore if we are to understand our world. This is the subject matter of Chapter 6.

The second relevant branch of physics is fluid mechanics; how a fluid moves, and why: that is, in response to various forces and pressures. Developed fully, this is a wonderfully rich, but deeply mathematical subject, and certainly beyond the scope of this book. Perhaps you will study it in a later course within your degree. In Chapter 7 we will introduce you to the 'broad brush' picture of atmospheric motion, which is fundamentally driven by temperature gradients, which produce both an afternoon sea breeze, and also the global circulation of the atmosphere. It is this that dictates where the monsoon rains occur, and where deserts are found.

5   The Dry Atmosphere

We turn now to the second of the key branches of science that are vital to an understanding of climate: physics. There are two branches of physics which are important here. Thermodynamics is the science which governs most of the basic, or 'static' properties of a system such as a gas. Fluid mechanics is the science of a fluid – liquid or gas – in motion. In this chapter we will be focusing on thermodynamics of the dry atmosphere. Water vapour is sufficiently important to deserve a separate chapter.

A thermodynamic system is characterised by a small number of **state variables**; primarily density, pressure, and temperature. After defining these, we will see how they vary as we move up through the atmosphere. Two key equations – the hydrostatic equation and the ideal gas equation – interconnect these variables and their profiles. The First Law of Thermodynamics tells us how the atmosphere responds to an input of heat, as well as a change in pressure, which may come about with elevation (vertical motion). This is the key to the atmosphere's vertical stability, and (ultimately) to cloud formation.

5.1   Thermodynamic Variables

When allowed to come to equilibrium, a (small) system such as a gas will have well-defined values of density, pressure, and temperature. (Clearly the atmosphere as a whole is not a 'small' system!) We need to start by defining these variables, even though they may seem obvious.

Density is the simplest: it is mass per unit volume; $\rho = m/V$. There are times, however, when it will be more convenient to turn this 'on its head'. In 'laboratory conditions', fixed volumes are the norm; however, in the atmosphere there are no obvious boundaries. By contrast, there will be times when we will be tracking the progress (in some sense) of a given parcel of air (to be formally defined below), with a certain mass. If our parcel rises, for example, it will expand; its mass remaining fixed, but not its volume. In such circumstances the quantity known as the **specific volume** is more useful, defined by $\alpha = V/m$. (The adjective *specific* is attached to any variable that is of the per kg form; e.g. specific heat.)

Pressure is defined as force per unit area. A gas exerts a force on any surface, including the ceiling, or another 'layer' of gas. This force comes about by virtue of the collisions of its molecules with that surface. A wall exerts a force on the molecules, causing them to rebound, and via Newton's Third Law the molecules exert a force on the wall. The SI unit of pressure is the Pascal (Pa): 1 Pa = 1 Newton per square metre. (The Newton is the unit of force; an obvious historical choice.) The surface pressure of our atmosphere is defined to be 101,325 Pa: known as standard atmospheric pressure, or 1 atmosphere. For historical reasons, meteorologists have chosen to use the hectopascal (also known as the millibar) as their unit in discussions of the weather. Thus, standard atmospheric pressure is usually quoted as 1,013.25 hPa (or 1,013.25 mb).

Temperature is about our sensation of hot and cold, and how we then quantify this in some way. While it seems simple enough – just pick up a thermometer – a fully scientific definition of temperature is not so straightforward. For example, how do we know that the mercury in our thermometer really expands linearly with temperature before we have both a rigorous set of definitions, and an equally rigorous set of tools with which to calibrate such an instrument, or at least verify that mercury does expand as assumed?

To define a temperature scale, we need two fixed points, which are usually chosen to be the ice point, and the steam point, both at a pressure of 1 atmosphere. (The ice point is largely independent of pressure, but not so the steam point, as will be discussed in Chapter 6.) In the **Celsius scale** they are assigned the values of 0°C and 100°C. (In the Fahrenheit scale they are 32°F and 212°F.) In physics, the absolute, or **Kelvin scale**, is central, and is defined by

$$K = 273.15 + °C.$$

Note that the 'step size' of both scales is the same.

Temperature is central to the thermodynamic concept of equilibrium. If thermodynamic system A is in equilibrium with both system B and system C – meaning that they do not exchange heat – then system B will be in equilibrium with system C. We then say that they have the same temperature. For historic reasons, this is known as the **Zeroth Law of Thermodynamics**.

5.1.1 Air Parcel

The atmosphere is, of course, a continuum, with no boundaries. Nevertheless, it will often make sense to focus on a small segment, known as an air parcel. The following properties are assumed for an air parcel:

- It is small enough for T and P to be uniform, but large enough – i.e. containing enough molecules – for T and P to be statistically meaningful.
- It is thermally insulated from its surroundings: that is, heat transfer by molecular conduction, etc., is negligible (at least on timescales relevant to any situation we might be focusing on).
- It is always at the same pressure as its surroundings: it can expand or contract rapidly in response to any pressure difference.

Because there will be times when we want our air parcel to move, particularly vertically, the following conditions are usually also assumed:

- The parcel does not mix with its environment.
- If the parcel rises or falls, the surrounding environment remains undisturbed.

- Frictional forces between the parcel and its environment are negligible (so our parcel moves 'smoothly').
- The parcel always moves slowly, so that its kinetic energy (as a 'body') is small compared with its internal energy – the kinetic energy of its molecules. (You will learn below that the speeds at which molecules travel are far greater than any wind speed.)

5.1.2 Atmospheric Profiles

Pressure, density, and temperature are not constant in the atmosphere, with the key variation being with respect to altitude. The way they vary with altitude is referred to as their (vertical) profiles. While the details do vary somewhat in time and space (an important factor in the daily weather we experience), we can make the following general statements.

Pressure and density both fall steadily, dropping by a factor of about 10 for every 16 km of altitude. In mathematics this is an exponential fall-off, which is best displayed via a so-called log–linear plot: Figure 5.1 (note that the horizontal scale is in powers of 10). This figure also shows the mean free path between molecular collisions, which we will return to later. (Note the units of all three plots, chosen for ease of display on a single graph.)

By contrast, temperature usually drops more or less linearly, up to roughly 10 km, before steadying, then rising, up to 50 km. After that it again drops steadily, before rising again above ~100 km (Figure 5.2). Note that below 50 km, temperature generally lies between 210 K and 310 K (–63°C to +37°C); or 260 ± 50 K; a 20% variation. By contrast, both pressure and density fall by a factor of 1,000.

The rate at which temperature *decreases* with altitude, in degrees per km, is known as the **lapse rate**, Γ: i.e. $\Gamma = -\Delta T/\Delta z$. (Note that local variability, such as temperature inversions, frontal activity, converging air masses, and the effects of water vapour, can have significant effects on the lapse rate within the boundary layer; the lowest ~1 km or so of the atmosphere.)

Figure 5.2 is the key to the standard layer names of the atmosphere. The lowest layer, where the temperature is steadily dropping, is the **troposphere**, where our weather occurs. As we will discuss below, this layer is convectively well mixed. The 'lid' on the troposphere is known as the

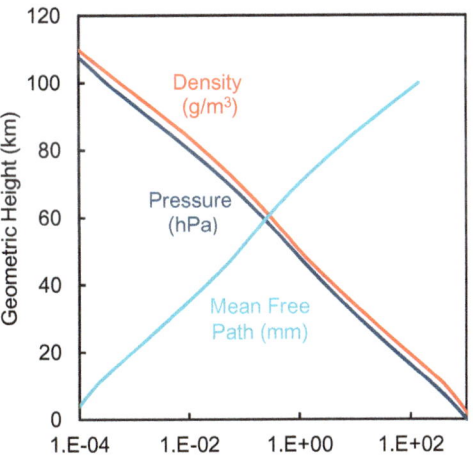

Figure 5.1 Vertical profiles of atmospheric pressure and density, and molecular mean free path (note the horizontal scales, and units for each). *Source: Physics of Radiation and Climate*, figure 1.6, Box and Box, CRC Press, 2016, reproduced by permission of Taylor & Francis Group.

Figure 5.2 Standard atmospheric temperature profile. *Source*: *Physics of Radiation and Climate*, figure 1.7 (modified), Box and Box, CRC Press, 2016, reproduced by permission of Taylor & Francis Group.

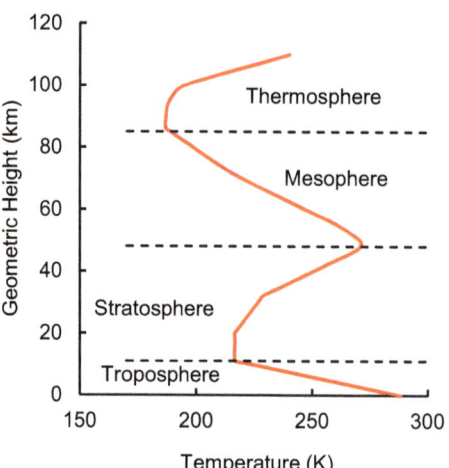

tropopause. Its height varies quite a lot: ~17 km in the tropics; ~11 km in the mid-latitudes; and ~8 km in polar regions. This variation is not uniform, but more step-wise, a result of the general circulation of the atmosphere, as will be discussed in Chapter 7.

Above the troposphere is the **stratosphere**. Its rising temperature profile can only be the result of an input of energy to that region. This comes about from the processes forming the ozone layer, which involve the absorption of solar UV radiation. (This will be discussed in Chapter 10.) The **stratopause** is quite uniform, at ~50 km, and a pressure of ~1 hPa. Above that is the **mesosphere**, and above that the **thermosphere.** Again, the rise in temperature in the latter is the result of the absorption of solar radiation, this time forming the ionosphere.

5.1.3 Hydrostatic Equilibrium

Pressure at any point/elevation in the atmosphere is a reflection of the weight of air above. Hence, pressure must always decrease with altitude. What can science tell us about the rate of decrease? Consider a thin layer of air, of density ρ and thickness Δz, with an area of 1 m^2. Then the mass of air in this layer is $m = \rho V = \rho \Delta z$, and its weight is $W = mg = \rho g \Delta z$. What stops this weight of air from falling to the ground? The answer is air pressure.

The pressure of the air below this layer, pushing up, will be some pressure, p. The pressure of the air above this layer, pushing down, will be given by $p + \Delta p$. (Δp must, of course, be negative: this way of expressing it is standard mathematical convention.) These two pressures, combined, must balance the weight of this layer of air (remember it has an area of 1 m^2, so that weight, which is a force, is now also pressure): Figure 5.3.

The pressure (force) from below must balance the weight plus the pressure from above:

$$p = (p + \Delta p) + \rho g \Delta z$$

Rearranging this we quickly find that

$$\frac{\Delta p}{\Delta z} = -\rho g = -\frac{g}{\alpha} \tag{5.1}$$

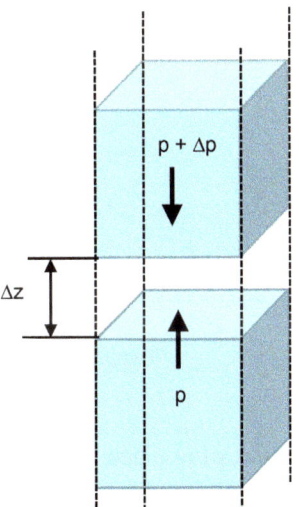

Figure 5.3 Schematic diagram to illustrate hydrostatic equilibrium. *Source*: *Physics of Radiation and Climate*, figure 2.1, Box and Box, CRC Press, 2016, reproduced by permission of Taylor & Francis Group.

This is known as the hydrostatic equation, and tells us the rate at which pressure falls with height. Finally, we may allow Δz to approach zero, and obtain the calculus version

$$\frac{dp}{dz} = -\rho g = -\frac{g}{\alpha} \qquad (5.1')$$

5.2   The Gas Law

Pressure, temperature, and density are not independent variables. Whenever we take a gas (in fact, any thermodynamic system), and confine it in such a way that it has certain values for density and temperature (for example), we find that it has a *unique* value for its pressure. The connection between these three variables is known by the general title the **Equation of State**. For most systems this is a very complicated relationship, and can often only be presented in tabular form, not as a mathematical expression.

5.2.1   Ideal Gas Equation

For gases, there are a number of mathematical forms for this equation, depending on density and/or temperature. Fortunately, our atmosphere has a sufficiently low density, even at ground level, that the simplest of these formulae suffices; namely, the so-called ideal gas equation:

$$p = \rho RT \text{ or } p\alpha = RT \qquad (5.2)$$

The gas constant, R, which appears in this equation, depends on the gas, or the mix of gases, which comprise the system. The composition of our atmosphere is sufficiently close to 'fixed' – with the exception of water vapour – that we may use a single value for R.

Figure 5.4 Illustration of Charles'
Law.

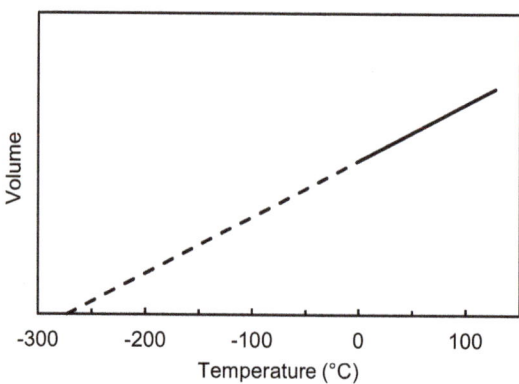

This equation occupies an interesting place in the history of science. In 1662 Robert Boyle (a contemporary of Newton) formulated his law: that pressure and volume are inversely proportional, whenever temperature is kept fixed. A century later, Charles (and others) formulated his law, which states that, for fixed pressure, the volume of a gas is proportional to its temperature, *provided* that temperature is measured via the absolute scale, defined above (Figure 5.4). Thus was absolute zero discovered. The ideal gas equation is the synthesis of these two.

Some readers may have encountered a slightly different version of this equation, used, for example, by chemists. If we replace density – i.e. mass per unit volume – by moles, or moles per unit volume, we may use a gas constant that is independent of the actual gas involved: referred to, logically, as the universal gas constant, $R* = 8,314.3$ J/kmol/K. This works well in the laboratory, but not so well in the atmosphere. For a gas of molecular weight M, $R = R*/M$. The average molecular weight of our atmosphere is just under 29 (four-fifths nitrogen, M = 28, and one-fifth oxygen, M = 32), so that the gas constant for dry air is 287 J/kg/K.

The one key compositional variable is, of course, water vapour. It also has a somewhat lower molecular weight (M = 18) than nitrogen and oxygen. This leaves us two choices. The more obvious one is to allow R to vary with the amount of water vapour in the part of the atmosphere we are currently focused on. The second is to 'tweak' the temperature in such a way that we may retain a single value of R whenever we make use of the ideal gas equation. The key is, we need to ensure that the product, RT (the right-hand side of Equation (5.2)), is correct. Because temperature is inherently variable throughout the atmosphere, and it would be nice to use a single value of R in our computer codes, for example, this is the way things are usually done. The resulting 'virtual temperature' rarely differs from the actual temperature by more than a degree or so: we shall ignore this technicality.

Technology has provided us with good instruments to measure both pressure and temperature in the atmosphere. By contrast, measuring density is much more of a challenge. For this reason, it is 'standard practice' to use the ideal gas equation to replace density in all equations in this field of science. An alternate version of the hydrostatic equation then becomes

$$\frac{dp}{dz} = -\frac{gp}{RT}$$

(5.3)

5.2.2 Constant Temperature Atmosphere

We now have two equations that connect our three variables of pressure, temperature, and density. If we had a third, we could 'solve' these, in combination, to derive the vertical profiles of these variables, from the surface on up. This is an exercise we have set our students in the past, and will set you at the end of the chapter, based on certain ad-hoc assumptions. One of the more realistic is the assumption made by the International Civil Aviation Organisation – namely, that temperature decreases linearly from its surface value, at a constant rate (i.e. a constant lapse rate), meaning $T = T_0 - \Gamma z$, at least up to aircraft cruising altitudes. (This one proved a challenge for some students, as the calculus is far from trivial.)

By far the simplest of the assumptions we might explore to provide a third equation is that of a constant temperature atmosphere: i.e. $T = $ const. This may seem crazy at first, because we are all very well aware that temperature decreases when we go up a mountain. However, when we remember that in the lowest 50 km of the atmosphere, temperature varies by only 50° (in fact, just 20% either side of the 'average' of 260 K), while both pressure and density vary by a factor of 1,000, maybe it's not quite so crazy after all? (Temperature changes of this magnitude can have significant impacts on the rates of chemical reactions, such as the metabolic processes which power our bodies, while having comparatively minor impacts on most physical processes.)

With this assumption, the hydrostatic equation (the version with density removed) can now be integrated directly, to give a simple exponential fall-off:

$$p = p_0 \, exp(-z/H) \tag{5.4}$$

where p_0 is the surface pressure, and $H = RT/g = 29.3T$ is known as the scale height. For typical atmospheric temperatures it is around 8 km. (You should check this result by differentiating with respect to z, to see that it gives you the hydrostatic equation, above.) As we have assumed that temperature is constant, the ideal gas equation tells us that density will have the same mathematical form as pressure.

This simple expression presents as a perfectly straight line on a log–linear graph, such as Figure 5.1. Note how close to straight are the graphs of pressure and density? That tells us that our 'crazy' assumption was far from crazy. The minor deviations above and below 'straight' are simply a reflection of the relative warmth, or 'coldness' of the air at that altitude.

5.3 Thermodynamics of Gases

Physics does, in fact, provide a third equation connecting our state variables, although it is not as simple to use as the other two, and that is the **First Law of Thermodynamics**. This is, in fact, one of the most important laws in all of science, as it is the fundamental Law of Conservation of Energy. This law says that a body (or a system, like a gas) contains internal energy, U, another state variable, which can be altered by either heat, H, or work, W. (Thus, we cannot talk about how much heat a body has, or how much work, only how much energy; heat being just one of the many forms of energy.) Formally we may write

$$\Delta U = H - W$$

for the change in internal energy, U, where H is the heat added *to* the system, and W is the work done *by* the system. In our usage we will replace the upper case symbols with lower case, to denote their per kg values (i.e. their *specific* values).

5.3.1 Specific Heat

There is one additional concept we need to introduce at this point: specific heat. The specific heat of a given substance is the amount of heat (energy) needed to raise its temperature by 1°C, per kg. In the case of a gas, this can be done in a range of ways, and we actually need to make two definitions. The specific heat at constant volume, c_v, is the amount of heat needed to raise the temperature of 1 kg of that gas by 1°C, while its volume is kept fixed. The specific heat at constant pressure, c_p, is the amount of heat needed to raise the temperature of 1 kg of that gas by 1°C, while its pressure is kept fixed. In this case, extra energy will be needed, as the gas will need to expand; i.e. push back on its surroundings, thus doing work.

5.3.2 The First Law

Rather than work with the fully general expression for this law, we will focus on the most logical form it takes in the atmosphere. As we will discuss in more detail below, for a dilute gas we may ignore any weak attractive forces between molecules, so that the only contribution to the internal energy of our atmosphere will be the kinetic energy of its molecules. From this fact, it may be shown that $u = c_v T$, where u is the internal energy (per kg) and c_v is its specific heat at constant volume. This implies that a change in internal energy will be reflected in a proportionate change in temperature.

A 'parcel of air' may do work on its surroundings by pushing back, and thus expanding. If the surrounding air pushes in on our air parcel, causing it to contract, this would represent work done *on* it: just a change of sign. In physics, work is defined as the force applied to something, times the distance displaced (pushed). As pressure is force per unit area, force (for a gas) may be expressed as pressure times area. So, force times distance equals pressure times area times distance, which can be expressed as pressure times volume. In symbols:

$$W = Fd = pAd = p\Delta V$$

where ΔV is the *change* in the volume of our air parcel. Also remember that if our air parcel expands by pushing back on its surroundings, then that can only come at the expense of some of its internal energy: it takes energy to do work. However, as we mentioned earlier, we will express many quantities in per kg form: so, we will replace ΔV by $\Delta \alpha$ (and W by w).

That only leaves the heat term. An air parcel can gain heat by absorbing solar or infrared radiation, and lose heat by emitting infrared energy. It can also exchange heat with the surface, assuming it is in contact. This time there is no simple expression to use, as these terms vary significantly. For now we will simply use the symbol h (for the heat input per kg). Thus, the First Law of Thermodynamics may be written

$$\Delta u = c_v \Delta T = h - p\Delta \alpha \tag{5.5}$$

5.3.3 Adiabatic Processes

The input of heat to a thermodynamic system, such as an air parcel, is called diabatic heating. Sometimes processes take place where there is no diabatic heating: such processes are referred to as adiabatic. For an air parcel that is not in contact with the surface, diabatic heating is via absorption and emission of radiant energy, both solar and infrared. On short timescales – say an hour or so – these are often (but not always) close to being balanced: that is, absorption equals emission. This means that we are, to a good approximation, in an adiabatic situation. In such a situation, the First Law simplifies to

$$c_v \Delta T = -p \Delta \alpha \qquad (5.5')$$

We are now going to need a little bit more mathematics (feel free to skip to the result, Equation (5.6), if you prefer): we need to ask what are the implications for the ideal gas equation when one of its terms is changing? Clearly this requires the other terms to change. So that equation becomes

$$(p + \Delta p)(\alpha + \Delta \alpha) = R(T + \Delta T)$$

$$p\alpha + p\Delta\alpha + \alpha\Delta p + \Delta\alpha\Delta p = RT + R\Delta T$$

Now we use the ideal gas equation to cancel off the terms which don't contain Δ's, and ignore the $\Delta\alpha\Delta p$ term, which is assumed to be very small ('second order of smallness' in calculus-speak) to give us

$$p\Delta\alpha + \alpha\Delta p = R\Delta T$$

When we combine this result with the previous equation we obtain

$$(c_v + R)\Delta T = \alpha\Delta p$$

so that

$$\Delta p = \frac{R + c_v}{\alpha} \Delta T$$

Finally, we combine this with the hydrostatic equation (in the form containing α) to obtain

$$\frac{\Delta T}{\Delta z} \equiv -\Gamma = -\frac{g}{c_v + R} = -\frac{g}{c_p} \qquad (5.6)$$

where the Kinetic Theory of gases (next section) tells us that the specific heats are connected via the relation $c_p = c_v + R$. Once again, the calculus limit is obvious.

So, what have we just shown? We have shown that, if we take a parcel of air, and adiabatically lift it, it should expand and cool at this rate. We can now put in numerical values for both the constants: they are 9.8 m s^{-2} and 1,000.5 J/kg/K. Thus, the **adiabatic lapse rate** is 9.8 degrees per km (10°C/km will be more than adequate most of the time).

5.3.3.1 Potential Temperature

Air temperature drops with altitude, as a result, primarily, of the adiabatic processes we have just outlined. How might we intercompare air temperatures at different altitudes, taking this into account? We do so by defining the potential temperature of an air parcel as the temperature it would have if it were brought adiabatically down to a reference pressure level of 1,000 hPa. As an air parcel moves adiabatically around in the atmosphere, its temperature may vary, but its potential temperature is conserved (i.e. remains constant).

5.3.4 Vertical Stability

The adiabatic lapse rate is central to the vertical stability of the atmosphere. Temperature falls through the troposphere: that is, the lapse rate is usually positive. (The lapse rate is negative in a temperature inversion.) However, the actual value varies quite a lot. At any given place and time, we may talk about the environmental lapse rate (more generally, and more accurately, the environmental temperature profile). What happens if, for some reason, an air parcel rises (or is somehow forced to rise) in such an environment? It will cool at the adiabatic lapse rate, Γ: 10°C per km. It will very quickly adjust to the local pressure, so its density relative to its surroundings will depend on its temperature relative to its surroundings.

Consider the situation in Figure 5.5 (left). If our parcel is moved up from height O to height A-B it will cool at the rate Γ, to the point A, where it will find itself colder than the 'environment', i.e. the local temperature (at B). Because its pressure will be equal to that of its surroundings, a lower temperature equates to a higher density: a direct result of the ideal gas equation. Hence it will be denser than its surroundings, and fall back down to O. This is stable equilibrium, which often leads to oscillations. This may be observed in the atmosphere as mountain or lee wave clouds. Note that mountain clouds have been 'identified' as UFOs (e.g. Mt Shasta in northern California)!

Now consider the situation in Figure 5.5 (right). This time our parcel ends up warmer, and hence less dense than its surrounds, and will now be pushed up even further. This is unstable equilibrium (i.e. buoyancy). Unstable equilibrium usually cannot survive for long, as enough parcels will move, some up, some down – convection – adjusting the temperature profile, and restoring the environment to stability. Unstable conditions are created on hot sunny days when solar energy is absorbed at the surface, warming the air above it; 'dragging out' the base of the temperature profile. This is the reason we often see those big convective (cumulus) clouds develop on summer afternoons. Very intense fires can also achieve this, leading to pyrocumulus clouds.

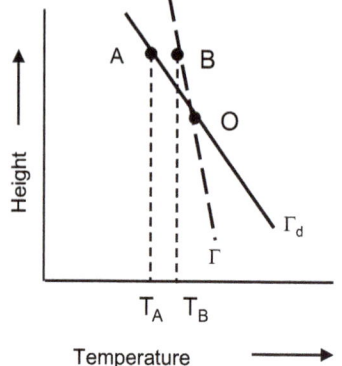 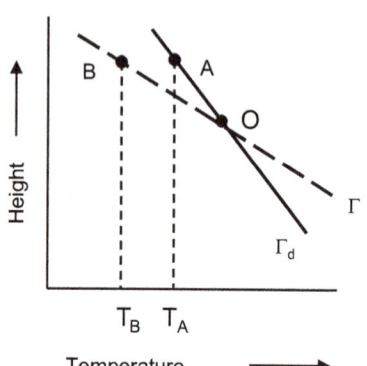

Figure 5.5 Vertical stability conditions: stable (left); unstable (right). *Source: Physics of Radiation and Climate*, figure 3.1, Box and Box, CRC Press, 2016, reproduced by permission of Taylor & Francis Group.

5.4   Kinetic Theory of Gases

The material in this last section is at a somewhat higher level than the rest of the chapter, and can be skimmed if felt appropriate.

The ideal gas equation – indeed the entire science of thermodynamics – makes no formal concessions to the fact that matter is discrete, rather than continuous: in reality, it is composed of atoms and molecules. To incorporate this reality, the next 'step' is known as Kinetic Theory. By assuming that the numbers of such components are sufficiently large, we may then use statistical tools to extract the information of importance to our interests.

Kinetic Theory makes the following assumptions:

1. A gas consists of particles called molecules.
2. These molecules are in random motion, and obey Newton's laws of motion.
3. The total number of molecules is large. This makes it possible, and valid, to do some statistical averaging.
4. The volume occupied by the molecules themselves is small, when compared to the total volume occupied by the gas.
5. No forces act between the molecules, except during collisions, which are assumed to be elastic: i.e. there is no loss of energy.
6. Collisions with the walls are also elastic.

Before we go further, we should ask ourselves how reasonable these assumptions are, in relation to our atmosphere. At standard temperature ($0°C$) and pressure (1 atmosphere) the number of molecules in one cubic metre is 2.687×10^{25} (known as Loschmidt's number), so it would be fair to say that assumption 3 is valid (even at the stratopause, where density is 1,000 times lower).

The density of air at ground level is ~1.2 kg m^{-3}. By contrast, the density of water is 1,000 kg m^{-3}. This factor of ~1,000 is a direct measure of the empty space between air molecules: to a reasonable approximation we may say that air molecules occupy only 1 part in 1,000 of the available space – even at ground level. So, assumption 4 is also valid: looking good so far!

On Venus this assumption is questionable, and the ideal gas equation would need to be modified to reduce the volume, V, by the volume actually occupied by molecules, as this volume is unavailable to another molecule. (The van der Waals equation of state incorporates this effect.)

Finally, a comment on assumption 5 is in order. Molecules do, in fact, exert weak, attractive forces on one another, but only when they are close to touching. Given the low density of our atmosphere, this can safely be ignored; however, it becomes vital when a gas is close to condensing. (The van der Waals equation of state also incorporates this effect.)

5.4.1   Application

Applying Kinetic Theory ideas necessarily becomes somewhat mathematical, so we will avoid those details. However, some of the results it produces are valuable for our understanding of the physical properties of our atmosphere. Pressure comes about from the collisions of the molecules with the walls of an appropriate container (which could, of course, be another layer of molecules): more specifically, by the change in momentum during these collisions. (The momentum of a moving object is the product of its mass, m, and velocity, v.) We now need to average over all possible speeds, and all possible angles of incidence, to arrive at

$$P = \frac{Nm\langle v^2\rangle}{3V} = \frac{2N}{3V}\langle KE\rangle$$

where N is the number of molecules in the container of volume V, m their mass, v their velocity, KE their kinetic energy ($\frac{1}{2}mv^2$), and we have used angled brackets to denote average values.

We may now compare this with the ideal gas equation above, remembering that in the kinetic picture the density is the product of the number of particles per unit volume – N/V – times their molecular weights, m. What we find is a direct connection between the (average) kinetic energy of the molecules, and the absolute temperature:

$$\langle KE\rangle = \frac{3}{2}k_B T \tag{5.7}$$

where $k_B = mR$ is known as Boltzmann's constant, one of the fundamental constants of nature. (This derivation is a little clearer if we had chosen the molar version of the ideal gas equation, as mentioned above: throw in Avogadro's number and stir gently.) The key point to note is that there is a direct connection between the kinetic energy of the molecules, and the absolute temperature, and this gives us a clearer picture of what absolute zero actually represents: when all molecular motion ceases.

One more technical point now needs to be made. In the case of diatomic molecules such as the nitrogen and oxygen that dominate our atmosphere, the factor of 3/2 becomes 5/2. The reason is easy enough to understand. A single atom has 3 '**degrees of freedom**': it can move in any of three directions. Each degree of freedom contributes a factor of ½. Linear molecules like nitrogen and oxygen (99% of our atmosphere) can also rotate about either of the two axes at right angles to the molecular bond axis (both translational motion and rotational motion involve kinetic energy): two further degrees of freedom; hence 5/2.

What else does Kinetic Theory tell us? The first thing of note follows immediately from the connection between temperature and kinetic energy, and hence velocity. While there is a dependence on the molecular weight of the particular gas, oxygen and nitrogen are sufficiently similar that an average, or typical, value can be assumed. For a room temperature of ~300 K, this works out at close to 500 m/s. (By contrast, for hydrogen it is 1,900 m/s, and for helium it is 1,350 m/s.) Kinetic Theory can also tell us about the mean free path of molecules between collisions, and, in particular, how this varies with density (Figure 5.1). At ground level this is about 0.1 μm, which, when combined with the average velocity, means that the average time between collisions is somewhere around 10^{-10} of a second!

James Clerk Maxwell (1831–1879), the physicist who brought together the known laws of electricity and magnetism (the laws due to Coulomb, Ampere, and Faraday), and then showed that light is 'just' an electromagnetic wave, also made valuable contributions to Kinetic Theory. In particular, he was able to derive an expression (along with Ludwig Boltzmann)

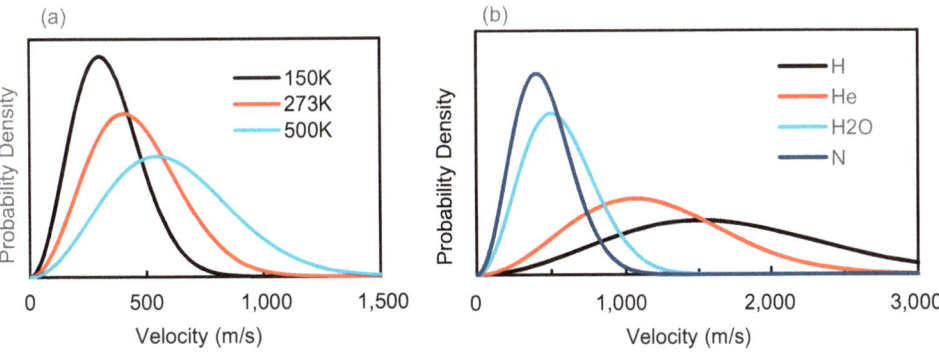

Figure 5.6 Maxwell–Boltzmann distributions of molecular speeds: left panel N_2 for 150 K, 273 K, and 500 K; right panel N_2, H_2O, He, and H_2 for 273 K.

for the *distribution* of the speeds of the molecules in a gas, not just their average speed. Figure 5.6a shows the **Maxwell–Boltzmann distribution** for nitrogen (N_2) for temperatures of 150 K, 273 K, and 500 K; Figure 5.6b shows the distributions for N_2, H_2O, He, and H_2 at 273 K. Note that these are probability density distributions, so that the vertical axis has no real scale. Figure 5.6b indicates that the tails of the distribution for He and H_2 extend sufficiently far that some will have speeds greater than the escape velocity, so that both gases will slowly escape Earth's gravity.

Finally, Kinetic Theory actually tells us ratios of the two specific heats, along with the gas constant. For a monatomic gas, $c_p:c_v:R = 5:3:2$; for a diatomic gas such as dominates our atmosphere, $c_p:c_v:R = 7:5:2$. In either case $c_p = c_v + R$.

5.4.2 Sensible Heat

Let's return to Equation (5.5), and turn it around, writing it in the form

$$h = \Delta u + p\Delta\alpha = (u_2 - u_1) + p(\alpha_2 - \alpha_1)$$
$$= (u_2 + p\alpha_2) - (u_1 + p\alpha_1) = e_2 - e_1$$

where e is known as the sensible heat (per unit mass), one component of the Earth's energy budget which we will examine in Chapter 8. Taking this a step further,

$$e = u + p\alpha = u + RT$$
$$= (c_v + R)T = c_p T \tag{5.8}$$

Note that in 'standard' thermodynamics texts this quantity goes by the name *enthalpy*, and is usually given the symbol h. However, we have used h for heat. The symbol q is often used for heat, but we will need it when we discuss water vapour. (When you spend enough time in physics you soon realise that there aren't enough letters/symbols to cover all your needs, so some doubling-up is inevitable.) Note that, from its definition, we see that enthalpy is derived from state variables, and so is, itself, a state variable.

Summary

The three key variables in our (dry) atmosphere are pressure, temperature, and density. In this chapter we have provided our readers with the necessary science that interconnects these: the hydrostatic equation, and the ideal gas equation. From these we gain a basic understanding of how these variables change with altitude, although the finer details require a knowledge of the energy absorbed and emitted by an air parcel: the First Law of Thermodynamics. While this can be complicated when investigated in full detail, much of the time we can get away with the adiabatic assumption: no exchanges of energy.

The most important piece of knowledge that we gain from all this relates to the vertical stability of the atmosphere: under what circumstances does convective motion take place? In the next chapter we will add water vapour to this, to understand cloud formation.

FURTHER READING

The material in this chapter is standard fare in every book on atmospheric science/physics, climate science, or meteorology. Among the many, we note the following:

- *Atmospheric Science, An Introductory Survey* by J. M. Wallace and P. V. Hobbs (Academic Press, 2006).
- *Meteorology Today* by C. D. Ahrens (Thomson Learning, 2007).
- *An Introduction to Atmospheric Physics* by D. G. Andrews (Cambridge University Press, 2000).
- *The Weather and Climate of Australia and New Zealand* by A. Sturman and N. Tapper (Oxford University Press, 2006).
- *Atmosphere, Weather and Climate* by R. G. Barry and R. J. Chorley (Routledge, 2003).
- *Atmospheric Thermodynamics* by G. R. North and T. L. Erukhimova (Cambridge University Press, 2010).

REVIEW QUESTIONS

1. Define specific volume.
2. Define pressure.
3. How is the Kelvin scale related to the Celsius scale?
4. What are the four standard layers of the atmosphere, from lowest to highest?
5. Explain the hydrostatic equation.
6. Explain the ideal gas equation.
7. How was absolute zero (i.e. 0 K) discovered?
8. Why is a 'constant temperature' atmosphere a reasonable approximation?
9. What is an adiabatic process?
10. Why might the adiabatic approximation be valid in the atmosphere?
11. How might the adiabatic approximation be useful in the atmosphere?
12. Does a steep atmospheric temperature imply stable or unstable conditions?

13. What is the connection between temperature and molecular kinetic energy?
14. By making reference to the Maxwell–Boltzmann distribution, explain why the gases hydrogen and helium are able to leak away to space.

EXERCISES

1. The Earth has a radius of 6,370 km. At the surface, the average pressure is 1,013.25 hPa. Use this information to compute the mass of air above each 1 m^2 of surface, and also the total mass of the atmosphere. (Ignore mountains; assume $g = 9.8$ m s^{-2} throughout.)
2. The average mass of an air molecule is 29 atomic mass units. Determine the number of molecules per cubic metre at ground level, where the pressure is 1,000 hPa, and the temperature is 20°C. (You may need to look up some data.)
3. Compute the total number of molecules in our atmosphere.
4. What is the average kinetic energy of a nitrogen molecule in the middle troposphere where the temperature is 250 K? What is its average speed?
5. Consider a constant density model of an atmosphere: that is, assume that the density is everywhere 1.25 kg m^{-3}. How thick would such an atmosphere be to account for the observed surface pressure? Use the ideal gas equation, and the hydrostatic equation, to determine the temperature profile. Find the lapse rate, Γ, and the temperature at the top of this atmosphere.
6. Assume that the temperature of the atmosphere decreases uniformly with height, with a profile given by $T = T_0 - \Gamma z$. Determine the relation between pressure, p, and height, z, in such an atmosphere. (Note: this requires good calculus skills.)
7. Integrate Equation (5.3) under the assumption that T is constant, to check Equation (5.4). If this is a bit of a challenge, differentiate Equation (5.4) to get back to Equation (5.3).
8. Calculate the scale height of the Martian atmosphere, assuming an average temperature of 222 K, a gravitational acceleration of 3.8 m s^{-2}, and that the atmosphere is mainly CO_2.

6 The Moist Atmosphere

Water is one of the most important, and fascinating, substances on the planet. It is, of course, essential for life, and so, of course, is the hydrological cycle. That cycle only exists because water can exist in all three phases in the atmosphere. Water also has large latent and specific heats, and the transfer of latent heat from the oceans to the atmosphere is a key component of the global energy budget (Chapter 8). Finally, water is most unusual in that it expands on freezing, which actually contributes to geological processes via its ability to split rocks.

Water vapour is the most variable substance in the Earth's atmosphere (measured in gross amount), so the first step we need to take is to study how and why it varies. This comes down to saturation vapour pressure, which is strongly temperature-dependent. Clouds form when moist air is lifted, a very important meteorological process. This causes some of the water vapour to condense to form water droplets, while also releasing latent heat. It is this latent heat which is, ultimately, the source of power to all storms. We will conclude this chapter with the surprisingly complex but very important story of how individual cloud droplets form.

6.1 Water in the Atmosphere

Although the distribution of water in the atmosphere is quite variable on many scales, we can say that, on average, it makes up 0.25% of its total mass: equivalent to a layer of liquid 2.5 cm deep. (By comparison, the oceans represent an average layer over 10^5 times thicker.) Of this mass, over 99% is in the form of vapour, with the rest either liquid droplets or ice crystals. Water vapour itself is a key atmospheric component, and its thermodynamic effects must now be added to our evolving story. We will start with the most basic of its effects, and introduce some of the additional notation we will need, before considering the variation of saturation vapour pressure with temperature.

We need to remind readers of a couple of definitions. The specific heat of a substance is defined as the amount of energy (e.g. heat), per kilogram, needed to raise its temperature by 1°C. (In reality it often varies, but only slightly, with temperature.) The **latent heat** of a substance is the amount of heat/energy needed to convert 1 kg from one phase to another: solid (ice) to liquid, L_{il}; liquid to gas (vapour), L_{lv}; or solid (ice) to gas (vapour), L_{iv} (known as sublimation: a good example is dry ice, solid CO_2). Latent heat is slightly dependent on temperature. In the case of water, at 100°C, L_{lv} = 2,250 kJ/kg. More useful, however, are the values at 0°C:

$$L_{il} = 334 \text{ kJ/kg}$$

$$L_{iv} = 2834 \text{ kJ/kg}$$

$$L_{lv} = 2500 \text{ kJ/kg}$$

Note that $L_{iv} = L_{il} + L_{lv}$ (of course).

When water evaporates, whether from the ocean, lake, or land surface, latent heat is required. When that water vapour subsequently condenses, that latent heat is released to the surrounding air. Thus, the hydrological cycle necessarily involves a transfer of energy from the surface to the atmosphere.

6.1.1 Moisture Parameters

A number of parameters are used to characterise, and quantify, the amount of water vapour in the atmosphere. Some of these may be plotted on thermodynamic charts used by meteorologists, or extracted from information plotted on such charts.

1. Mixing ratio, r: mass of vapour/mass of dry air.
2. **Specific humidity**, q: mass of vapour/mass of moist air.

$$q = \frac{m_v}{m_v + m_d} = \frac{r}{1 + r} \approx r$$

3. **Vapour pressure**, e: that part of the atmospheric pressure exerted by water vapour: the 'partial pressure' of water vapour (remember Dalton). Note that

$$r = \frac{m_v}{m_d} = \frac{M_v}{M_d}\frac{e}{p - e} = \varepsilon\frac{e}{p - e} \approx \varepsilon\frac{e}{p} = 622\,e/p$$

Note that we usually express r in g/kg, and both e and p in hPa.

4. Saturation mixing ratio, r_s: mixing ratio in the case of air that is saturated with water vapour.
5. Saturation specific humidity, q_s: specific humidity of saturated air. $q_s \approx r_s$.
6. **Saturation vapour pressure**, e_s: vapour pressure in the case of air that is saturated with water vapour. Again, we note that

$$r_s = 622\ e_s/p$$

7. **Relative humidity**, RH: a measure of humidity relative to saturation.

$$RH = (q/q_s) \times 100\% \approx (r/r_s) \times 100\% = (e/e_s) \times 100\%$$

8. **Vapour pressure deficit**, VPD: $e_s - e$.
9. **Dew point**, T_d: the temperature to which a sample of moist air must be cooled at constant pressure for it to become saturated. This is a good measure of human (dis)comfort. Note that

$$e(T) = e_s\,(T_d)$$

$T - T_d$ is known as the **dew point depression**.

10. Wet-bulb temperature, T_w: the lowest temperature to which air may be cooled, adiabatically and isobarically (i.e. at constant pressure), by evaporating water into it. This is much easier to measure than T_d. Remember that, unlike for T_d, we are adding extra water vapour to the (local) atmosphere, so that

$$T_d \leq T_w \leq T$$

A wet-bulb thermometer is (or at least once was) part of the standard set of meteorological instruments included inside a **Stevenson screen**. When performing a wet-bulb measurement, we evaporate water into the air, which has two effects: the vapour pressure of water in the air increases, while the temperature of the air decreases, as it must supply the latent heat needed for the evaporation.

6.1.2 Saturation

At any temperature below the boiling point, water vapour can attain equilibrium with respect to the liquid or solid phase, when equal numbers of molecules are moving from the liquid/solid into the vapour, and vice versa (Figure 6.1). Technically, this equilibrium is defined with respect to a *flat* surface of *pure* liquid (or solid): the importance of this will be seen below. The maximum vapour pressure that can be maintained with respect to such a surface is the saturation vapour pressure, e_s, which varies strongly with temperature. In general, of course, the actual vapour pressure at a temperature, T, will be less than the saturation vapour pressure, meaning that relative humidity is less than 100%.

Exactly how e_s varies with T is given by a general result known as the **Clausius–Clapeyron equation**, which can be derived using the Second Law of Thermodynamics. The result is given by the somewhat complicated equation

$$e_s = 6.11 \, \exp\left\{\frac{L}{R_v}\left(\frac{1}{273} - \frac{1}{T}\right)\right\}$$

Figure 6.1 Schematic representation of a liquid and vapour in equilibrium.

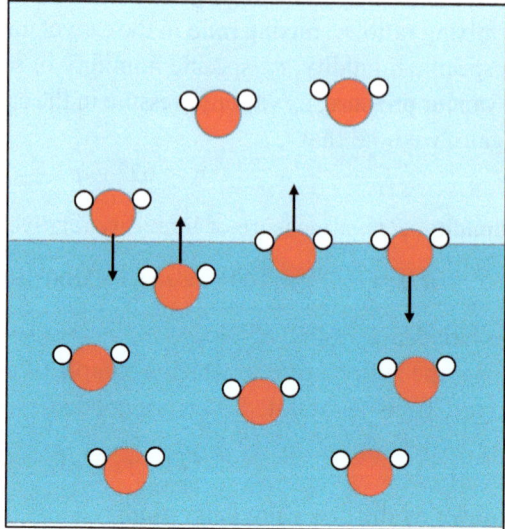

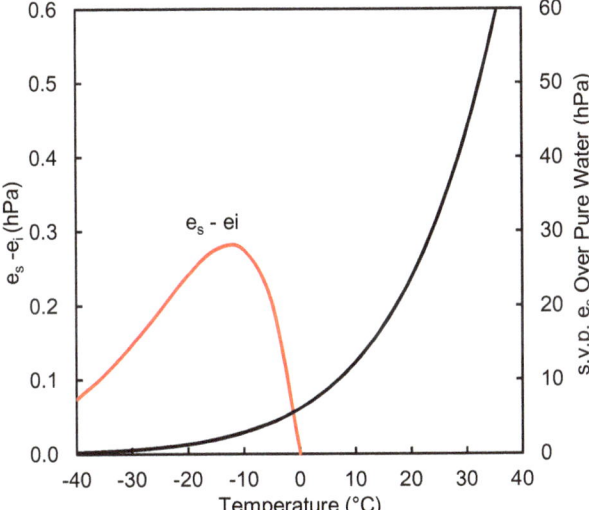

Figure 6.2 Saturation vapour pressure (SVP) over liquid water as a function of temperature (right axis); and difference between SVP over liquid water, and over ice (left axis). *Source: Physics of Radiation and Climate*, figure 2.6, Box and Box, CRC Press, 2016, reproduced by permission of Taylor & Francis Group.

where 6.11 hPa is the saturation vapour pressure at 0°C, and we will use the latent heat value at 0°C. (Strictly, this result assumes that the latent heat does not vary with temperature. In fact, it does, but slowly. Assuming the value quoted above is usually sufficiently valid for a range of ±20 or so degrees, adequate for most purposes.) The gas constant for water vapour has the value 461.5 J/kg/K. The results from this relationship are presented in Figure 6.2 (so don't worry if you're not comfortable with an equation like the one above).

At temperatures below 0°C, water can exist both as solid crystals and supercooled liquid droplets, as well as vapour. The saturation vapour pressure over ice is given by the Clausius–Clapeyron equation in exactly the same form as for the liquid case, except that we must use the appropriate latent heat value. Because $L_{iv} > L_{lv}$, we may easily show that for temperatures below 0°C, the saturation vapour pressure over ice is lower than that over liquid water. Figure 6.2 also shows the difference between the two saturation vapour pressure values at temperatures below 0°C. The consequences of this in cold clouds will be explored below.

Tropospheric temperatures range from around 30°C in the tropics near the ground, to as low as –80°C at the tropopause. The Clausius–Clapeyron equation shows that, over this range, the saturation vapour pressure of water varies significantly, from as high as 40 hPa (equivalent to 4% of atmospheric surface pressure) to less than 0.1 hPa. As a consequence, the water vapour distribution in the atmosphere is far from uniform, being strongly concentrated in the lower altitudes and latitudes. Yet storms are quite common at high latitudes. Clearly, the amount of moisture in the tropics (e.g. the monsoon rains) is substantial!

Time for another technical point. It is often said that 'air can hold a certain amount of water vapour, at a given temperature'. Strictly speaking, saturation vapour pressure has no direct connection to 'the air', except that it is the air temperature which is the variable in the equation above. If we placed some water in a container with nothing else – i.e. with a vacuum above the surface – some water would evaporate into that space, with the amount depending on the temperature of this 'system' of liquid and vapour (and container walls).

Global warming is increasing atmospheric temperatures, not just the surface temperature. Thus we would expect that the atmosphere is likely to contain more water vapour (which happens to be a greenhouse gas). The available data suggest that, in fact, specific humidity, q, is increasing, while relative humidity, RH, is actually decreasing. The implications of this for both drought and flood conditions will be examined in Chapter 19.

6.1.3 Boiling Point

The definition of boiling point is that temperature at which $e_s = p$ (i.e. ambient pressure). For a pressure of 1,013.25 hPa (standard sea-level pressure), the boiling point for water is 373 K, or 100°C. (Note that if you set $e_s = 1,013.25$ hPa in the equation above, and solve for T, you obtain a lower value: a direct consequence of the variation of L between 0°C and 100°C, which we have chosen to ignore.)

On a high mountain, where pressure is lower, the boiling point is also lower. By assuming an exponential fall in pressure with height (which we know to be a reasonable assumption), combined with the exponential nature of the Clausius–Clapeyron equation, we may obtain the following approximation for the boiling point as a function of altitude:

$$T_{boiling} = \frac{L/R_v}{a + z/H}$$

where H is the scale height, and a is a constant that may be obtained in these manipulations (Exercise 5). However, a more accurate (and hence more useful!) value may be determined by requiring that we get the right answer at sea level (i.e. $z = 0$). You may check that this yields

$$a = \frac{L/R_v}{373}$$

$$\therefore T_{boiling} = \frac{373}{1 + z/H^*}$$

where

$$H^* = \frac{L}{R_v} \frac{H}{373} = 14.523\,H.$$

6.2 Elevation of Moist Air

In Chapter 5 we considered the elevation and vertical stability of dry air. Will water vapour change any of our conclusions? The elevation of moist air is the central process leading to the formation of clouds. When a parcel of moist air rises it will initially cool at the dry adiabatic lapse rate, Γ_d. However, it will eventually reach a point (altitude) where its temperature, T, has dropped to its dew point, T_d. That means the air parcel is now saturated. This is known as its **lifting condensation level**, LCL.

We saw in Section 5.3 that when air is lifted (dry) adiabatically, T falls at a rate of 9.8°C/km. It is relatively easy to show that the dew point, T_d, falls at a rate of close to 1.8°C/km

[$\pm0.1°$C/km]. This means that they approach each other at a rate of $\approx8°$C/km. From this we may quickly estimate the height of the LCL (i.e. cloud base) from measured data:

$$LCL = (T - T_d)/8 \text{ km}.$$

6.2.1 Release of Latent Heat

If our parcel is lifted further, some of the water vapour must condense, and a cloud will start to form. As the parcel rises further, it can still be assumed to be an adiabatic process (that is, no *external* input of heat): but there's one key difference. As water vapour condenses, it releases latent heat: this is converted into molecular kinetic energy, which means temperature; hence T must be decreasing more slowly than it otherwise would, meaning that the lapse rate will be less than Γ_d. There is no simple expression for the **saturated adiabatic lapse rate**, Γ_s, but a value of about 5°C/km is often assumed/reasonable for the lower troposphere. The path followed by an air parcel that is forced to rise is known as its **forced ascent curve**.

If a moist air parcel is forced to rise far enough, perhaps to go over a mountain, much of its moisture will condense, releasing latent heat. *If* most of the condensate precipitates, that latent heat will stay with the air parcel as it descends on the other side. Now it will descend (largely) dry adiabatically, and end up distinctly warmer than at the same level on the other side (i.e. latent heat has been converted to 'warmth'). These descending winds are sometimes known as **Chinooks** ('snow eaters'), or a foehn (or a zonda).

6.2.1.1 Equivalent Potential Temperature

An air parcel contains three forms of 'internal energy'. The first is thermal energy – sensible heat – reflected directly in its temperature, T. The second is potential energy, due to its elevation. The third is latent heat, reflected in its water vapour content. Unfortunately, all three are quantified using different scales, making comparison effectively impossible. (We previously defined potential temperature, which is a measure to combine the first two.)

Now consider the following. If we were to raise our air parcel to the top of the troposphere, so that virtually all the water vapour had condensed, *and* it all precipitated out, we could then bring back down to the ground the warmest possible air parcel, where all its latent heat has been converted into temperature: known as the equivalent potential temperature: temperature (sensible heat), plus elevation, plus latent heat.

6.2.2 Conditional Stability

We have already looked at the question of vertical stability, and concluded that if the local environmental lapse rate was less than the (dry) adiabatic lapse rate, the atmosphere was stable – that is, convective motions would not occur. However, if the environmental lapse rate was greater than the adiabatic lapse rate, the atmosphere was unstable, and convective motions would ensue. Now we have two adiabatic lapse rates: dry and saturated. So, we have some more complex questions to ask: Figure 6.3.

If the environmental lapse rate is less than the saturated adiabatic lapse rate, then it must, of course, be less than the dry adiabatic lapse rate, and so the atmosphere is said to be absolutely

Figure 6.3 Stability conditions for a moist atmosphere.

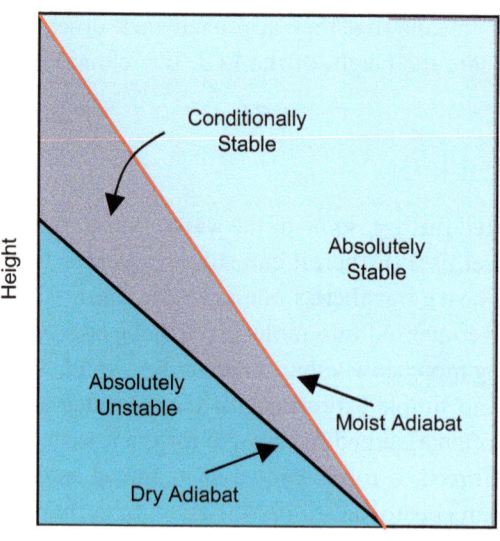

stable. If the environmental lapse rate is greater than the dry adiabatic lapse rate, then, again, it must be greater than the saturated adiabatic lapse rate: this time the atmosphere is said to be absolutely unstable. But what about those cases when the environmental lapse rate lies between the dry and saturated lapse rates? This is referred to as conditionally unstable.

Consider now what happens when an air parcel is lifted up a little. Assuming it is initially unsaturated, it will fall back; stable conditions. But what happens if, somehow, our parcel is raised to its lifting condensation level? Now, if pushed further, it must follow a saturated adiabat, and (relative to its original starting point) our parcel is still stable. A convective cloud will be produced. Its vertical ascent may be followed in Figure 6.4. It required energy, of some form, to lift our parcel to its lifting condensation level, and it is still liable to fall back, due to this energy debt.

If, however, our parcel does continue to rise, it may cross over the environmental temperature profile and become absolutely unstable: at its **level of free convection**, or LFC. Now it will continue to rise due to its buoyancy. Kinetic energy is being generated by this buoyancy. Somewhere near the top of the troposphere it must again meet the local temperature profile – the **level of neutral buoyancy**, LNB. This places a lid on the cloud we have been forming – often seen as an anvil – although convective overshoot may occur. All of these steps can be followed on Figure 6.4.

From the level of free convection up to the level of neutral buoyancy, energy is being released – known as **convective available potential energy**, or CAPE. This is the energy released from the condensation of water vapour – latent heat – and it is this that 'powers' thunderstorms. The energy required to lift our original air parcel up to its level of free convection is known as **convective inhibition energy**, or CINE. This energy must be supplied from somewhere: a trigger. That could be the convergence of two air masses due to terrain, or it could be supplied from the CAPE of a nearby convective cloud, as part of a storm system. Meteorologists charged with forecasting local thunderstorms talk about a tripod: sufficient water vapour; upper-level instability; and a trigger.

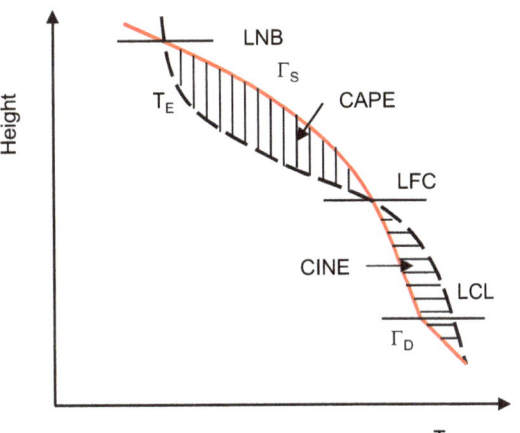

Figure 6.4 Forced ascent curve. *Source: Physics of Radiation and Climate*, figure 3.1, Box and Box, CRC Press, 2016, reproduced by permission of Taylor & Francis Group.

6.2.3 Cloud Formation

A cloud forms when air becomes supersaturated, usually by cooling a moist air parcel/layer. The primary way this comes about is by lifting: for example, as just discussed. A number of questions need to be addressed:

- What other mechanisms are available?
- When air is supersaturated, do droplets (or ice crystals) just form 'automatically' (in midair, so to speak)?
- Why do clouds 'exist'? That is, why don't their droplets simply fall to the ground?
- How do cloud droplets/crystals grow?

We have just seen that the elevation of moist air is the key factor in cloud formation. What are the macro-scale physical processes that might lead to this? Textbooks usually focus on four processes, which we now outline, along with a less-discussed cloud type.

- Instability, as just discussed. Note that while some air is rising, other air must be falling, leading to scattered clouds, with diameters from 0.1 to 10 km. Updraft velocities can be a few m/s, and liquid water content ~1 g/m^3.
- Orographic lifting: air is forced to rise to go over terrain. For high mountains we may have heavy rain on one side and a rain shadow on the lee side. For lower hills we get lenticular clouds, and possibly lee wave clouds. Details depend on wind speed, slope angle, etc.
- Frontal lifting: one air mass pushes under or over another. Updraft speeds are a few cm/s, and liquid water content a few tenths of a gram per cubic metre. Such clouds can be extensive (layers), and last many hours.
- Low pressure centres: here we find that air 'spirals into' the centre of a low and is forced to rise (slowly): a few cm/s. Converging air masses (such as a sea breeze), and troughs, also require air to rise.
- Marine boundary layer clouds (which are ubiquitous): they are a complex mix of dynamics and radiative cooling.

6.2.3.1 Cloud Classification

Clouds are generally classified as one of

- *Stratiform* ('layered') clouds develop through large-scale lifting, such as from the passage of a cold front.
- *Cumuliform* ('piled') clouds develop through the buoyant motions just discussed (including orographic lifting).
- *Cirriform* ('fibrous') clouds are found at high altitudes (7 km and up), and are mainly composed of ice crystals.

Terms like *nimbo* and *alto* are added to indicate 'rain-bearing' and 'mid-level', respectively. The altitude ranges of mid- and high-level clouds vary with latitude. Meteorologists use two-letter codes: Sc, Cb, Ci, etc.

6.3 Cloud Microphysics

6.3.1 Cloud Droplet Formation

So far, we have talked as if the production of cloud water droplets is automatic when air reaches saturation. Well, it ain't necessarily so! We said earlier that saturation is an equilibrium, defined 'with respect to a *flat* surface of *pure* water'. There are no such surfaces in the atmosphere.

It turns out that saturation vapour pressure over a curved surface/droplet is (much) higher than over a flat surface. The details depend on the radius of curvature. If we try to start a droplet with, say, 100 molecules, we find it will very quickly re-evaporate. However, even in the most pristine environments, there are always large numbers of aerosol particles – sea salt, dust, organic compounds, etc. (**condensation nuclei**, CN). Many are **hygroscopic**, and take up large amounts of water when the *RH* gets close to 100%. Thus, they grow. We now have some droplet solutions with sizes around 0.1 µm or larger. Saturation/equilibrium with respect to such a drop (sphere) occurs at around *RH* = 100.2% (referred to as a supersaturation of 0.2%). This embryonic droplet can now absorb more water vapour, and start to grow.

We know that anthropogenic emissions of SO_2 have led to an increase in sulphate aerosols, excellent nucleating agents. This should lead to more, but smaller, cloud droplets. Such a cloud will be more reflective of solar radiation, increasing the planetary albedo. This is known as the **Twomey effect**, or **aerosol–cloud interaction**, and can be observed when a ship cruises underneath cloud cover. Its emissions (low-grade marine diesel is a poorly regulated, dirty fuel) seed the clouds above, producing a bright line where it has sailed, known as a ship track, which may be observed by a NASA satellite overpass (Figure 6.5).

This concept has actually been suggested as one of a number of measures to counter global warming, by actively and directly reducing the inflow of solar radiation, under the general heading of **Geoengineering** (we'll say more in later chapters). As you will get a bigger effect from 'polluting' clouds in a clean environment, marine clouds, especially in the more remote oceans, would be the likely target: referred to as **marine cloud brightening**.

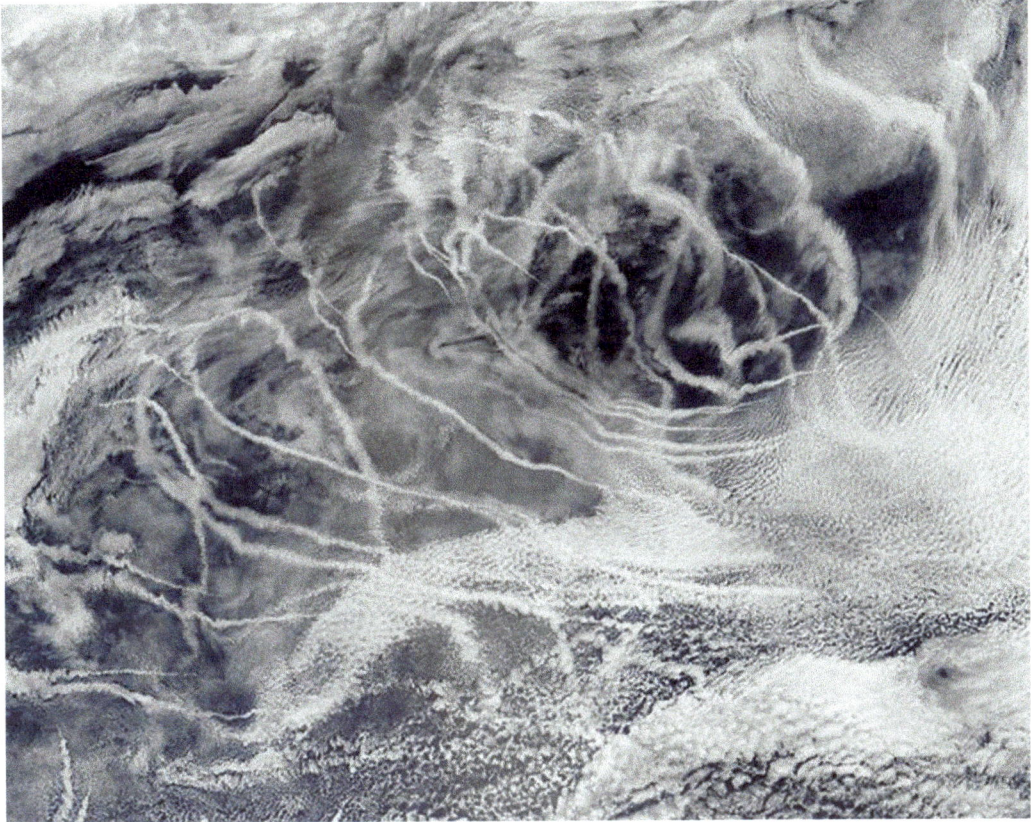

Figure 6.5 Ship tracks observed by satellite (NASA image).

6.3.2 Cloud Droplet Growth

A (small) object falling in air is subject to air resistance, which increases with fall speed. If released, its speed will increase until resistance equals the force of gravity: at this point the particle has reached terminal velocity. Both measurements and theory for particles of this size are far from straightforward. Table 6.1 gives what we might describe as 'reasonable values' for some drop diameters. Because clouds mostly form in rising air, terminal velocity must exceed this before a droplet actually descends.

Cloud droplet growth is both slow and complicated. Initially, our embryonic droplet will grow by **condensation**: simply absorbing water vapour from its supersaturated environment. This will allow it to grow to, maybe, 20 µm in diameter, but beyond this it really slows down. However, by this stage in a cloud's development, we are likely to have a range of drop sizes, including a few that are distinctly larger than the rest. They will have larger terminal velocities, and so will 'fall through' the cloud of smaller droplets, sweeping them up, and growing quite rapidly. This is called **coalescence**.

Table 6.1 Terminal velocities of water droplets in still air at 1 atm and 20°C.

Droplet diameter (μm)	Terminal velocity (m s^{-1})
10	0.003
30	0.025
50	0.070
100	0.255
300	1.15
1,000	4.00

In many parts of the world temperatures within clouds are below freezing. It turns out that water can remain liquid in these clouds – 'supercooled' – down to as low as −40°C. (They need a 'freezing nucleus', with the right crystal structure, which are comparatively rare.) Such cold clouds will generally have a mix of a lot of liquid droplets and a small number of ice crystals. As the latent heat for ice-to-gas is larger than for liquid-to-gas, saturation over ice is lower than over liquid (Figure 6.2). What this means is that we can have an environment that is subsaturated with respect to liquid drops, but supersaturated with respect to ice crystals.

In such an environment, water molecules can leave the droplets, and condense on the crystals: that is, the crystals will grow at the expense of the droplets. This is known as the Bergeron–Findeisen or cold cloud process. Ice crystals often grow quite large before falling. If they melt on the way to the ground we have (cold) rain. This process is also the basis of many cloud-seeding endeavours. A substance is introduced to the clouds (dry ice or AgI) with the right crystal structure to act as ice nuclei, in the hope that they will then grow and precipitate. Note that this is only possible with the right (cold) clouds.

6.3.3 Precipitation

Water in solid or liquid form (or indeed a mixture) precipitates in a range of physical manifestations, collectively known as hydrometeors.

6.3.3.1 Rain

Any falling liquid drops of water; however, the diameter really must be greater than 0.5 mm. Smaller droplets are classed as *drizzle* (usually falling from stratus clouds). Note that raindrops may partially evaporate on the way down, and so end up as drizzle. If they fully evaporate it is called *virga*. *Showers* occur when a sudden downdraft forces down droplets that had been previously held aloft within a convective cloud environment. A *cloudburst* is an especially heavy shower. Continuous rain falls from layered clouds.

6.3.3.2 Snow

Snow is any ice crystal, formed via the Bergeron–Findeisen process, that has not melted on the way down. Because ice crystals are such effective light scatterers, falling snow looks far more distinctive than falling rain. In fact, it is often possible to see where falling snow melts to raindrops by the change in scattered light (this may easily be confused with virga).

6.3.3.3 Sleet

If falling snow melts while travelling through a warmer air layer, but then encounters a subfreezing air layer, and freezes into an ice pellet, it is called sleet. (Unlike snow, such pellets bounce, and make a distinct noise when they hit something.) If the subfreezing air layer is too thin, the droplets will not have time to freeze, but will freeze as soon as they hit anything (if it, too, is below freezing). This is called *freezing rain* and can be very damaging (for example, by bringing down power lines).

When supercooled droplets strike an object below freezing, they freeze, forming *rime.* This can happen when ice crystals are falling through supercooled cloud layers. When the ice particle has accumulated so much rime that it is no longer identifiable as a snowflake, it is called *graupel.* The ultimate example of this process is *hail,* where ice crystals/graupel have been swept up and down in a cumulonimbus cloud many times, collecting more and more water.

6.3.4 Fog

Fog is ground-level cloud: the result of cooling without lifting. There are two major types, or processes. **Radiation fog** occurs when the ground cools radiatively, usually during the night, cooling the air above. Slow cooling allows fog droplets to grow quite large. **Advective fog** occurs when warm moist air moves over a cold surface. This time, cooling is more rapid, which leads to higher supersaturation and more nucleation. This means more but smaller droplets (for the same amount of liquid water). It turns out this scatters light more efficiently, so an advective fog usually looks 'thicker'.

Summary

Water (vapour) plays multiple roles in weather and climate. It is essential for life. It is the rain we sometimes curse, and sometimes bless. Evaporation and condensation are a significant component of the Earth's energy budget. And as we will see in later chapters, it is also a major greenhouse gas.

In this chapter we have covered three important pieces of science. Firstly, we saw how the amount of water vapour the atmosphere 'can hold' depends strongly on temperature. Because we know that the world's temperature is rising, this implies that there is likely to be more of this greenhouse gas in the atmosphere as time goes on: and heavier rainfall. We'll return to this, many times.

The second piece of science concerns the role of any latent heat that might be released by convection and subsequent condensation of moisture in the air. It is this that powers all storms. Will they get worse? This is a subject of considerable scientific interest right now.

Finally, we looked at cloud droplet formation, and the role of aerosol seeds. What role might changes in either the numbers, or types, of aerosols have on clouds? Again, that is a topic for later chapters.

FURTHER READING

All of the references at the end of Chapter 5 will cover the material in this chapter, to a greater or lesser extent. In addition, there are specialist cloud physics books, such as

- *A Short Course in Cloud Physics* by R. R. Rogers (Pergamon, 1976): quite readable.
- *Microphysics of Clouds and Precipitation* by H. R. Pruppacher and J. D. Klett (D. Reidel, 1978): not for the faint-hearted!
- *Atmospheric Thermodynamics* by G. R. North and T. L. Erukhimova (Cambridge University Press, 2010).

REVIEW QUESTIONS

1. What is the difference between specific humidity and relative humidity?
2. Which is easier to understand? And why?
3. What does boiling point actually mean?
4. Why does boiling point drop when you go up a mountain?
5. How does the condensation of water vapour affect the adiabatic lapse rate?
6. What is CAPE? Why is it important?
7. Why does an advection fog look 'thicker' than a radiation fog?
8. What does 'hygroscopic' mean?
9. Why is a cloud condensation nucleus – i.e. a seed – necessary for droplet formation?
10. Explain the Twomey effect; and its implications.

EXERCISES

1. As we noted in Section 6.1, the evaporation and subsequent condensation of water vapour is a major component of the Earth's energy budget. One estimate gives this as 80 W m^{-2}, globally averaged. Use this number, and the latent heat of water (at 0°C) to calculate the globally averaged rate of evaporation, which must equal globally averaged precipitation. Your answer will, of course, be in SI units (metres per second), and rather small! Convert this result first to mm/h (severe weather units), then to mm/day (standard weather units), and finally to mm/year (climatology units).
2. The Clausius–Clapeyron equation may be used to determine the boiling point of water, which is defined as that temperature where saturation vapour pressure equals ambient pressure. Determine the boiling point at standard pressure, 1,013.25 hPa, using the value of latent heat at 0°C. Your answer will not be what you expected, as we have ignored how latent heat varies with temperature.
3. Repeat Question 2 for Denver, and for the top of Mt Everest. (In both cases you will have to obtain a 'reasonable' estimate of the relevant pressure.)

4. A person perspires in order to cool. How much liquid water would you need to lose in order to reduce your temperature by 1°C? (The result will be too large to make sense: in reality we only reduce our skin temperature, not our core temperature.)

5. In Section 6.2.2 we presented an expression for how boiling point might vary with altitude. This equation may be obtained by combining the Clausius–Clapeyron equation with equation (5.4), to give $1013.25 \exp(-z/H) = 6.11 \exp\left\{\frac{L}{R_v}\left(\frac{1}{273} - \frac{1}{T_b}\right)\right\}$, where we have used 1,013.25 hPa for surface pressure, and 6.11 hPa for saturation vapour at 0°C. Now you need to 'undo' the exponentials. Your manipulations will give you a value of the constant a that differs from the suggested alternative. However, using your value, you should be able to re-check your answers to questions 2 and 3.

6. A cloud of cross-sectional area 10 km^2, and a height of 3 km, has a supercooled liquid water content of 2 g m^{-3}. This nucleates to ice crystals present in a uniform concentration 1 l^{-1}. Calculate the mass of each ice crystal, the total number in the cloud, and the total rainfall eventually produced.

7   Atmospheric Circulation: Weather and Climate

One of the fundamental facts about the physical environment of our planet is that there is a major geographic gradient in the inflow of solar radiant energy. This imbalance is manifested in the large temperature difference between equatorial and polar regions. The tilt of the Earth's rotational axis modulates this temperature difference, producing the seasonal cycles we all experience.

However, this latitudinal temperature contrast would be significantly greater if it were not for the fact that there is a considerable transport of energy by the Earth's two fluid components. While the majority of this energy transport takes place in the atmosphere, the oceans also make an important contribution. In the process, both the atmosphere (the focus of this chapter) and the oceans (Chapter 13) flow in well-defined circulation patterns that strongly influence our day-to-day weather.

We will start with a semi-qualitative discussion of the forces that act on a fluid, remembering that we live on a rotating Earth. While it is pressure differences/gradients that move the air, it is temperature gradients that sustain these pressure gradients. The general circulation of the Earth's atmosphere, the dominant influence on the weather and climate most of us experience, is the result. It governs the transport of heat from the tropics to higher latitudes, and also explains why the world's deserts are where they are.

7.1   Forces on Fluids

Just like solid objects, fluids obey Newton's Laws of Motion, which means his Second Law in the form $F = ma$, where F is the force imposed on the 'object' of mass m, and a is the resulting acceleration. This is effectively a two-step exercise. Firstly, we need to identify what forces are acting; for example, the force of gravity. Secondly, we need to 'do the mathematics': this might be easy (textbook stuff), or it might be challenging. In the case of a fluid, which is effectively continuous, we need to focus our attention on a small segment, an additional complication.

Atmospheric motions are largely created by pressure differences, and then 'steered' by the Coriolis force, at least if we are above the frictional effects of the boundary layer (the lowest ~1 km). Pressure is force per unit area, and acts in all directions. Hence differences in pressure imply differences in force – or an unbalanced net force – acting on the fluid in between. In the vertical, the hydrostatic equation tells us that gravity balances the pressure gradient force, to a very good approximation. In the horizontal, pressure differences – or pressure gradients – exert forces on a segment of the atmosphere. We need to ask a number of questions:

1. Exactly how does a pressure gradient lead to fluid motion?
2. Is the pressure gradient the only force involved?
3. How do (horizontal) pressure differences/gradients arise?

7.1.1 Pressure Gradient Force

Consider a horizontal 'tube' (segment) of fluid, of cross-section A, and length Δx. At each end, the fluid next to it will exert an inward pressure (Figure 7.1). Let the difference between these two pressures be Δp; then the net (horizontal) force on this tube of fluid is $A\Delta p$. The volume of fluid in the tube is $A\Delta x$, and so its mass is $\rho A\Delta x$, where ρ is its density. Hence Newton's Second Law becomes

$$F = A\,\Delta p = \rho\,A\,\Delta x\,a$$

As we don't have fixed masses in the atmosphere, we usually talk in terms of the acceleration, or 'force per unit mass'. So, this equation becomes

$$a = (-)\frac{\Delta p/\Delta x}{\rho}$$

Pressure gradient *force* points from higher pressure to lower pressure, and so must the acceleration, whereas the pressure gradient points from lower to higher, hence the minus sign.

What other forces act on a fluid parcel? Firstly, there is *gravity*, which is balanced by the vertical pressure gradient force, as shown in the hydrostatic equation. Because this force is so strong, geophysical fluids mostly travel horizontally (actually *isobarically*: on surfaces of constant pressure) far more easily than vertically. Then there are **frictional forces**. Near the surface, the air is trying to move over the stationary Earth. This produces frictional forces, which (almost) always occur whenever there are variations in velocity. These effects are mainly felt in the lowest kilometre of the atmosphere, and will be considered shortly.

7.1.2 Coriolis Force

The Earth is a rotating object: known to physicists as a 'non-inertial reference frame'. If we want to do Newtonian mechanics in such a reference frame, we need to include a fictitious

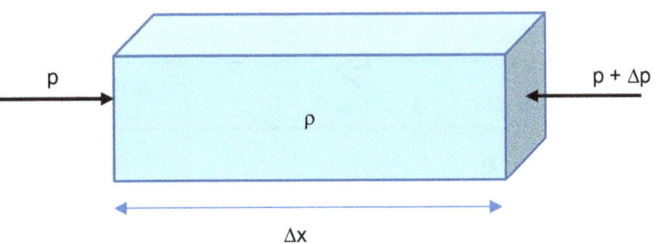

Figure 7.1 Pressure gradient force.

force – the Coriolis force (per unit mass) – to account for this rotation. A full analysis shows that

$$\text{Coriolis force/unit mass} = 2\Omega\, V \sin\varphi$$

where Ω is the Earth's rotation rate (once in 24 hours), V is the object's velocity (every moving object 'feels' this force), and φ is the latitude. This force points at right angles to the velocity vector: in the Northern Hemisphere, objects are deflected to the right; in the Southern Hemisphere, to the left.

The Coriolis force, or Coriolis effect, may be understood as follows. Consider a mass m at the equator, which is stationary with respect to the Earth's surface. Because the Earth is rotating at an angular velocity Ω radians per second, the mass has a linear velocity $u = R\Omega$ (in an inertial reference frame) in the easterly (or x) direction, where R is the Earth's radius. Now assume that our mass moves northward, without friction, along a meridian of longitude (at least to begin with): it will still have the same eastwards velocity u, but the underlying surface will have a smaller velocity of $R\cos\varphi\Omega$ (where φ is the latitude). Thus, to a stationary observer on the Earth, it will appear to be moving eastwards with a velocity $u = R(1 - \cos\varphi)\Omega$. The further it moves northwards, the faster it appears to be moving: it appears to be accelerating.

7.1.3 Geostrophic Winds

We will concentrate initially on the atmosphere above the boundary layer, so that we can ignore any frictional forces. Horizontal air motions are then governed by the two forces we have just introduced: the pressure gradient force, and the Coriolis force. Consider a small air parcel that is initially stationary: it won't feel the Coriolis force, only the pressure gradient force. Thus, it will accelerate, in a direction from high to low pressure. As soon as it does, it will have velocity, and from then on it will feel the Coriolis force. It will then be deflected sideways – to the right in the Northern Hemisphere, and to the left in the Southern Hemisphere.

As a consequence, the velocity vector will veer from its original direction and start to move more and more parallel to the isobars (lines of constant pressure), not across them. Before long we will reach an 'equilibrium' situation, where the pressure gradient force is pushing it in one direction and the Coriolis force is pushing it in the opposite direction, and these forces exactly balance each other. Now acceleration, but not velocity, will cease. This is shown in Figure 7.2.

Figure 7.2 Motion of an air parcel in the Northern Hemisphere under the pressure gradient and Coriolis forces.

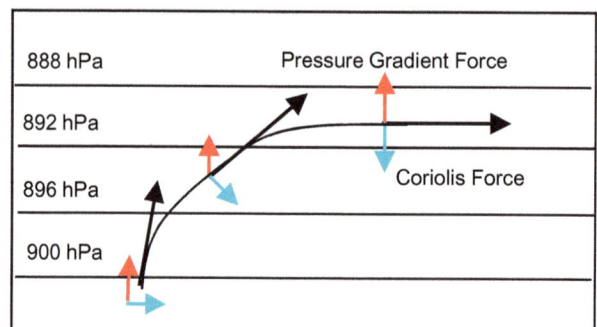

We can now calculate the wind speed, which is given by balancing the two forces:

$$\frac{1}{\rho}\frac{\Delta p}{\Delta x} = 2\Omega\,V\,\sin\varphi$$

Hence

$$V = \frac{\Delta p / \Delta x}{\rho\,2\Omega\,\sin\varphi}$$

This is known as the **geostrophic wind**.

The key point is that the wind speed is proportional to the pressure gradient. Where the isobars are more closely packed together – that is, the larger the pressure gradient – the faster will be the winds. We may understand this as follows. When the isobars are closer together, the pressure gradient is larger, and so is the pressure gradient force. The only way we can adjust the Coriolis force to balance this is to increase the wind speed. Note also that at higher altitudes, where air density is lower, a given pressure gradient will produce faster geostrophic winds. (This is one of the factors in jet stream formation.)

7.1.4 Friction and Surface Winds

Near the surface, frictional forces need to be considered. The lowest layers of air blow across the surface, and are slowed by the resulting frictional forces. Faster-moving layers above these layers experience friction from these slower layers, up to a height of roughly 1 km. Because of friction, the velocity is reduced, reducing the Coriolis force, but not the pressure gradient force. As a result, surface winds tend to blow partially across the isobars, from high to low pressure. The crossing angle varies with terrain, but is roughly 30°.

This situation is illustrated in Figure 7.3. The pressure gradient force (PGF) must point from high pressure to low pressure, and the Coriolis force (CF) must point at right angles to the velocity vector. Finally, the frictional force (FF) is always directed opposite to the velocity. The only way that these three forces may balance is for the velocity vector to have rotated through an angle α from its geostrophic direction of parallel to the isobars.

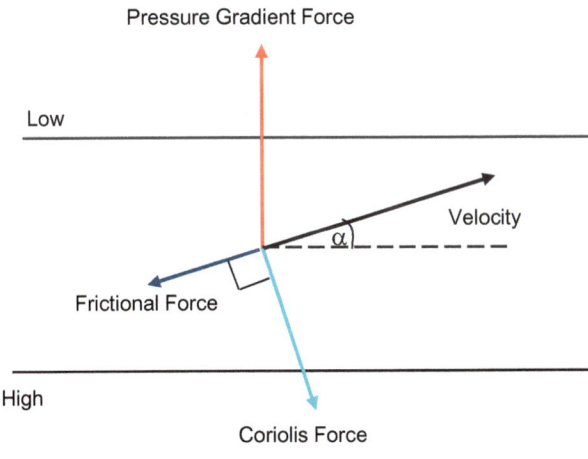

Figure 7.3 Balance of forces in the boundary layer. *Source*: *Physics of Radiation and Climate*, figure 6.1, Box and Box, CRC Press, 2016, reproduced by permission of Taylor & Francis Group.

Buys-Ballot's law states that *in the Northern Hemisphere*, if you stand with your back to the wind, Low pressure will be on your Left. The opposite applies in the Southern Hemisphere, of course. This law applies (more or less) strictly aloft. At the surface, however, we need to account for the effects of friction. You should stand with your back to the wind, but then rotate yourself 30° clockwise before pointing to high- and low-pressure centres. How would you apply this rule in the Southern Hemisphere?

When the full physics of fluid mechanics is developed in more advanced treatments, combining the various forces/effects we have discussed above – pressure forces, nonlinear viscous forces, and the Earth's rotation – the result is known as the Navier–Stokes equation, often referred to simply as the **momentum equation**. This is way beyond the level of our book!

7.2   Thermal Circulation

7.2.1   Charting Pressure

Before we go further, we need to think a little about how pressure varies with altitude. In Chapter 5 we studied the more or less exponential drop in pressure with height: by a factor of around 10 for every 16 km. How do meteorologists study this, in ways that can assist their endeavours? For example, what sort of graphing might we use to 'picture' the situation in, say, the mid-troposphere? As we shall discuss briefly below, this information can help them predict the progress of storm systems into the (near) future.

You might think that the answer is to plot a chart of pressure at an altitude of, say, 5 km, which we might agree is mid-troposphere: at least it's a nice round number. In fact, meteorologists reverse this. Rather than plotting pressure at a series of altitudes, they plot the height, in metres, of, say, the 500 hPa pressure surface, among others. Now we remember the hydrostatic equation, in the form where density has been removed, but temperature appears: let's turn it upside down:

$$\frac{\Delta z}{\Delta p} = -\frac{RT}{gp}$$

i.e.

$$\Delta z = -\frac{RT}{gp}\Delta p$$

Suppose we choose to plot, say, an 800 hPa chart, and a 700 hPa chart. What might be the *difference* in altitude? Well, we have fixed Δp to be –100 hPa, so what determines Δz? The key variable in the above equation is temperature, T. If T is comparatively large, this 'gap' will be larger, lifting the higher (in altitude) pressure surface (the 700 hPa surface) compared to other parts of the globe with lower temperature. What we have outlined by this somewhat round-about argument is something we assume our readers intuitively know: a warmer air column expands, upwards.

7.2.2   Thermally Driven Circulation

If it is pressure differences which drive atmospheric motion, what creates – and more to the point, sustains – pressure differences? Ultimately, atmospheric pressure differences are the result of temperature differences, and the most obvious cause of temperature differences is

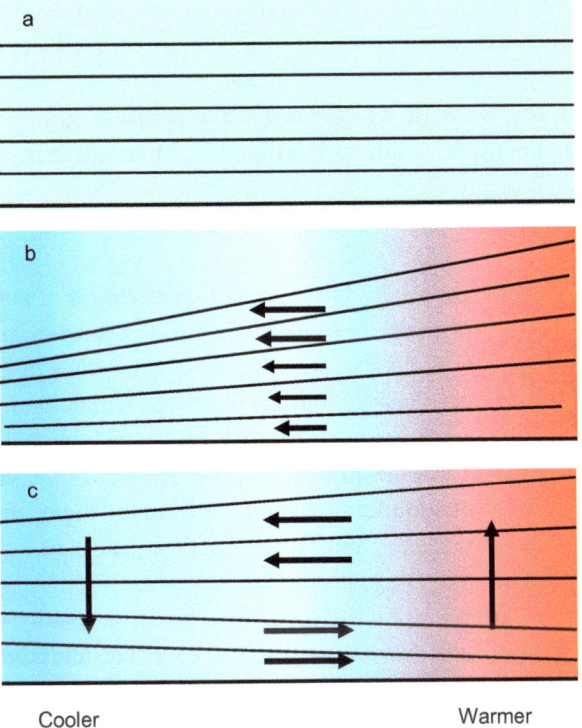

Figure 7.4 Development of a thermal circulation. *Source*: *Physics of Radiation and Climate*, figure 6.3, Box and Box, CRC Press, 2016, reproduced by permission of Taylor & Francis Group.

latitude. At a more local level, differences in the thermal properties of land and ocean surfaces can also be important. Let us examine the consequences of thermal imbalances between two air columns, and see how this can lead to a 'thermal circulation'.

Consider a 'slice' of atmosphere, which initially has horizontally homogeneous pressure and temperature profiles, as represented in Figure 7.4a. Some representative isobars have been included to indicate this. (The 'values' of these isobars will, of course, decrease with altitude. We have not labelled them, as this depends on the 'scale' of the motion being considered.) Now let us assume that, for whatever reason, the right-hand end of this slice is heated, while the left-hand end is cooled. Assuming this is done steadily, and over a sufficiently large area (minimising sharp gradients), both ends should remain more or less in hydrostatic equilibrium, and the argument we outlined above will apply. In particular, the warm end will expand vertically, while the cool end will contract: this is shown in Figure 7.4b. Because the total mass in each column remains unchanged – so far – the surface pressure remains unchanged, but the isobars at higher elevation must become progressively more tilted.

Now the (horizontal) pressure gradient force operates at all altitudes, not just ground level. This implies that, in the upper levels, there will be transport of air from the warm end of the region to the cool end, as also indicated in Figure 7.4b. Now the column masses are no longer equal, as the cool end has acquired some mass from the warm end, which means that its surface pressure is now higher. The isobars at lower elevations must now be sloping downwards from the cool end to the warm end, as shown in Figure 7.4c. Thus, at lower elevations there is a mass flow from the cool end to the warm end, creating a circulation pattern, including ascending and descending air motions, also shown in Figure 7.4c. Provided that we maintain the temperature difference/gradient between the two ends of this atmospheric slice, this thermal circulation will continue indefinitely.

There are two key ways in which such a temperature gradient may be maintained: the equator-to-pole temperature gradient, and land–sea thermal inertia differences.

7.2.3 Sea Breeze

During the daytime, and especially in the longer, hotter days of summer, the sun heats a land surface to a higher temperature than it does a neighbouring sea surface, due to the much lower heat capacity of land compared to ocean. (This is a product of their considerably different specific heats, and also their densities and the depths to which the heat penetrates.) As a result, some of this heat is transferred to the air column above the land, making it the warm end of our atmospheric slice. Thus, at low elevations, air flows from over the sea to over the land – the sea breeze many enjoy on a hot summer afternoon. During the night, the situation will be reversed, as the land will cool down faster than the sea surface, leading to a reverse circulation: a land breeze. Both of these situations are illustrated in Figure 7.5.

These circulation patterns are clearly going to be more readily observed in summer, when the insolation is greater, and for longer, leading to larger temperature gradients. Those who live close to the ocean are, of course, well aware of the sea breeze, and of its cooling effects on a hot afternoon. In Perth, Western Australia, this is known as the Fremantle Doctor, because of the welcome relief it can bring in summer; Fremantle being the coastal port for the city of Perth.

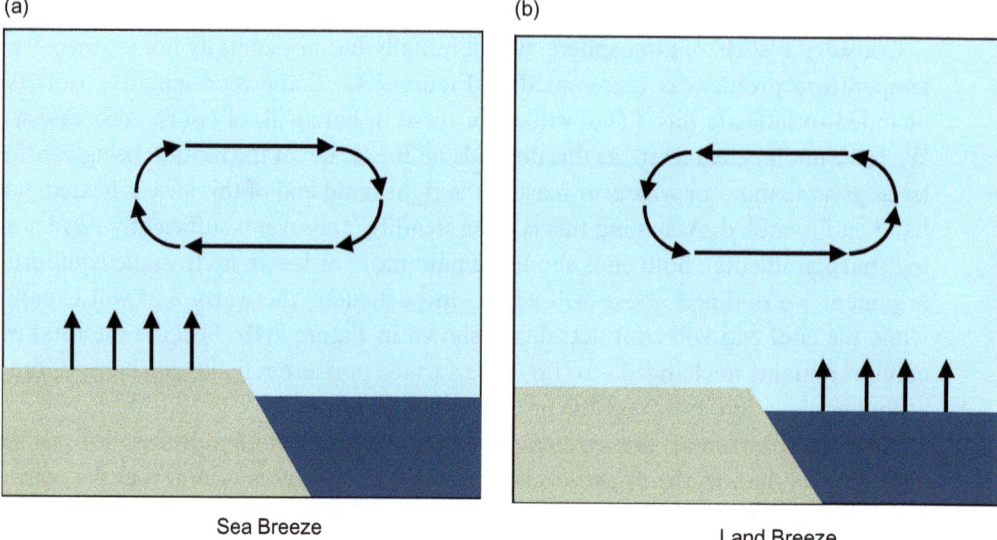

Figure 7.5 Sea and land breezes. *Source: Physics of Radiation and Climate*, figure 6.4, Box and Box, CRC Press, 2016, reproduced by permission of Taylor & Francis Group.

There is one other situation where the land–sea thermal difference plays an important role, and that is at the continental scale. During summer, the interior of a continent will heat up to higher average temperatures than the surrounding (or neighbouring) oceans, creating a local low-pressure centre, known as a 'heat low'. This will have implications for the synoptic-scale circulation in the region.

Global warming appears to be creating an 'interesting' variation on this pattern. Ice and snow are highly reflective, so that ice-covered regions such as the Arctic Ocean can be very cold. When that ice melts, the ocean now absorbs solar radiation and will, in general, be warmer. This may be affecting weather patterns in the higher northern latitudes, such as the wave pattern of the jet stream (below), assisting with occasional outbreaks of arctic weather down into parts of the USA, seeming to contradict what we have come to expect of a warming planet.

7.2.4 Hadley Circulation

The first attempt to explain atmospheric circulation was by the English lawyer and amateur meteorologist George Hadley, in 1735. He proposed a simple, direct thermal circulation in each hemisphere, with warm air rising near the equator, and moving poleward at higher elevations, and a surface flow back to the equator (see Figure 7.6(left)). This flow pattern would be fine on a slowly rotating planet, but strikes problems on Earth. (For a planet that has one face locked to the Sun, as our Moon has to the Earth, the thermal circulation would actually resemble Figure 7.6(right), as the thermal gradient would be from the sunlit side to the dark side.)

A thermal circulation as described by Hadley would require a surface low at (or near) the equator, and a surface high at the poles (which happens to be true), with a pole-to-equator pressure gradient and resultant airflow, in the lower troposphere. However, the Earth's rotation leads to flows parallel to the isobars as just discussed, which would lead to easterly winds throughout the lower troposphere. Due to frictional forces at the surface, such winds would be in a fight with the Earth's surface, attempting to transfer westerly angular momentum to the Earth, and easterly angular momentum to the atmosphere, a battle the solid Earth would obviously win. Such a situation could not persist in a steady state.

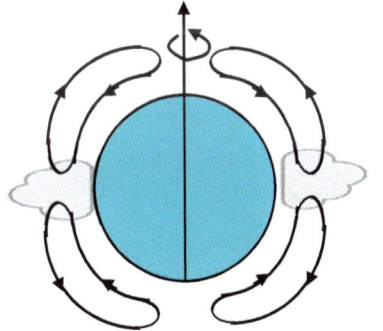 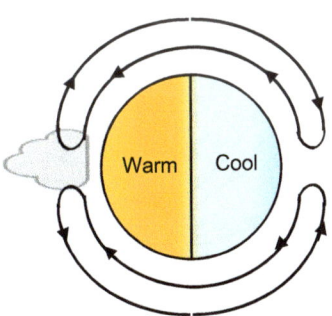

Figure 7.6 Simple circulation patterns on a slowly rotating and a non-rotating planet. *Source*: *Physics of Radiation and Climate*, figure 6.5 a and b, Box and Box, CRC Press, 2016, reproduced by permission of Taylor & Francis Group.

7.3   The General Circulation

While we do generally observe low pressures at, or near, the equator, and high pressures at the poles, the observed surface pressure patterns are somewhat more involved than the simple equator-to-pole pressure gradient of Hadley's imagining. What we actually see is a low-pressure belt near the equator (the **Inter-Tropical Convergence Zone**, or ITCZ), which migrates north and south with the solar seasons, and then a band of high pressure at about 30°N and 30°S, known as the subtropical high-pressure belts (or ridges). Between these and the equator, there are northeasterly winds in the Northern Hemisphere and southeasterly winds in the Southern Hemisphere: these are the trade winds used by the early sailing ships. Around 60°N and 60°S are two low-pressure belts, and between them and the subtropical highs are the prevailing westerlies, including the 'roaring forties' and the 'furious fifties'. At the poles are usually highs, and the winds there are usually easterlies.

What this means is that if we take a vertical slice through this pattern, we see not one but three cells (Figure 7.7). Between the equator and the subtropical highs, we have a thermally driven direct circulation, with warm air rising at the equator and sinking in the subtropical ridges, in accordance with the mechanism outlined above. This cell is known as the **Hadley cell** (in his honour), and it is clearly capable of transporting energy (heat) from tropical to subtropical latitudes. Between 30° and 60°, where temperature gradients are often strongest, the meridional (i.e. north–south) circulation is opposite to what would be expected of a thermal circulation: it is said to be **thermally indirect**. It actually has to fight against the forces we have been discussing. This is called the **Ferrel cell**, and clearly cannot be a heat transport mechanism.

Figure 7.7 Global circulation of the Earth's atmosphere. *Source: Physics of Radiation and Climate*, figure 6.5c, Box and Box, CRC Press, 2016, reproduced by permission of Taylor & Francis Group.

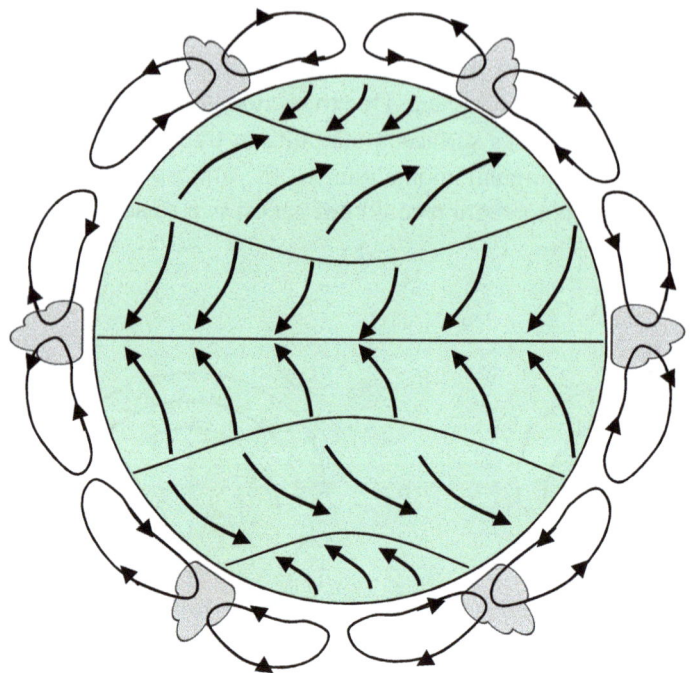

At the highest latitudes, a third **polar cell** exists, although its circulation is relatively weak: there isn't much of the globe's surface area left by those latitudes. Like the Hadley cell, this cell is also thermally direct.

7.3.1 Mid-Latitude Transport

Although the poleward transport of energy in the tropics and subtropics is produced by a direct, thermal circulation – the Hadley cell – this mechanism is not available in the mid-latitudes. In this region the dominant transport mechanism is eddy motion, on the synoptic to planetary scale. Eddies develop when air is forced to veer around slow-moving, or even stationary, high- and low-pressure regions. When the stream lines approach closest to the poles, we refer to **ridges**, while places where the stream lines move closest to the equator are called **troughs**. Whilst the air is in a trough region, it is likely to pick up energy, perhaps by passing over warmer land or ocean surfaces (depending on the season). This energy may be lost, both by radiation and by thermal contact/transfer, when the air is in a ridge.

What causes the necessary troughs and ridges – the high- and low-pressure zones? One major cause is the seasonal difference in thermal properties of land and sea surfaces, as we noted above. During summer, land surfaces, with their smaller thermal inertia (i.e. heat capacity) heat up more quickly than the surrounding seas. This will establish a thermal circulation, with a surface low over the land, and a surface high over the sea. In winter, this situation is reversed. Eddies produced by this mechanism tend to have continental-scale dimensions. A second mechanism produces seasonal, migrating high- and low-pressure cells, of synoptic scale. This mechanism is basically dynamic, and is associated with the fronts which appear between contrasting air masses. Finally, mountain ranges also play a role (although mainly in the Northern Hemisphere), as might be expected.

7.3.2 Vertical Air Motions

The circulation of the atmosphere is, of course, three-dimensional, so examining two-dimensional maps of pressures and winds is only half the story. While the horizontal component of the wind vector is invariably much larger than the vertical component, vertical motion is an essential feature of the circulation. Near the surface, air moves across the isobars, from high to low pressure, as outlined above. Thus, air is spiralling into the centre of a low, converging. Where does it go? The only answer is up. Above a height of around 6 km, air will slowly spiral out again, diverging. Similarly, air is spiralling out of a high-pressure area at the surface, and can only have come from aloft. The rate of air rise or descent above a low- or high-pressure centre is quite small – about 1.5 km per day, or several cm per second (~1 m per minute). Nevertheless, this is quite high enough to elevate typical cloud droplets, as we have seen in Chapter 6.

If the upper-level divergence (convergence) exactly balances the surface-level convergence (divergence), then there will be no change to those pressure systems. If not, there will be a weakening or strengthening of those systems, as the case may be. This is one of the reasons meteorologists study upper-level charts.

7.3.3 Upper Air Motion

Because equatorial regions are warmer than polar regions, there is a strong upper tropospheric pressure gradient from equator to pole in both hemispheres. An air parcel starting out near the equator, in either hemisphere, will start to move directly to the pole, before being deflected by the Coriolis force: to the right in the Northern Hemisphere, and to the left in the Southern. Thus, in *both* cases, the winds end up blowing basically west to east: the term for this is **zonal.** Note also that in the winter hemisphere the temperature gradient is greater, and so must be the pressure gradient. That is why winds tend to be stronger in that hemisphere (i.e. that season). At the upper boundaries of our three cells the slope in the isobars tends to be greatest: this leads to the strong winds known as the **jet streams**. Again, the larger temperature gradient in the winter hemisphere increases the strength of the jet.

In the middle troposphere, say a pressure altitude of 500 hPa, the influence of surface inhomogeneity is greatly reduced compared to the surface, and air flow is much smoother: more zonal. Day-to-day weather patterns often show a reasonably strong wave structure, but seasonally averaged this is masked. Ridges are places where a height line has arched poleward, indicating warmer air, while troughs are places where a height line has arched equatorward, indicating colder air. The Southern Hemisphere mean flow is particularly smooth, as this hemisphere is ocean dominated: the polar vortex, for example, is stronger, which has direct connections to the Antarctic ozone hole, as will be seen in Chapter 10.

Any daily weather map will show a well-developed wave structure, caused by thermal patterns and orographic features, that travel more slowly than the winds which actually blow through them. There are generally 3–6 in the Northern Hemisphere and 3–4 in the Southern Hemisphere. The Northern Hemisphere mean flow in January (winter) shows a ridge over the eastern Atlantic (caused by the warm Gulf Stream) and Pacific oceans, and troughs over the east of North America (due to the Rockies), as well as over East Asia and eastern Europe. The July (summer) pattern is less intense.

The position and intensity of the waves influences surface weather patterns. Air travelling through a ridge actually moves faster than geostrophic speed, due to the centripetal force – 'supergeostrophic' – whereas air travelling through a trough will be subgeostrophic. As a consequence, air will converge ahead (east) of a ridge, subside, and diverge at the surface, producing a surface high (anticyclone). Similarly, air will diverge ahead of a trough, inducing convergence at the surface, and rising air: a surface low.

7.3.4 Hydrology

The circulation of the atmosphere has enormous implications for the circulation – and especially precipitation – of water vapour. Consider the rising arm of the Hadley cell, near the equator. Air flowing in to this region is largely flowing over warm oceans, especially over the western Pacific Ocean and the Indian Ocean, and will thus accumulate very large quantities of water vapour. As this air rises, it cools adiabatically, as we discussed in Chapter 6, and is no longer able to hold all of this moisture. Thus, the ITCZ is the site of intense rainfall.

When this air flows north or south in the mid- to upper troposphere, to the region of the subtropical highs, it is thus comparatively dry. As it descends at these latitudes it warms adiabatically, so that its *relative* humidity decreases even further, and thus produces little or no rainfall. It is precisely for this reason that the world's deserts are located at these latitudes, not, as might have been naively guessed, around the equator. Not only do these latitudes receive very little rainfall, dry air means few clouds, which means strong solar radiation, especially in the summer months.

The ITCZ moves north and south with the seasonal movement of the sun. Thus, from December to February it is found over northern Australia. In July, the Indian subcontinent and South-East Asia experiences its monsoonal rain. Due to the differing heat capacities of land and sea, the ITCZ is attracted to the continents. When coupled with differing amounts of land in the two hemispheres, we find that the ITCZ moves further to the north than it does to the south. It is for this reason that India actually receives two monsoons each year, as the ITCZ sweeps across it, and back again. Both the Hadley and Ferrel cells migrate with the seasons to some extent, with the Southern Hemisphere subtropical highs moving north and south by $\sim 10°$. The situation in the Northern Hemisphere is more complicated due to the distribution of the continents.

Summary

Our atmosphere acts, in many ways, as a machine, driven by the difference in temperature between low and high latitudes. It is these differences – gradients – that create the circulatory system which governs much of our weather and climate. Thus, the fundamental take-home message we expect our readers to remember is the concept of a thermal circulation.

This circulatory system is also the reason we have monsoons (with their seasonal cycle) near the equator, and deserts around 30° north and south: the descending arm of the Hadley cells; a.k.a. the subtropical ridges. In Section 18.2.3 we will see that the IPCC's AR6 has found evidence that the northern Hadley cell appears to be expanding northward. This, of course, has the potential to increase rainfall in some locations and reduce it in others, possibly heightening tensions between neighbouring countries and peoples.

FURTHER READING

There are many books that take a broad, general look at weather and climate, including

- *Fundamentals of Weather and Climate* by R. McIlveen (Chapman Hall, 1992).
- *Meteorology Today* by C. D. Ahrens (Thomson Learning, 2007).
- *The Weather and Climate of Australia and New Zealand* by A. Sturman and N. Tapper (Oxford University Press, 2006).
- *Atmosphere, Weather and Climate* by R. G. Barry and R. J. Chorley (Routledge, 2003).
- *Atmospheric Science for Environmental Scientists* by C. N. Hewitt (Wiley-Blackwell, 2020).

A fully detailed fluid mechanical treatment of atmospheric circulation is one of the most highly mathematical branches of atmospheric science. For readers who might wish to delve into this topic, we suggest

- *An Introduction to Atmospheric Physics* by D. G. Andrews (Cambridge University Press, 2000)
- *Atmospheric Science, An Introductory Survey* by J. M. Wallace and P. V. Hobbs (Academic Press, 2006)

REVIEW QUESTIONS

1. Explain the pressure gradient force.
2. Explain the Coriolis force.
3. Explain how frictional forces alter atmospheric flow in the boundary layer.
4. Outline the principles of a thermal circulation.
5. Explain why the original single-cell model of atmospheric circulation does not work on Earth.
6. Why do we say that the Ferrel cell is 'thermally indirect'?
7. What happens when air spirals into a low-pressure centre?
8. Explain why the world's deserts are where they are.

EXERCISES

Collect your local synoptic chart, either from a newspaper, or online, for two weeks, and note the progress of features such as high- and low-pressure centres, fronts, etc. Make a note of any unusual weather during this period. You might also like to take careful note of the forecasts provided in your area. (Note that weather presenters in the media do not always understand some of the subtleties of the 'official' weather service forecasts, or may not be afforded sufficient airtime to cover such details, especially where they involve statements around likelihood: try to get hold of these as well.)

Part III

Radiant Energy

Apart from a tiny amount of energy generated in the core by radioactive decay, and tidal energy from the gravitational interaction of the Moon, the Earth generates no energy of its own. Yet this is a planet full of life, both biological, and physical – assuming we may stretch the language, and call atmospheric circulation, and the hydrological cycle, forms of 'physical life'. (James Lovelock's Gaia hypothesis treats the entire planet as a living organism.)

Fortunately, not only is our planet part of a solar system, with the Sun at the centre, but it is also in the so-called 'Goldilocks zone', where liquid water can exist. That is to say, a region where surface temperatures lie between 0°C and 100°C. The primary determinant of a planet's inclusion in that zone is the solar energy – insolation – it receives, which in turn depends on solar output (luminosity) and orbital distance. We look briefly at these in Chapter 9.

This incoming energy must be balanced by energy emitted by the planet (or else it would melt!), the amount of which is strongly temperature-dependent. The key science that we need for this is the work of two giants of early twentieth-century physics, Max Planck and Albert Einstein (completing some earlier work by Stefan and Boltzmann). When balance between the inflow and outflow of radiant energy is established, it also establishes the (radiative) temperature of that planet.

However, when we do the numbers for our planet, we find a temperature that is well below 0°C. The 'solution' to this obvious 'problem' is to be found in our atmosphere, and in certain of the gases it contains. These act much like a blanket on a bed at night, keeping the surface temperature (either yours, or the Earth's) at a comfortable level. This is what we refer to as the greenhouse effect.

In Chapter 8 we present the 'big picture stuff', starting with the work on thermal emission of electromagnetic radiation we mentioned above. We need to know how much, and over what wavelength range, the emission occurs. The greenhouse effect is all about the exchanges of energy between the planetary surface, the atmosphere, and space. We will illustrate the basic science of the greenhouse effect with a very simple model.

To understand how and why our climate is changing, we need to fully understand both the inflow and outflow of energy through the atmosphere, which separates where we live from the void of space, and how any changes to the atmosphere might be impacting on those flows. So, we start in Chapter 9 with the simpler of the two, the inflow of solar radiation. How much might be absorbed in the atmosphere, and why? How much might be scattered back to space? Might either of these be changing, as a result of the changes we have noted in previous chapters?

The key to the habitability of our planet, as we noted above, is certain gases found in 'just the right amounts' in our atmosphere. These greenhouse gases trap outgoing radiant energy, and re-emit some of it back to the ground, raising its temperature to a comfortable level. Because it is changes in these gases that are causing so much concern among climate scientists, and more and more community leaders, we need a thorough understanding of how this works. This will be covered in Chapter 11. This is somewhat complex material, involving the details of how radiation interacts with individual atoms and molecules, requiring a qualitative excursion down into the quantum world.

Like Parts I and II, this part has three science chapters. But we have also included two chapters that make use of that science: applications chapters. The first of these, Chapter 10, looks at the vital ozone layer, without which terrestrial life would be impossible. Ozone exists in the stratosphere as a result of a series of photochemical reactions, not dissimilar to those you saw in Chapter 2, again creating a photochemical balance. Unfortunately, that balance has been under threat from certain man-made chemicals. Fortunately, this is one threat that was recognised in time, and the world community has taken what we hope are the right actions to avoid disaster.

A great deal of the data which are fed into our numerical weather models come from satellite observations. In parallel with this, scientists are finding more and more ways to use satellite technology to 'measure' changes in our environment, whether in air quality or the climate. How can an instrument 500 km (and more) above the Earth provide so much information about the lowest few kilometres of the atmosphere? That is where all of the science that we have covered in this part comes together, as we will show in Chapter 12.

8   Radiant Energy, the Primary Climate Driver

We turn now to the third of the three branches of science which underpin our changing climate: radiant energy. Although clearly the subject matter of physics, radiant energy, or the energy carried by light and other electromagnetic waves, and the ways it interacts with the gases and particles in our atmosphere, is so central to our understanding of climate that we feel it makes more sense to place it on its own pedestal.

In this chapter we will outline the necessary science, and especially the laws of thermal emission. It turns out that we are bathed in electromagnetic radiation that we can't see. We will then show how these laws lead to the counter-intuitive outcome that the 'temperature' of our planet is actually well below freezing!

Fortunately, however, our Earth has a 'good' atmosphere, containing some very important gases – greenhouse gases – which act as a blanket, keeping the surface at a suitable temperature for liquid water, and hence for life. A simple model will help you gain a basic understanding of the greenhouse effect, in preparation for the heavy stuff in Chapter 11.

8.1   Electromagnetic Radiation

Visible light is part of the **electromagnetic spectrum**, which is comprised of waves of oscillating electric and magnetic fields, in accordance with the equations obtained by James Clerk Maxwell 150 years ago. The story of Maxwell, and his famous equations, is fascinating. Many of his contemporaries refused to accept it because, based largely on mathematical logic, he added an extra piece that had not been found by experiment. It is this term that allowed him to obtain a wave solution to his equations, and determine that the speed of this wave was, in fact, the speed of light. So now we understand that light – indeed, all electromagnetic radiation, from radio waves to X-rays – consists of oscillating electric and magnetic fields, with their electric and magnetic field vectors at right angles to the direction of propagation, and to each other (this allows light to be polarised). The whole of optics is governed by this theory: reflection, refraction, scattering; *everything*.

Any wave may be characterised by its wavelength, λ, the distance between successive peaks, and its frequency, f, the number of peaks that pass by any given point per second (the unit here is Hz, or cycles per second). The product of wavelength times frequency is thus the speed of the wave, v. In the special case of light and other electromagnetic waves, this speed is a fundamental constant of Nature, and so is given its own symbol, c (now *defined* to be 299,792,458 m s^{-1}). We may express this connection in the form $c = \lambda f$. Among many other important properties, all waves carry (i.e. transport) energy, with their **intensity** (power per unit area) usually measured in watts per square metre: W m^{-2}.

The electromagnetic spectrum spans (in principle) all wavelengths from zero to infinity. We prefer to think in terms of wavelength (more concrete than frequency), which we usually measure in micrometres (microns), and some of the time in nanometres. **Visible light** spans the range from 0.4 to 0.7 µm (400–700 nm). (Note that the frequencies of visible light are of the order of 10^{14} Hz.) **Ultraviolet** covers shorter wavelengths (down to X-rays and gamma rays), while **infrared** covers longer wavelengths, out to 1,000 µm (or 1 mm). Beyond that lie microwaves, used for radar, and radio waves, used for various forms of communication.

In 1887, Heinrich Hertz put Maxwell's ideas to the test, using electrical circuitry to generate electromagnetic waves in the radio frequency range, effectively ushering in the whole of modern communication technology. His name lives on, of course, in our unit of frequency.

With this confirmation, some physicists convinced themselves that this work, plus Newton's mechanics, signalled the completion of our attempts to understand the physical world. How wrong they were! Lurking in the background were one or two little niggles that many chose to ignore. Yet just a decade later, struggling to sort these out, a couple of unexpected discoveries (X-rays, radioactivity, the electron) would open up the atomic, and sub-atomic, world and lead on to quantum theory. While most of that can wait for Chapter 11, we must now look at one of these puzzles, and its solution.

8.1.1 Thermal Emission

All bodies with temperatures above absolute zero (0 K) emit electromagnetic radiation; hence, this is often referred to as thermal or heat radiation. A **black body** is an ideal object that *absorbs* all radiation incident on it, and reflects none; hence, it should appear black. A black body is also the most efficient *emitter* of radiation. This can be thought of as Nature's way of obtaining a balance, or equilibrium. A black body at temperature T(K) emits radiation from its surface at the rate (intensity) given by the Stefan–Boltzmann equation $I = \sigma T^4$ W m^{-2}. σ is the Stefan–Boltzmann constant: its value is $\sigma = 5.67 \times 10^{-8}$ W m^{-2} K^{-4}. As you can see, the emission increases rapidly with temperature.

The emission spectrum (energy as a function of wavelength) of black-body radiation follows Planck's law. This law is one of those puzzles just alluded to, and has a very interesting history. Nineteenth-century physicists tried and failed to solve the problem, although they did produce some good science, including the Stefan–Boltzmann equation. Planck 'solved' this problem in 1900 by assuming that light can only be absorbed and emitted in discrete amounts, or quanta. The Planck function may be written in the form

$$E(\lambda, T) = \frac{2hc^2}{\lambda^5[exp(hc/\lambda kT) - 1]}$$

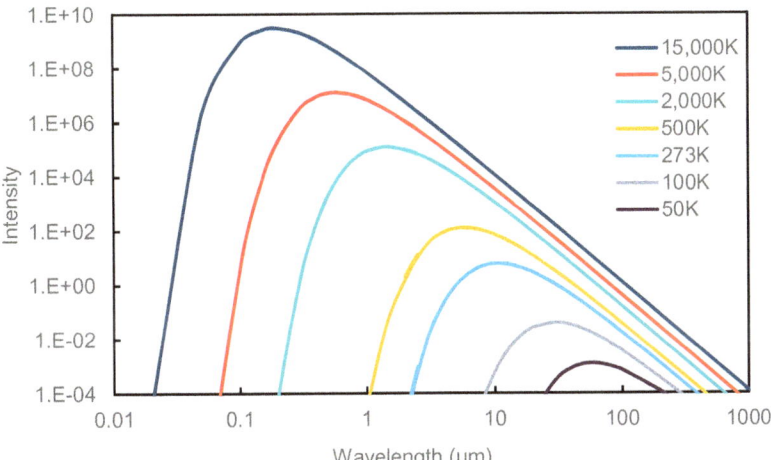

Figure 8.1 The Planck function for a range of absolute temperatures.

Here, T is the absolute temperature, λ is the wavelength, c is the speed of light, k is Boltzmann's constant which we met earlier, and $h = 6.634 \times 10^{-34}$ Js is Planck's constant.

Figure 8.1 shows the Planck function for a range of temperatures. The wavelength at which this spectrum peaks is inversely proportional to temperature (Wien's law). As the temperature increases, thermal emission increases at all wavelengths, but especially at the shorter end.

For a body which is not 'black' we may 'invert' the Stefan–Boltzmann equation to obtain a temperature, which we may interpret as its effective (or radiative) temperature. That is to say, if we can measure I, then we can use the Stefan–Boltzmann equation to work back to a temperature, T, which 'effectively' radiated the measured intensity, I.

A body that is 'non-black' will both absorb, and emit, less radiation than a black body. We'll need a gut feeling. A body which is not 'black' will absorb a fraction, a_λ of the radiation incident upon it: its **fractional absorptivity**. This varies (strongly in the case of a gas) with wavelength, λ, hence the subscript. Such a body will also emit a fraction, e_λ, of the radiation that a black body would emit at that wavelength (and temperature): its **fractional emissivity**. **Kirchhoff's law** then tells us that $a_\lambda = e_\lambda$ for all wavelengths, λ. So, a good absorber is (potentially) a good emitter: it will obviously depend on both temperature and wavelength; how large is the Planck function, E, for that particular T and λ?

8.1.2 Light Quanta

In 1905 Einstein published four remarkable papers. The two best known relate to his Special Theory of Relativity (solving another of the niggles referred to above). The third related to his PhD thesis topic, Brownian motion. In the fourth, he solved the photoelectric effect/problem by proposing that, while light usually behaves as a wave, under certain circumstances it can behave as a particle, called a **photon**. In doing so he justified the assumption that Planck had introduced in 1900. (It was for this work, *not* for either Special or General Relativity, that he received the Nobel Prize in 1921.) This property is central to the way in which light – indeed, all electromagnetic radiation – interacts with atoms and molecules (photochemistry; Chapter 2):

and especially how it is absorbed, including by the photoreceptors in our eyes. (This is the basis of colour vision, for example.)

Each photon carries an exact amount of energy, which is proportional to its frequency: $E = hf$, where h is Planck's constant. The value of h is a very small number, a reflection of the fact that quantum theory deals with 'the very small': that is the key reason it took science and its tools so long to become aware of its existence. (Similarly, of course, special relativity deals with 'the very fast': speeds approaching c.) The frequency dependence of this formula means that we become more aware of the particle properties of light as the frequency gets higher: for example, gamma rays. By contrast, the photons in a radio wave carry so little individual energy that we need so many that you would never be able to observe the 'lumpiness'.

8.1.2.1 The Electron Volt

When we enter the quantum world, we are entering the realm of individual atoms and photons: a world in which the macroscopic units of mass and energy lose their meaning, so we need to introduce new ones. For mass, the atomic mass unit, standard throughout chemistry, is appropriate. We now need to introduce a new unit of energy.

When an electrically charged particle is placed in an electric field it will be accelerated, and gain (kinetic) energy, equal to qV, where q is its electric charge, and V is the potential difference through which it is accelerated. We define the electron volt to be the energy gained by an electron which has been accelerated through a potential of 1 volt. Because the electric charge carried by an electron is 1.6×10^{-19} coulombs (the unit of electric charge), that means 1 eV = 1.6×10^{-19} J. An alternative value for Planck's constant is $h = 4.135 \times 10^{-15}$ eVs.

As we stated above, we prefer to think in terms of wavelength, not frequency, in which case the Planck/Einstein formula becomes

$$\begin{aligned} E = hf &= hc/\lambda \\ &= 4.135 \times 10^{-15} \times 3 \times 10^8 /\lambda \\ &= 1.24/\lambda \, (\lambda \text{ in } \mu m, E \text{ in eV}) \end{aligned}$$

Visible light spans the range 0.4–0.7 μm, with a photon energy range of 3.1–1.8 eV. UV radiation (i.e. photons) below ~0.31 μm, which have energies above ~4 eV, can break (bio) chemical bonds in cells and cause skin cancer, etc. [The UV is generally subdivided according to: UVA, 0.32–0.4 μm; UVB, 0.29–0.32 μm; UVC, <0.29 μm; different authors may use somewhat different cut points.]

8.2   Radiation and Climate

The Earth's climate is governed by the energy balance between energy coming in to the planet, and energy going out; specifically (see Figure 8.2):

- incoming *solar* radiation, of intensity F, minus the fraction, α, which is reflected (referred to as the planetary **albedo**); both are measured by satellite; and

• the emission of thermal (a.k.a. '*terrestrial*') radiation: *but* what is it that determines this emission?

If we *assume* that the Earth is a black body with an unknown effective/radiative temperature T, then we can determine T by demanding this balance. So, we equate these two terms:

$$(1 - \alpha)F\pi R^2 = \sigma T^4 4\pi R^2$$

$$\therefore T = \sqrt[4]{\frac{(1 - \alpha)F}{4\sigma}} = 255K = -18°C$$

where we have used $F = 1{,}364$ W m^{-2} and $\alpha = 0.3$. Note that the Earth intercepts solar radiation as a disc, of area πR^2, but radiates as a sphere, of area $4\pi R^2$.

Seems a bit cold! (Average surface temperature is ~15°C.) But satellite observations confirm this result, at least in terms of the incoming and outgoing energies. Is the physics wrong? No: it's just incomplete.

8.2.1 Greenhouse Effect

An examination of the Planck spectra for $T = 5{,}750$ K (the sun's approximate temperature), and 255 K (the Earth's effective temperature), shows quickly that (see Figure 8.3)

• 99% of sunlight has wavelengths of less than 4.0 μm: we speak of **short-wave radiation**;
• 99% of earthlight has wavelengths more than 4.0 μm: we speak of **long-wave radiation**.

Our atmosphere contains a number of gases that absorb in the long-wave spectral region (i.e. $\lambda > 4.0$ μm): these are known as greenhouse (or **radiatively active**) gases. These include water

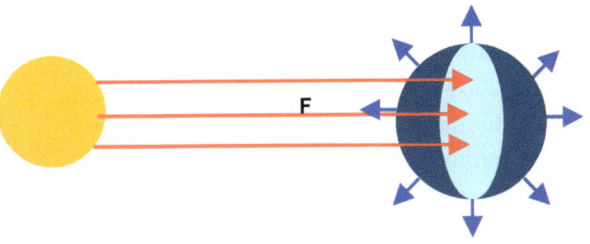

Figure 8.2 Schematic diagram of the inflow of solar radiation and the outflow of thermal radiation.

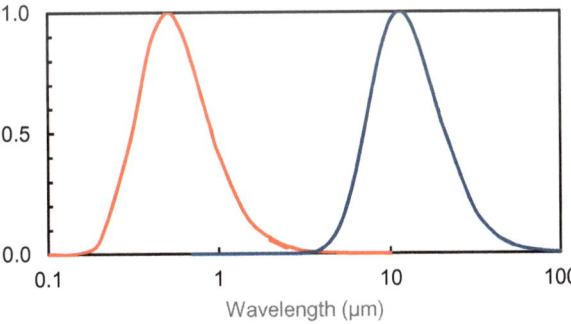

Figure 8.3 Normalised Planck function for 5,750 K (left) and 255 K (right).

vapour (H_2O), carbon dioxide (CO_2), methane (CH_4), nitrous oxide (N_2O), ozone (O_3), plus some man-made gases such as the CFCs, etc.

Because of its temperature, the Earth's surface emits radiation in the 4.0–100.0 μm region (and beyond), as we see from Figure 8.3. Most of this is absorbed by the greenhouse gases. But the atmosphere is at a 'comparable' temperature to the surface, so by Kirchhoff's law these gases will re-emit (much of) this radiation; some to space, but more back to the surface, making the surface warmer. This is the greenhouse effect, or more pedantically, the 'atmosphere effect'.

8.2.2   Atmospheric Energy Budget

Having introduced the most basic ideas of the Earth's energy balance, and the gases which modulate long-wave radiation flows, it is time to go into more quantitative detail. If we were to venture to the top of the atmosphere (TOA) with a suitable detector aimed directly at the Sun, we would discover that it intercepted approximately $F = 1,364$ W m^{-2} of radiant energy, as noted above: F is known as the **solar constant**. (This value varies a little, for a number of reasons which we will discuss later, and has recently been revised down a tad, but will do for now.)

The solar beam effectively sees the Earth as a disc of area πR^2, while the Earth is, of course, a sphere of area $4\pi R^2$. Thus, the average downward flux at the top of the atmosphere will be a quarter of 1,364, or 341 W m^{-2}. However, that is only the start. If we now turn our detector to point downwards, we would find that, on average, ~30% of this energy is reflected back by the atmosphere – especially clouds – and the planetary surface below. This reflected component is the albedo, a. Again, you have already seen this number. Thus the (average) *net* inflow of solar energy is $341 \times 0.7 \approx 239$ W m^{-2}.

A suitable downward-looking detector would also measure upwelling thermal radiation from the Earth/atmosphere of (on average) 239 W m^{-2}. (All of these measurements are, in fact, routinely performed by space-based instruments.) So, what we have seen is that energy inflow equals energy outflow. This is, effectively, the basis of the calculation above.

If we were to repeat this set of measurements at the Earth's surface, we might be surprised at just how different the numbers were. Figure 8.4 displays what is known as the Earth energy budget (or, less correctly, radiation budget). On the left we have the flows of incoming solar radiation; on the right, the flows of outgoing thermal radiation; and in between are two non-radiative contributions. We will examine these in turn.

On the incoming (solar/short-wave) side, we start with 341 W m^{-2}, globally averaged. Of this, 79 W m^{-2} are reflected by the atmosphere, mostly by clouds, but also by gas molecules and aerosol particles. A further 23 W m^{-2} are reflected by the surface, especially by ice and snow. By contrast, both oceans and forests are very dark, while deserts are relatively bright. Dark surfaces, by definition, strongly absorb incident radiation, and this leads to a net absorption at the surface of 161 W m^{-2} (close to half the incident TOA flux). The remaining 78 W m^{-2} are absorbed in the atmosphere, mostly by water vapour in the lowest ~1–2 km.

Skipping across to the outgoing (thermal/long-wave) side, we see that 396 W m^{-2} are emitted by the surface. However, of this only 40 W m^{-2} escape directly to space, while 356 W m^{-2} are absorbed by gases in the atmosphere. The key factor here is that the 40 W m^{-2} are found in a

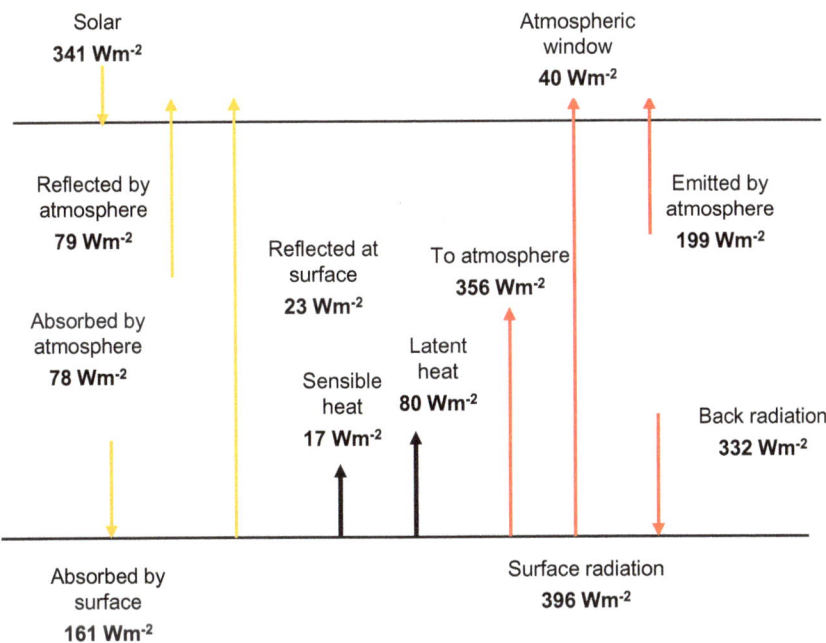

Figure 8.4 Schematic diagram of the Earth's energy budget, based on data from Trenberth et al., *Bull. Am. Meteor. Soc.* 90, 311, 2009. *Source*: *Physics of Radiation and Climate*, figure 1.1, Box and Box, CRC Press, 2016, reproduced by permission of Taylor & Francis Group.

spectral region that is not absorbed (known as the window region), for reasons which will have to wait for the moment. The atmosphere also emits radiant energy in this spectral region (i.e. wavelengths greater than 4 μm): 199 W m^{-2} eventually escape to space, and 332 W m^{-2} return to the surface. Note that some photons are absorbed and re-emitted multiple times before their ultimate fate is achieved. It is the job of physics to explain, and *quantify*, these processes.

Finally, we turn to the two terms in the middle of the diagram. The Earth's surface temperature is, in general, higher than that of (the lower part of) the atmosphere, and as a result, heat energy flows from the former to the latter in accordance with the laws of thermodynamics. This flux of 17 W m^{-2} is known as *sensible heat* (Section 5.4.2). Whenever water evaporates, it absorbs latent heat; whenever it condenses, it releases that latent heat. Thus, the evaporation of water vapour from the oceans and its subsequent condensation in the atmosphere, followed by precipitation in condensed form, represents a flux of *latent heat* of 80 W m^{-2} from the surface to the atmosphere.

To complete this story, we now examine the energy budgets of the surface, the atmosphere, and the planet as a whole. You should do this for yourselves: make sure that the total flux of energy absorbed by the surface equals the total flux emitted; do the same for the atmosphere; and finally at the top of the atmosphere, or TOA. Note that radiation that is reflected is neither absorbed nor emitted. (A quick, technical point: the First Law of Thermodynamics makes it clear that all inputs/outputs of energy to a system are additive, regardless of their previous 'nature'. To paraphrase Orwell: all energies are created equal.)

8.2.2.1 Quantification

These are the types of numbers that prompt scientists to ask questions. Here is the first, and most basic of all: how much of this radiation trapping can we assign to each of the radiatively active gases in our atmosphere? This question, like so many good questions, turns out to be easier to ask than to answer, but that is one of the reasons some of us choose careers in science. However, to a rough approximation we can say that water vapour is responsible for about 60% of the total, carbon dioxide around 35%, with the minor gases such as methane, nitrous oxide, and some man-made chemicals accounting for the remaining 5%. These numbers are subject to a range of assumptions, and are also changing slightly as we change the composition of our atmosphere, but they do provide a reasonable guide to the situation.

8.2.3   The Role of Clouds

Clouds, which typically cover half the globe, are one of the dominating factors in the Earth's radiation budget; hence, some brief comments are in order. Firstly, as noted above, clouds – and especially thick, layer clouds and towering monsoon clouds – are highly reflective of solar radiation, contributing more than two-thirds of the planetary albedo.

In the long-wave spectral region, clouds are very efficient absorbers. Low-lying clouds will absorb most of the radiation emitted by the surface. As their temperature is likely to be only slightly lower than the surface temperature, they will then re-emit a comparable amount of radiation. Therefore, their effects on long-wave radiation are quite minor.

Thin, high-altitude cirrus clouds are an interesting subject. They scatter comparatively little incoming solar radiation, and a substantial fraction of that is scattered downwards. Hence their contribution to the albedo is quite small. By contrast, they are reasonably good absorbers of long-wave radiation. And because of their altitude, they are cold, so their re-emission is much less than for boundary layer clouds. Put simply, they are greenhouse (i.e. warming) agents. One suggested method of countering global warming is to find some way to reduce such clouds, perhaps by cloud seeding.

8.2.4   Two Misconceptions

We will now take a moment to address two misconceptions that have formed the basis of some misguided attacks on the science underpinning climate science, and more pertinently, climate change.

The first of these concerns the radiant energy that flows from the atmosphere down to the Earth's surface: the 'back radiation' of 322 W m^{-2}. We know that the temperature of the troposphere is lower than the surface temperature. And we also remember that one of the basic claims of the Second Law of Thermodynamics is that heat does not flow from a cooler body to a warmer body. So, don't we have a violation? No. The Second Law only demands that the *net* flow is always from a warmer to a cooler object. So, let's add up the numbers. The flow from the surface to the atmosphere is 356 (long-wave radiation), plus 80 (latent heat), plus 17 (sensible heat): for a total of 453 W m^{-2}. As this is significantly higher than the 322 W m^{-2} back radiation, the Second Law has not been violated. One final point: scientists make use of an instrument known as a pyrgeometer to actually measure this radiation; it is certainly real.

The second misconception stems from the fact that water vapour absorbs more long-wave radiation than carbon dioxide, making it a more potent greenhouse gas. True; but that is only half the story. Think what would happen if all the atmosphere's CO_2 was suddenly removed. That would mean reduced absorption of the long-wave radiation emitted from the surface. This reduction in input energy would, naturally, lead to a cooling of the atmosphere. Now we remember the Clausius–Clapeyron equation: the amount of water vapour the atmosphere can hold is strongly temperature-dependent, so some would be forced to condense and precipitate. That, in turn, would mean a further reduction in the amount of long-wave radiation absorbed, and a further lowering of the temperature. We have clearly entered a downward spiral, leading inexorably to a very cold, dry atmosphere, with much lower greenhouse gas concentrations, and a corresponding reduction in the greenhouse effect. CO_2 is the key to the greenhouse effect.

8.3   Grey Atmosphere Models

A thorough treatment of the flow of radiation into and out of the Earth's atmosphere requires serious mathematics. Instead, we will now introduce you to some quite simple models which, although definitely just approximations, still provide valuable insights. They are known as grey atmosphere models because they ignore the spectral details of the greenhouse gases, which will be explored in Chapter 11.

In Figure 8.3 we outlined the energy balance of the Earth and its atmosphere, with the key being the absorption and emission of long-wave radiation by the atmosphere. This suggests a suite of very simple models to get a handle on these energy flows. The simplest model consists of just a single atmospheric layer.

So, we'll start by treating the atmosphere as a layer with uniform properties – primarily its temperature – much like the glass roof of a greenhouse. The flows of solar and terrestrial radiation are then treated with their own sets of rules. This very flexible model has many useful pedagogic applications, which will be explored (by you, the reader). Finally, we will extend the model to many layers, which will help us understand the greenhouse effect on Venus.

8.3.1   Single-Layer Model

In the simplest of these models, we assume that the layer is transparent to incoming solar radiation, but absorbs all terrestrial radiation: that is, we will treat it as a black body with an unknown temperature T_a (The lack of atmospheric short-wave absorption does not agree with Figure 8.3; however, the strong convective coupling of the troposphere to the surface means that this is not as serious a defect as it might appear.) We also treat the surface as a black body, with an unknown temperature T_g. Figure 8.5 shows the radiation flows, with solar/short-wave

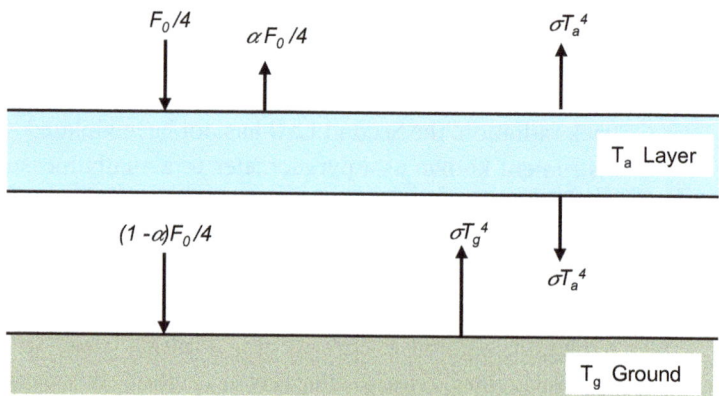

Figure 8.5 Schematic diagram of the simple one-layer model. *Source*: *Physics of Radiation and Climate*, figure 13.1, Box and Box, CRC Press, 2016, reproduced by permission of Taylor & Francis Group.

radiation on the left, and terrestrial/long-wave radiation on the right. Remember that a planet intercepts solar radiation as a disc, but emits it as a sphere, so the average incoming solar flux at TOA is given as $F_0/4 = 341$ W m^{-2}.

We will treat the planetary albedo, a, as known. Now we may examine these energy flows and, by demanding equilibrium, arrive at the following equations for energy balance:

$$\text{Above the atmosphere} \quad F_0/4 = a F_0/4 + \sigma T_a^4 \tag{8.1}$$

$$\text{In the layer} \quad \sigma T_g^4 = 2\sigma T_a^4 \tag{8.2}$$

$$\text{At the ground} \quad (1 - a) F_0/4 + \sigma T_a^4 = \sigma T_g^4 \tag{8.3}$$

It might appear that we now have three simultaneous equations; however, if we add the last two, we obtain the first. Thus, we have two simultaneous equations, which we may solve for two unknowns: the two temperatures. Equation (8.1) is, in fact, identical to the situation we studied above, so we already know the solution:

$$T_a = 255\,\text{K} = -18\,°\text{C} \tag{8.4}$$

If we now substitute this into Equation (8.2) we find

$$T_g = \sqrt[4]{2}\, T_a = 1.189 \times 255 = 303\,\text{K} = 30\,°\text{C} \tag{8.5}$$

We have gone from being 33°C too cold with no atmosphere to being 15°C too hot. That's an improvement, but it's still not much of a model.

8.3.1.1 Extending the Model

The simple model we have just considered had three major deficiencies when compared with the details of the Earth's energy budget as presented in Figure 8.4. These are:

1. We did not include absorption of solar radiation in the lower atmosphere.
2. We did not allow for the partial transparency of the atmosphere to terrestrial radiation (the window region).
3. We did not include the non-radiative exchanges between the surface and the atmosphere.

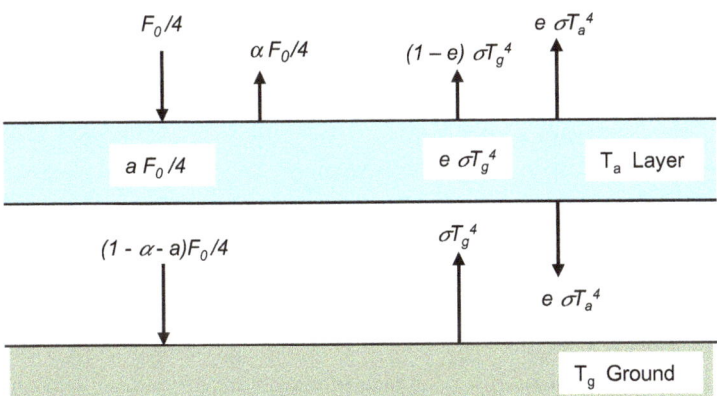

Figure 8.6 Schematic diagram of the extended one-layer model. *Source: Physics of Radiation and Climate*, figure 13.2, Box and Box, CRC Press, 2016, reproduced by permission of Taylor & Francis Group.

We will now extend our model to account for the first two of these. In Figure 8.6, α is the planetary albedo, a is the short-wave absorptivity (fractional absorption) of the atmosphere, e is its long-wave absorptivity/emissivity (i.e. fractional absorption of long-wave radiation, which equals the fraction of the long-wave emission of a black body), T_a is the temperature of the atmospheric layer and T_g is the ground temperature. The incoming solar radiation is $F_0/4$.

Again, solar radiation is incident on the top of the atmosphere: some is reflected back to space; some is absorbed by the atmospheric layer; and the balance reaches the ground, where it is absorbed. The ground emits radiation, as a black body. This radiation then passes up through the atmospheric layer where a fraction, e (in reality the majority), is absorbed, and the remainder is transmitted directly to space. The atmospheric layer, which has absorbed both solar and terrestrial radiation, emits long-wave radiation ($e\sigma T_a^4$) in both directions: to space and back to the ground. We will assume again that these emissions are equal, although this is not in fact the case, as Figure 8.4 shows.

For radiative equilibrium we require that incoming radiation is balanced by outgoing radiation at each level (absorbed radiation/energy equals emitted radiation). This gives us three equations, although again only two are independent:

$$\text{Above the atmosphere} \quad F_0/4 = \alpha F_0/4 + (1-e)\sigma T_g^4 + e\sigma T_a^4 \tag{8.6}$$

$$\text{In the layer} \quad a F_0/4 + e\sigma T_g^4 = 2 e\sigma T_a^4 \tag{8.7}$$

$$\text{At the ground} \quad (1 - \alpha - a) F_0/4 + e\sigma T_a^4 = \sigma T_g^4 \tag{8.8}$$

Once again, we can only solve for two unknowns: this is simply a fact of mathematics. However, our model now has two additional unknowns: a and e. Thus, we are forced to make some choices, as we will need to specify two of our four potential unknowns, and solve for the other two.

The most logical choice is to allocate values for the short-wave absorptivity, a, and the surface temperature, T_g. Then we may solve for the long-wave emissivity, e, and the atmospheric temperature, T_a. Perhaps the simplest 'philosophical' argument for these choices is that the two unknowns are actually 'effective' values, involving some sort of averaging procedure (over altitude, for example) – this is certainly true of T_a. It may seem that this model is somewhat unsatisfactory, as we need to specify some of the values we might have hoped to solve for. However, that is the best we can do.

Here is an exercise: in the model above, assume $F_0 = 1,364$; $\alpha = 0.3$; $a = 0.15$; and $T_g = 15°C = 288$ K. Now solve this set of equations for e and T_a. [Answers: 0.904; 250.5 K.] Why has the atmospheric temperature dropped from the previous version?

8.3.2   A Multilayer Model

We have seen that a simple single-layer model produces a surface temperature that is somewhat too warm for the Earth. However, Venus has a very much larger greenhouse effect than Earth, with a radiative temperature of ~240 K, but a surface temperature of ~700 K. No single-layer model can reproduce that!

However, we may extend our single-layer model by simply placing an additional layer on top of the first layer, and then another and then another (like putting more blankets on your bed). Assume now that we have n layers (counting from the top, down), with temperatures T_i (Figure 8.7). We assume, of course, that energy emitted by one layer is absorbed by the layers either side (or the ground, or escapes to space, as appropriate).

We may then obtain the equations for the energy balance in each layer, plus at the ground. The top layer receives energy from layer number 2, and emits energy both up and down. Layer i receives energy from layers $i - 1$ and $i + 1$, and again emits both up and down. The ground receives both the transmitted solar energy, and energy from layer n, and emits radiation upwards (only), at a temperature T_g. Thus, we obtain the following set of equations:

Figure 8.7 Schematic diagram of the multilayer model.

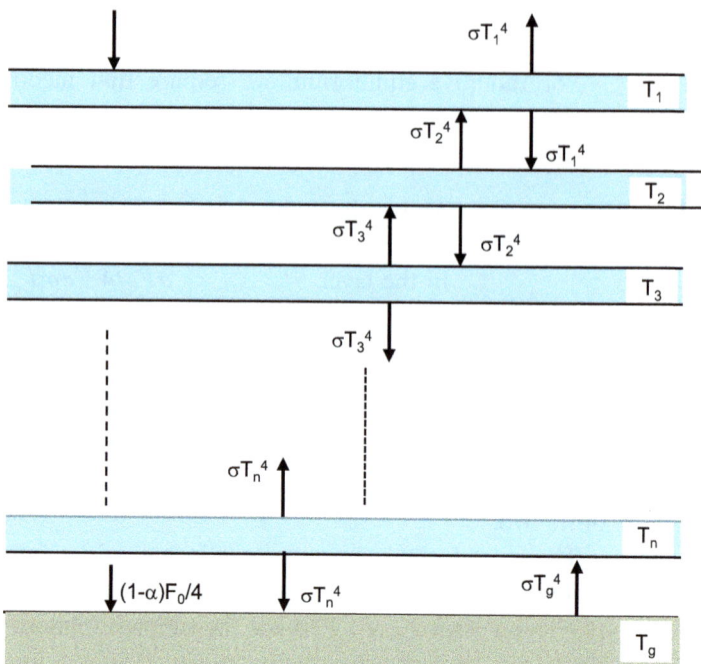

$$\text{Top layer} \qquad \sigma T_2^4 = 2\sigma T_1^4 \qquad\qquad (8.9)$$

$$\text{General layer} \qquad \sigma T_{i-1}^4 + \sigma T_{i+1}^4 = 2\sigma T_i^4 \qquad\qquad (8.10)$$

$$\text{The ground} \qquad (1-\alpha)F_0/4 + \sigma T_n^4 = \sigma T_g^4 \qquad\qquad (8.11)$$

How do we go about solving this set of equations, especially when we may not even know how many layers to use? (Note that these equations are linear, as all the temperatures appear only as their fourth power: the equations are thus linear in this set of unknowns.) In fact, in analogy with our very first model, the temperature of the top layer is just the radiative temperature of the planet. We may then solve Equation (8.9) for T_2, and then iterate downwards to the bottom of the atmosphere, and the surface temperature. Alternatively, if we know the surface temperature, we can iterate until we have sufficient layers to 'model' our atmosphere.

Summary

You have just finished reading the pivotal chapter of the book, covering the science at the heart of climate, and especially climate change. While we will develop some of this material more fully in Chapters 9 and 11, it is essential that you thoroughly grasp the ideas presented here.

Firstly, we introduced you to thermal radiation: the Stefan–Boltzmann equation and the Planck function. Every object in the universe (with a temperature above absolute zero) emits such radiation.

Secondly, we have taken you through the Earth's energy budget. While you can always check back for the numbers (which are likely a little out of date), you should understand *all* of the physical processes involved.

Finally, we have introduced you to a suite of simple models of the greenhouse effect. Although they grossly oversimplify much of the reality of radiation flows through the atmosphere (especially the long-wave flows; Chapter 11), they should help focus your thinking.

FURTHER READING

The material covered in this chapter is quite basic, and can be found in numerous books, such as those by Andrews, Barry and Chorley, Neelin, Taylor, and Wallace and Hobbs, listed in previous chapters. Some specialist texts on atmospheric radiation, which will be relevant to the more advanced material of Chapters 9 and 11, include

- *An Introduction to Atmospheric Radiation* by K. N. Liou (Academic Press, 2002).
- *Radiation in the Atmosphere* by W. Zdunkowski, T. Trautmann and A. Bott (Cambridge University Press, 2007).
- *Atmospheric Radiation* by R. M. Goody and Y. L. Yung (Oxford University Press, 1989).
- *Physics of Radiation and Climate* by M. A. Box and G. P. Box (CRC Press, 2016).

- *Atmospheric Science for Environmental Scientists* by C. N. Hewitt (Wiley-Blackwell, 2020).

Once again, you will find many interesting ideas and exercises relevant to the material in this chapter in one of our favourite references

- *Consider a Spherical Cow* by J. Harte (University Science Books, 1988).

REVIEW QUESTIONS

1. How does radiant energy emission increase with temperature (for a black body)?
2. How does the spectral distribution of the emission vary with temperature?
3. Explain Kirchhoff's law.
4. It has been stated that the flow of radiant energy from the colder atmosphere to the warmer ground violates the Second Law of Thermodynamics. Why is this not true?
5. What would happen to planetary temperatures if all of the CO_2 was removed from the atmosphere?

EXERCISES

1. Venus orbits the Sun at a distance of 108 million km, and has an albedo of 0.71. Find the solar constant at Venus, and its effective temperature. Does this result surprise you?
2. Repeat question 1 for Mars, which orbits at 228 million km from the Sun, and has an albedo of 0.17.
3. In another 5 billion years (give or take a few), the Sun is expected to become a red giant star, with a photospheric temperature of 4,000 K, and a radius of 3.5 million km. Calculate the new solar constant, and effective temperature, for Earth, assuming no albedo change.
4. Show that adding Equations (8.2) and (8.3) gives Equation (8.1). Also show that adding Equations (8.7) and (8.8) gives Equation (8.6).
5. Consider the extended model in Section 8.3.1.1. Suppose that the long-wave emissivity, e, that we found there is increased by 5%. Find the new values for the two temperatures, T_a and T_g. You might also like to try other variants on this question: increase the planetary albedo, a, to account for an increase in aerosol scattering; increase the short-wave absorptivity, a, to account for aerosol absorption.
6. Venus has an effective temperature of 240 K, and a surface temperature of 720 K. How many layers of the multilayer model would be needed to account for this effect? To help you do this exercise, try to find a general solution for the temperature of the nth layer.

9   Solar Radiation and Its Atmospheric Interactions

Virtually all of the energy at the surface of the Earth comes to us from the Sun. By contrast, the heat flux from the Earth's interior, due to the residue of the radioactive isotopes that were present at the planet's birth, plus the cooling of the interior from its formation, is smaller by a factor of several thousand. This is the energy that, in combination with the greenhouse gases in the atmosphere, maintains our current surface temperature in the range where liquid water is present. As we saw in Chapter 7, this is also the source of all the energy contained in atmospheric motions.

In this chapter we will examine solar radiant energy and its most important interactions with the various constituents of the Earth's atmosphere. The highest-energy photons are absorbed first, creating the ionosphere, while the next-highest-energy range creates the ozone layer. Most of the rest reaches the lower troposphere, some to be absorbed by water vapour, and some to be scattered by clouds, molecules, or aerosols. The rest (roughly half) reaches the ground.

Human activity has the power to alter these inflows, in a number of ways. We devote one section to the potentially disastrous **Nuclear Winter** scenario, and another to the potentially useful idea, known as **Geoengineering**, of reflecting some of the incoming sunlight in order to re-balance the Earth's energy budget.

9.1   Solar Radiation

We will start by looking at the Sun, our source of energy: how much it emits, and how much arrives at the top of our atmosphere. While this may seem quite straightforward, there are some key sources of variability that we need to be aware of.

9.1.1   Nuclear Fusion

This subsection is 'for interest', and may be skipped.

The Sun, and all stars, create energy by the process known as nuclear fusion, with the dominant reaction being the conversion of four protons (hydrogen nuclei) into a helium nucleus: the overall process being represented by

$$4p^+ + 2e^- \rightarrow {}_2He^4 + 2\nu + \gamma's + 26.73 \text{ MeV}$$

where 1 MeV = 10^6 eV = 1.6×10^{-13} J.

Central to this is the conversion of two of the protons into neutrons, a fundamental Weak Interaction process. Because of the electrical repulsion of protons as they approach each other, this requires high kinetic energy, which is the result of the high temperatures in the Sun's core: around 10 million degrees. This high temperature came from the gravitational collapse of the original gas cloud that produced the Sun: gravitational potential energy was converted to kinetic energy, and we saw in Chapter 5 that there is a direct connection between KE and temperature.

The Sun's **luminosity** – total rate of emission of electromagnetic energy – is 3.86×10^{26} W. From this number, and the number above, we can calculate that 3.7×10^{38} hydrogen nuclei are converted into neutrons per second, or we may say that the Sun is 'burning hydrogen' at a rate of 6.1×10^{11} kg/s. (Central to all this is Einstein's famous $E = mc^2$.) Fortunately, the Sun still has lots of hydrogen!

9.1.2 Solar Spectrum and Solar Constant

The solar spectrum covers the complete wavelength range from gamma rays to radio waves, and is mostly emitted from its photosphere and chromosphere. The majority of the energy is in the form of a continuum, although considerable line structure – the Fraunhofer absorption lines – can be seen on close examination of the spectral details. To a reasonable approximation, the solar spectrum can be said to be that of a black body (Planck) distribution with an absolute temperature of ~5,750 K. However, as shown in Figure 9.1, there are significant variations about this spectrum.

The division of the electromagnetic spectrum into γ-rays, X-rays, ultraviolet (UV), visible, infrared, microwaves and radio waves is essentially arbitrary, and the boundaries do vary from author to author. In this book our focus is essentially confined to the UV (wavelengths mostly 100–400 nm), visible (400–700 nm), and infrared (700 nm to ~1 mm); on occasions we may choose to further subdivide these regions. Visible light is also known as **photosynthetically**

Figure 9.1 The solar spectrum along with the Planck function for 5,500 K and 6,000 K for comparison. *Source*: based on data downloaded from PVEducation: www .pveducation.org/pvcdrom/ appendices/standard-solar-spectra.

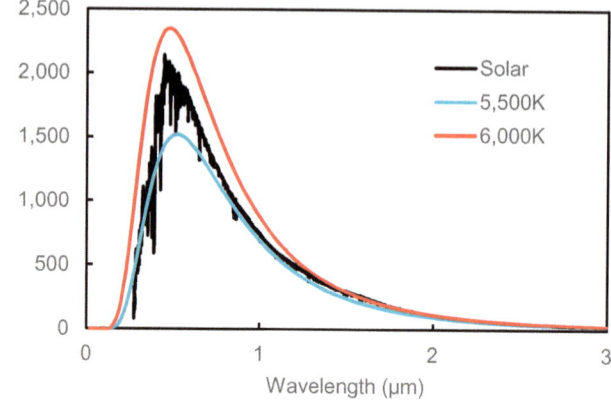

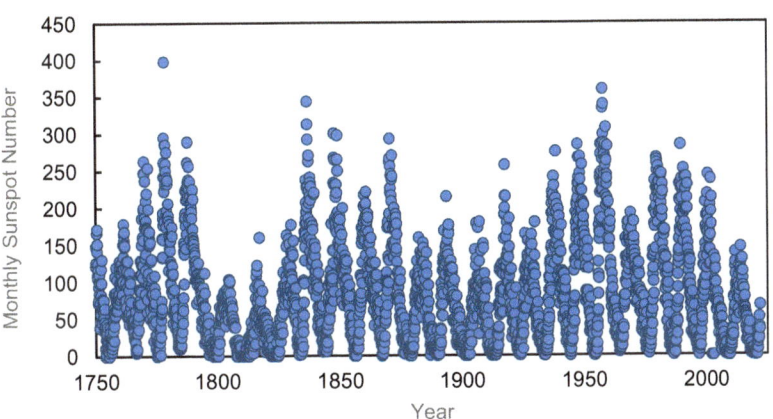

Figure 9.2 Sunspot numbers since 1750. *Source*: Data downloaded from NOAA data, https://psl.noaa.gov/gcos_wgsp/Timeseries, updated values to 2021. *Source*: WDC-SILSO, Royal Observatory of Belgium, Brussels.

active radiation (PAR), as it is the key to photosynthesis, i.e. plant growth; indeed, the entire biosphere.

The **solar constant** is defined as the total energy flux of solar radiation received at the mean Earth–Sun separation distance, and is clearly the key driver of our climate. Ground-based measurements were originally used to provide an estimate of the solar constant. These days, satellites provide the best source of data, although calibration has been an issue in the past. A recent analysis of instrumental errors, and particularly scattered light within each instrument, has led to a consensus that the total solar irradiance during the 2008 solar minimum period was $1{,}360.8 \pm 0.5$ W m^{-2}, slightly lower than previous values.

Our Sun undergoes an 11-year **sunspot cycle**: more correctly a 22-year cycle, as the Sun's magnetic field reverses from one 11-year subcycle to the next. Figure 9.2 shows sunspot numbers since 1750, when regular observations began. Sunspots range from the barely visible, to areas more than 150,000 km across, with an average around 10,000 km. Smaller sunspots persist for several days, while the largest may last for weeks, long enough to reappear in 27 days, after a solar rotation. Almost all sunspots appear at latitudes between 40° north and south of the equator, and never close to the poles.

Sunspots are cooler regions, with an average temperature around 4,000 K, which is why they tend to look 'black' against the brighter solar disc (remember Figure 8.1). However, sunspot activity is also accompanied by an increase from the magnetic regions surrounding the spots, involving faculae and plages, where emission is enhanced. As a result, total solar irradiance increases with increased sunspot activity, by ~1 W m^{-2}.

9.1.3 The Earth's Orbit and Solar Insolation

Our solar system is comprised of the four terrestrial planets (Mercury, Venus, Earth, and Mars), the four major or giant planets (the gas giants Jupiter and Saturn, and the ice giants Uranus and Neptune) and the dwarf planet Pluto, as well as asteroids (some of which are now also classed as dwarf planets) and other material. All of the planets revolve around the Sun in the same direction, with all planetary orbits lying in, or close to, the same plane; known as the ecliptic. The Earth takes 365.2422 mean solar days to complete its orbit around the Sun.

Figure 9.3 An elliptical orbit.
Source: *Physics of Radiation and Climate*, figure 12.4, Box and Box, CRC Press, 2016, reproduced by permission of Taylor & Francis Group.

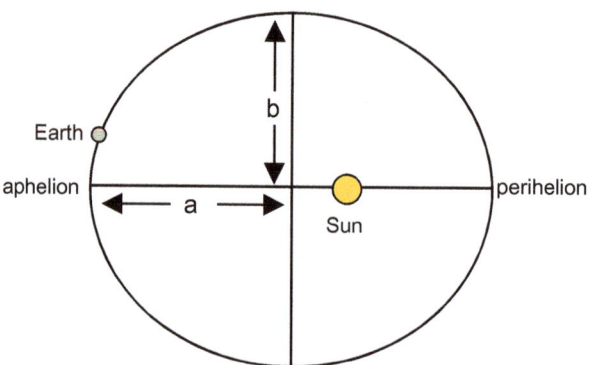

If the Earth's orbit were circular, it would always be at the same distance from the Sun, and the Earth would receive the same quantity of solar radiation all year round. However, this is not the case. Tycho Brahe (1546–1601) was the finest observational astronomer before the advent of the telescope. Upon his death, his observations of the planets passed to Johannes Kepler (1571–1630), who was as good a mathematician as Tycho was an astronomer. He set to work in the hope that these data would verify the highly controversial (at the time) proposal by Nicolaus Copernicus (1473–1543) that the Earth, and the other (known) planets, orbited the Sun, in 'perfect', circular orbits.

The trouble was, Tycho's data were just too accurate to reach such a conclusion. After a careful study of Mars' orbit – which fortunately happens to be one of the more non-circular – he was forced to a different conclusion, as stated in the first of his three laws: The Earth and planets move around the Sun in elliptical orbits, with the Sun at one of the two foci (not the centre). One of the early triumphs of Newton and his three Laws of Motion, plus his Law of Gravitation, plus his mathematics (calculus), was to show that elliptical orbits may be produced by an inverse square law for the attractive force between the Sun and planets.

A circle may simply be defined by its radius (half the diameter), which is constant. An ellipse may be defined by its **semimajor axis**, a, and its semiminor axis, b, being, respectively, half the largest and smallest 'diameters' of the ellipse (Figure 9.3). The **eccentricity** of an ellipse is then defined by $e = (a^2 - b^2)^{1/2}/a$. The current value of the Earth's eccentricity is 0.0167, making it very close to circular. (We will discuss how it varies in time, in Chapter 16.)

As a consequence of the orbital eccentricity, the Earth–Sun distance varies slightly throughout the year, with a minimum (**perihelion**) of 147.1×10^6 km on 3 January, and a maximum (**aphelion**) of 152.1×10^6 km on 4 July. The semimajor axis is the mean value: 149.6×10^6 km. (For readers who delight in precision, or trivia, the International Astronomical Union has recently redefined the Astronomical Unit, their name for the semimajor axis, to be 149,597,870,700 metres; so now you know: 149.6×10^6 km is more than adequate for most purposes.)

As the energy emitted from the surface of the Sun spreads out through the solar system and beyond, that energy will be spread out over the surface of a growing sphere of radius R, where R is the distance from its centre. Because the surface area of a sphere is $4\pi R^2$, the intensity – power per unit area – must decrease at a rate of $1/R^2$. This is what we refer to as an inverse square law fall-off.

The variation from closest distance to furthest amounts to 3.4%. Because the intensity of solar energy decreases as the square of the distance from the Sun, this equates to a variation in solar insolation of 6.8%. This should make Southern Hemisphere summers warmer than Northern Hemisphere summers, and winters cooler. However, a larger regional impact comes from the differing land/ocean ratio in the two hemispheres. Continental interiors (outside the tropics) experience much larger temperature ranges than coastal regions.

9.2   Absorption in the Upper Atmosphere

Shorter-wavelength radiation, UV and beyond, corresponds to photon energies, which are sufficiently high that they may cause a range of atomic and molecular processes such as electronic rearrangements, photodissociation, and photoionisation. The shortest wavelengths and highest photon energies tend to be so readily absorbed that they do not penetrate far into the atmosphere. In this section we will briefly look at the gases, and reactions, that absorb ultraviolet radiation in the upper atmosphere. The level of technical detail in some parts of this section is probably higher than most readers will need, but the concepts should be easily digested.

Ozone is of sufficient importance that we will devote the next chapter to its formation and destruction in the stratosphere, the threats to our ozone shield, and our response.

Radiation with photon energies greater than ~9 eV – wavelengths shorter than ~140 nm – are able to strip electrons off key molecules in the Earth's upper atmosphere, such as N_2, N, O_2, O, and NO, thus producing an ionised region: the ionosphere. The highest-energy photons are absorbed most readily, ionising the molecule that requires the most energy, N_2. Successively lower-energy ranges are progressively absorbed at lower altitudes, producing the various layers of the ionosphere:

- The F region, above ~170 km, created by photons with wavelengths less than ~80 nm.
- The E region, around 110 km, created by photons with wavelengths less than ~100 nm.
- The D region, from 60 to 90 km, is a region where somewhat lower-energy photons are able to penetrate.

Not surprisingly, the characteristics of the ionosphere undergo a diurnal cycle. The ions may recombine with their missing electron, although such processes depend on the air density, of course, so that this cycle is more pronounced in the D region.

Radio waves may pass through the ionosphere, or be reflected, depending on both the frequency and the electron density. Reflection from the E layer is important in long-distance radio reception. As such waves must pass twice through the D region, radio reception generally improves at night. Such transmission can be greatly affected by solar storms, something which comes under the broad heading of **Space Weather**.

9.3   Absorption and Scattering in the Troposphere

9.3.1   Absorption in the Visible and Near IR

The visible spectral region is largely transparent, apart from molecular, aerosol and cloud scattering, and this is the fundamental reason eyes have evolved to be sensitive to this wavelength range. There is weak absorption in the broad Chappuis band of ozone, which has a typical maximum optical thickness (at ~610 nm) of 0.03–0.035. Molecular oxygen has narrow absorption bands in the red and infrared (IR). The most important are the A band at 762 nm, the B band at 688 nm, and the γ band at 628 nm.

Water vapour absorbs a significant amount of radiation in the near IR. The most important feature for the absorption of solar radiation is the 2.7-μm band. There are also a number of bands in the near IR, which are centred at 0.94, 1.1, 1.38, and 1.87 μm, as well as weak bands at 0.72 and 0.82 μm, which contribute to the heating rate of the lower troposphere because of the high solar flux.

Carbon dioxide also absorbs in this region, mainly via weak bands at 1.4, 1.6, and 2.0 μm. Its stronger 2.7-μm band is overshadowed by the water vapour band. Nitrous oxide has bands at 2.87, 2.97, 3.90, and 4.06 μm. Methane has a strong band at 3.31 μm, as well as a number of weaker bands. Nitrogen dioxide absorbs weakly across the visible (and UV-A), with consequences we discussed in Chapter 2. Finally, some absorption by transitional dimer molecules such as O_2–O_2 (in the visible) and O_2–N_2 (at ~1.26 μm) has been reported.

9.3.2 Clouds

Water clouds are efficient scatterers of incoming solar radiation, and contribute about two-thirds of the planetary albedo. While they vary in time and space – especially latitude, as we saw in Chapter 7 – they typically cover roughly half of the globe at any one time. A very thick cloud might have an albedo close to 1.0, although such situations are rare. Stratus cloud albedos typically range from 50% to 85%, depending on thickness, while tall cumulus clouds can be a little higher. At the other extreme, thin cirrus (ice) clouds have albedos in the 15–20% range.

Liquid water has negligible absorption in the visible and near infrared range, out to about 2.5 μm. As we have just noted, this also happens to be the spectral region where water vapour is a strong absorber. Thus, clouds can be said to be pure scatterers of incoming solar radiation, but are very efficient absorbers of long-wave radiation.

9.3.3 Molecules and Aerosols

Molecular scattering (also known as Rayleigh scattering) of incoming solar radiation is strongly wavelength-dependent, with an optical thickness given by

$$\tau(\lambda) = 0.00888 \, \lambda^{-4.05} \quad \lambda \text{ in } \mu m.$$

To a good approximation, we may regard this as an inverse fourth-power dependence. It ranges from 0.36 at the blue end to 0.038 at the red end: this is why the sky is blue: what we see is scattered sunlight. Rayleigh scattering is highly polarised, the details of which are beyond the scope of this book. However, if you hold a sheet of Polaroid in your hands, while looking at right angles to the incoming solar beam, you can see this effect as you rotate the sheet.

We looked at aerosol scattering in Chapter 4. Aerosols are, as we noted, a highly variable component of our atmosphere. Their optical thickness in the visible spectral region can range from as low as 0.1 to as high as 1.0 or more in highly polluted environments, and even higher during a dust storm. Sunlight scattered by aerosols is only weakly polarised, and generally only shows a small wavelength dependence (compared to Rayleigh scattering). Also discussed in Chapter 4 is the increase in the amounts of anthropogenic aerosols, which is leading to a slight increase in the planet's albedo; sometimes referred to as solar dimming.

Finally, we should recall that soot – black carbon – can be quite absorbing of any radiation that penetrates far enough, which means the visible region, and the windows in the water vapour absorption in the near IR.

9.3.4 Surface Albedo

In those places where clouds are absent, a sizeable amount of solar radiation reaches the ground. What is its fate? That depends very much on the nature of the underlying surface.

A water surface has a very low albedo, typically around 5–10%. However, if the solar beam is coming from a low angle, known as grazing incidence, this increases to quite high values. Such angles mainly depend on latitude, so that reflection at high latitudes is generally higher than at low latitudes. Rough seas will, of course, present a range of angles to the incident radiation, leading to somewhat higher average reflection.

Forests are also dark, with albedos typically 10–15%. Crops and grasses have albedos that typically range from 15% to 25%. Dry sand dunes can have albedos up to ~40%. By contrast, ice and snow are quite reflective, with fresh snow having an albedo near 90%. Old snow, and sea ice, have albedos ranging from 30% to 70%.

9.3.5 Radiative Transfer

One of the tasks of atmospheric physics is to compute the flow – or flux (which is the quantification of such flows) – of solar radiation through the atmosphere, part of the subject known as radiative transfer. When the complexities of multiple scattering are fully accounted for (especially if polarisation is included), this can be a computational challenge. There are times, of course, when we must face these challenges. However, when we need to incorporate radiative transfer into a climate model, for example, we simply cannot afford the time. Instead, we are forced to turn to quite simple approximations, which – we hope – capture the energy flow sufficiently accurately. These approximations, known as two-stream models, are regularly benchmarked against our most accurate models, and any improvements incorporated.

9.4   Nuclear Winter

The release of large quantities of both CO_2 and CFCs (which posed a threat to the ozone layer: Chapter 10) into the atmosphere may be regarded as environmental engineering, and so, perhaps, might we regard the release of significant quantities of SO_2. What other situations might lead to unintended environmental consequences?

Our military capability is now sufficiently potent that its large-scale utilisation is certain to have at least a localised impact on the environment: but what of larger conflicts, and especially those which might involve significant numbers of nuclear warheads? Such questions were being asked in the late 1970s and early 1980s, and prompted a large effort from scientists (and others), much of it under the umbrella of the Scientific Committee on Problems of the Environment (SCOPE) of the International Council of Scientific Unions (ICSU).

SCOPE held a series of workshops and encouraged a range of research on a number of issues, before releasing a report on Environmental Consequences of Nuclear War, Volume I. Physical and Atmospheric Effects (SCOPE 28) in 1985. Its chapter titles give a good indication of the range of material covered:

1. Direct Effects of Nuclear Detonations;
2. Scenarios for a Nuclear Exchange;
3. Sources and Properties of Smoke and Dust;
4. Atmospheric Processes;
5. Meteorological and Climatic Effects;
6. Nuclear and Post-Nuclear Chemical Pollutants and Perturbations;
7. Radiological Dose Assessments;
8. Research Recommendations.

While much of this is beyond the scope of this book, we will briefly look at the possible atmospheric/climatic impacts that were canvassed, the most extreme of which have come to be referred to as 'Nuclear Winter'. Anyone interested in other aspects of this important subject will find that book quite readable, although we would not describe any of it as an 'enjoyable read'.

A low-altitude nuclear explosion releases considerable quantities of radioactive material, and also sweeps up surface material into its 'mushroom cloud', much of which subsequently comprises the fallout. So the nature of the lofted material depends on the local surface environment, which in turn depends on targeting scenarios.

Among the first scenarios to be considered in these investigations was a so-called 'counter-force attack', where one side targets the other's missile silos, in the hope of 'disarming' them. As these are invariably located in relatively isolated rural areas, the lofted material was primarily soil dust. While this resulted in a temporary loss of sunlight at the surface, the effects were considered both local and temporary.

However, the growing sophistication of satellite and related detection technology has come to render counterforce obsolete, as your opponent's missiles would almost certainly have been launched before yours reached their now empty silos. A more devastating scenario involves an attack on an opponent's cities and industrial centres (including command and control), which are likely to contain considerable quantities of petroleum, plastics, and other highly combustible materials. This would create the sort of fire-storm conditions that occurred in some conventional bombing raids in the Second World War, with Dresden and Tokyo being the best known. Under these circumstances a primary combustion product is black smoke, or soot.

When large quantities of smoke are lofted into the upper troposphere, the initial effects are much the same as for lofted soil: a significant reduction in surface insolation. There is, however, one important difference: unlike dust, smoke absorbs most of the blocked sunlight, rather than reflecting it. Thus, the smoke layer warms, becomes buoyant, and rises further, rather than beginning to settle out. Modelling suggested that this layer would rise into the stratosphere, which would then effectively 'decouple' from the troposphere, and the smoke layer would spread, at least throughout the Northern Hemisphere.

A number of important modelling studies were conducted, although we shall focus on just two. Turco, Toon, Ackerman, Pollack, and Sagan – known as the TTAPS study – used a globally and annually averaged one-dimensional climate model to simulate many exchanges of several thousand megatons, with the smoke (and dust) encircling the Earth for a week or so. Average light levels fell to a few per cent of ambient, and land temperatures dropped to around –20°C. Were such an event to occur in the key growing season of late spring/early summer it could easily destroy crops, leading to widespread food shortages, and in many parts of the world to starvation. They also suggested that horizontal and vertical temperature gradients between smoky regions and others might spread the layer to the Southern Hemisphere, subjecting those populations to similar effects (although we need to remember that the seasons are reversed). The TTAPS paper included 'Nuclear Winter' in its title.

Because of the controversy that this work caused, other modellers decided to use more sophisticated general circulation models to better examine the impacts, and to include the dynamical effects which would obviously ensue. The oceans are, as we have seen, a very significant heat store, so that we ought to expect a sea-breeze effect, but this time carrying warm air from over the oceans onto the land, or at least the coastal regions. This reduced the impacts somewhat, but temperatures in the continental interiors still fell well below freezing.

This field of study remains somewhat contentious, and is regularly revisited. Robock et al. used a 'modern climate model' (the first time a full atmosphere–ocean general circulation model (AOGCM) had been used for such a task) to study the consequences of a range of nuclear war scenarios, and found that there would be significant and long-lasting climatic impacts. (More recently they have also looked at some impacts on ozone photochemistry.) A scenario involving 150 Tg of smoke would still produce Nuclear Winter conditions, with globally catastrophic consequences. While it is probably correct to say that the term Nuclear Winter should only be applied to the most extreme scenarios and impacts, and some have suggested Nuclear Autumn as a more appropriate title, it is clear that the research to date gives us reason to pause and reflect.

9.5   Geoengineering

As we know, the Earth's temperature is rising. Why? Well, the short answer is that, for the past century or so, more energy has flowed into the Earth than has flowed out, with obvious consequences. The primary reason for this, of course, is that the steady increase in greenhouse gases has trapped more and more of the outgoing long-wave radiation, effectively choking the outflow. While ceasing greenhouse gas emissions is obviously essential to restore the balance, it is not happening fast enough. Are there any other options to slow, or even reverse, the temperature rise?

Geoengineering, or **Climate Engineering**, refers to a suite of initiatives designed to modify our environment in ways that might offset the effects of global warming: we hope. According to the Royal Society it is 'the deliberate large-scale manipulation of the planetary environment to counteract anthropogenic climate change'. While we should always prefer mitigation, we may one day need one or more of these ideas, in one form or another.

These initiatives may be conveniently divided into two main strategy directions:

- Carbon dioxide removal (CDR) looks for ways to remove carbon dioxide from the climate system (this is clearly the better option).
- Solar radiation management (SRM) aims to rebalance the radiation budget by increasing the planetary albedo, thus reducing the inflow of solar energy.

Because it is the increase in atmospheric CO_2 that has caused our planet to warm, removing some from the atmosphere is clearly a no-brainer. A number of ideas have been suggested, including

- growing special crops, harvesting and using them as biofuels, coupled with carbon capture and storage;
- increasing soil carbon (which is already included in a nation's carbon accounting);
- adding certain nutrients to the ocean, particularly iron, in the hope of increasing plankton growth, with some of it falling to the ocean floor (Section 3.3.4).

Unfortunately, modelling studies suggest that the rate of removal is likely to be distinctly less than our current emissions, so this is not a magic bullet. However, on longer timescales such measures will most likely be invaluable in helping to stabilise our climate. (We'll return to this in Section 20.4.)

9.5.1 Solar Radiation Management

As the thrust of this chapter has been solar energy inflow, and its fate within the Earth System, this is a suitable place to look at the second of these strategy directions, solar radiation management. A number of fanciful ideas have been advanced, which we'll quickly mention.

- Paint all roofs white, increasing their albedo. Almost certainly too little to be effective globally, but useful locally; e.g. reducing the heat island effect.
- Replace native grassland with genetically modified species which are more reflective. Or perhaps simply change the strain of wheat sown? But what might be the impacts on biodiversity? Or the water and carbon cycles?
- Orbiting space shields/umbrellas.

Currently the focus is on what might be characterised as 'using pollution to fight pollution'.

9.5.1.1 Stratospheric Aerosols

We know that reflection of sunlight by aerosols is one component of the planetary albedo. However, tropospheric aerosols have typical lifetimes of a week, so a deliberate plan to increase such aerosols would achieve little, apart from increasing air pollution, and reducing visibility. By contrast, material injected into the stratosphere will generally stay there for a year or two. For example, volcanic emissions of sulphur gases lead to a sulphate aerosol layer. The Mt Pinatubo eruption (one of the biggest of the twentieth century) cooled the Earth by ~0.5°C for a year or two (Figure 4.1).

Perhaps we could inject such aerosols, or their precursor gases, into the stratosphere. This would need to be ongoing, of course. Low latitudes would mean more sunlight to scatter. It has been estimated that 10 Mt S/year might produce enough cooling to counter the warming effects of doubled CO_2, at a possible cost of several hundred million dollars per year: a pittance, really.

9.5.1.2 Cloud Brightening

Clouds form when the amount of water vapour exceeds the (temperature-dependent) saturation vapour pressure. However, every cloud droplet requires an aerosol seed. Hence an increase in aerosols should mean the cloud liquid water is spread over *more*, but *smaller*, droplets. Known as the Twomey effect, or aerosol indirect effect, this should lead to an increase in the cloud's albedo. In addition, there are reasons to believe that such a cloud should take longer to precipitate.

Both of these effects would constitute negative forcings. Low clouds are the major contributor to the Earth's albedo. Could we increase this contribution? Because an increase in suitable aerosol numbers has the potential to make a cloud brighter, might there be a way to exploit this? Firstly, appreciate that relatively 'clean' clouds with low aerosol (CCN) numbers are the obvious candidates, as the effects are likely to be bigger.

Marine stratocumulus over the pristine Southern Ocean and South Pacific have been suggested, based on the direct injection of sea spray particles (or DMS) from below. (This has been suggested to cool Australia's Great Barrier Reef, which has suffered a number of recent episodes of coral bleaching.) The albedo increase would probably only be modest.

9.5.2 Important Questions

Before we seriously contemplate any such measures, there are some very important questions that must be addressed.

9.5.2.1 Key Issues

Can we do it? Should we do it? Should we even investigate it? Firstly, there are the **moral** issues. Have we the right? Do we own the planet? Well, we are already messing with the planet, so is it okay to try to 'undo' our damage?

Could geoengineering provide an **excuse** to carry on with 'business as usual'? What if it does, and then our endeavours to deploy geoengineering solutions fail? Should we do the research, but keep it secret? It's a bit late for that: the IPCC is now openly discussing it. On the other hand, if we see a dangerous 'tipping point' ahead we may need to be able to spring into action.

Could these technologies be weaponised, perhaps to damage enemy agriculture, or to modify a battlefield? Manipulation of cloud fields could certainly be used to favour the weapons available to one side. In 1976, 85 countries signed up to the UN Convention on the Prohibition of Military or any other Hostile Use of Environmental Modification Techniques. This clearly prohibits the use of geoengineering for hostile purposes, but that does not eliminate the risks, it only highlights them.

Who should govern global-scale interference? What are the rules; who are the rulers? Open disclosure is essential. We would need **three teams**: one to do the research; one to independently assess the results; and one to monitor progress, and be in a position to call a halt if danger becomes apparent.

9.5.2.2 Drawbacks

The biggest drawback of any SRM approach, in the absence of mitigation, is that it allows CO_2 emissions to keep increasing, so **ocean acidification** gets worse. This is just one of the reasons why CRM is preferred, assuming it can be made to work.

The stratospheric aerosol idea has its own question marks. Would it produce a rain of sulphuric acid from on high? The answer is yes, but far too small to be of any concern. The ozone hole is 'based' on chemical reactions that take place on the surface of aerosol particles (Chapter 10). So, such injection of sulphate aerosols could delay its closure by decades. Photosynthesis, and our plans to expand the use of solar energy, might be impaired by reduced sunlight.

No doubt there are other side effects that we now only dimly comprehend. As just one example, any changes in the radiation flows and related phenomena may alter aspects of the general circulation, with at least localised impacts on weather.

9.5.3   The Harvard Program

Harvard University is taking the stratospheric aerosols idea very seriously, with a total of 12 specific projects underway:

1. Decision theory and anomalies on solar geoengineering;
2. Laboratory studies of stratospheric aerosol;
3. Interdisciplinary research on solar geoengineering;
4. Governance for the stratospheric controlled perturbation experiment (SCOPEX);
5. Solar geoengineering and global climate governance under the Paris Agreement;
6. Heterogeneous chemistry and ageing of designer aerosol particles to assess the risk of solar geoengineering;
7. Stratospheric controlled perturbation experiment (SCOPEX);
8. Effect of stratospheric aerosol injection on global oxidative capacity;
9. The ethical and political dimensions of solar geoengineering;
10. Governance of solar geoengineering: advancing understanding and action;
11. Public attitudes of solar geoengineering;
12. Moral hazard.

As can be seen, only number 7 will focus on actual stratospheric modification, with 2, 6, and 8 helping to minimise any potential harm. The rest address moral and political issues. More information, including a short, informative video, can be found on their website: https:// geoengineering.environment.harvard.edu/

Summary

The Sun is the source of virtually all the energy that is available on Earth. It is the ultimate provider for the planetary temperature, its weather and climate, and via the process of photosynthesis, of (virtually) all life, including ours.

In this chapter we have looked at how much arrives at the top of the atmosphere, and the reasons for its minor variations. After that we have tracked its progress from the top of the atmosphere to the Earth's surface. One key point to remember is that higher-energy photons are invariably absorbed earlier – i.e. higher in the atmosphere. This is just 'logical' physics.

After that, we have introduced you to two important, although somewhat controversial, subjects: Nuclear Winter, and geoengineering.

FURTHER READING

For more information on processes in the upper atmosphere, one useful reference is

- *The Solar–Terrestrial Environment* by J. K. Hargreaves (Cambridge University Press, 1992).

Optical phenomena in the atmosphere can be both subtle and spectacular:

- *Rainbows, Halos and Glories* by R. Greenler (Cambridge University Press, 1989).
- *What Light Through Yonder Window Breaks?* by C. F. Bohren (Wiley, 1991).

For more information on Nuclear Winter, we suggest

- *Beyond Darkness* by A. B. Pittock (Sun Books, 1987; Barrie Pittock was a member of the SCOPE 28 team).
- Nuclear winter: Global consequences of multiple nuclear explosions by R. P. Turco et al., *Science*, vol 222, 1283–1292, 1983.
- 'Nuclear winter': A diagnosis of atmospheric general circulation model simulations' by C. Covey et al., *J. Geophys. Res.*, vol 90, 5615–5628, 1985.
- Nuclear winter revisited with a modern climate model and current nuclear arsenals: Still catastrophic consequences by A. Robock et al., *J. Geophys. Res.*, vol 112, D13107, 2007. doi:10.1029/2006JD008235.

REVIEW QUESTIONS

1. Why are the shortest wavelengths of solar radiation absorbed highest in the atmosphere?
2. What is the role of clouds on the flows of radiation through the atmosphere?
3. Which surfaces are most reflective? Which surfaces are least reflective?
4. Why is soot – black carbon – the key to Nuclear Winter?
5. Why might it be wise to have three teams when we start to undertake solar radiation management experiments in the atmosphere?

EXERCISES

1. Check the numbers in the last paragraph in Section 9.1.1. (You won't get quite the same numbers that are quoted, as there are one or two technicalities we have chosen to omit.)
2. Assume the Sun to be a black-body radiator, with a radius R_S and temperature T_S. How much power does it radiate? Remembering that this radiation (energy) per m^2 decreases as

the square of the distance from the Sun, derive an expression for the 'solar constant' for a planet at a distance R from the Sun. Assume that this planet has an albedo α, derive a general expression for the effective temperature of this planet.

3. The distance between the Earth and the Sun varies during the year, with a minimum in January and a maximum in July: the difference being 3.4%. By how much does the Earth's effective temperature change?

4. Check out the Harvard University geoengineering website, and write a report on one of the projects being undertaken. (By the time you read this, the material we have provided in Section 9.5.3 may well be out of date.)

10   The Ozone Layer and (Its) Health

Temperature rises in the stratosphere, between 20 and 50 km, due to the absorption of solar ultraviolet radiation by ozone. The ozone layer performs the vital role of protecting all terrestrial life from the damaging effects of UV radiation. In fact, terrestrial life could not appear until there was sufficient ozone to provide this protective shield. Because of the importance of this layer to humanity, and because of the threats to it that have been identified, and (perhaps narrowly) averted, we will now examine the ozone layer in some detail.

We will start by looking at the photochemical reactions that form the ozone layer. It turns out that these reactions, on their own, would produce an ozone layer distinctly 'thicker' than measurements indicate. The reason for the difference is that there is a suite of additional reactions that reduce the amount of ozone. We will then briefly look at the major biological impacts of the UV radiation that the ozone layer largely, although not completely, filters out.

The ozone hole will be discussed in the last two sections. The first will examine just what is the cause of the hole, and why it is largely confined to Antarctica. Finally, we look at the Montreal Protocol that has saved it; and us (fingers crossed).

10.1   Formation of the Ozone Layer

The formation of the ozone layer starts with the photodissociation of molecular oxygen:

$$O_2 + hf \rightarrow O + O \quad \lambda < 242 \, nm \tag{10.1}$$

Atomic oxygen may then attach itself to an oxygen molecule via a three-body reaction:

$$O + O_2 + M \rightarrow O_3 + M \tag{10.2}$$

Once again, the spectator molecule, M, is essential to ensure the balancing of both energy and momentum. The primary mechanism for the destruction/removal of ozone is also photochemical:

$$O_3 + hf \rightarrow O + O_2 \quad \lambda < 1,100 \, nm \tag{10.3}$$

Note that although this reaction is possible for wavelengths up to 1,100 nm, the rate drops significantly as wavelengths exceed ~300 nm.

Two recombination reactions should be included in the ozone 'budget' equations:

$$O_3 + O \rightarrow 2O_2 \tag{10.4}$$

$$O + O + M \rightarrow O_2 + M \tag{10.5}$$

These five reactions govern the concentration of ozone, at least in the absence of other species. Because the density of oxygen atoms is quite low, reaction (10.5) is usually neglected, by comparison with (10.2). Reaction (10.4) is also quite slow, so that the primary fate of the free oxygen atom created in reaction (10.3) is recombination via reaction (10.2). This means that reactions (10.2) and (10.3) equilibrate on short timescales, while reactions (10.1), (10.4), and (10.5) equilibrate on longer timescales. The upshot of this is a simple lesson: although ozone appears to be destroyed in reaction (10.3), that is really only temporary, as its constituents are likely to recombine.

One of the 'themes' of the ozone layer and its photochemistry is 'a story in odds and evens'. Here is the first instalment. Ozone is an odd-oxygen species (triatomic). Single oxygen atoms, another odd-oxygen species (monatomic), are 'potential ozone molecules', so they should clearly be seen as beneficial. To paraphrase George Orwell: 'odd oxygen good (in the stratosphere), even oxygen not so good'. (We won't go as far as 'bad', as without molecular oxygen there would be no ozone!)

10.1.1 Dobson Units

Ozone column abundance is usually measured/quoted in Dobson units. G. M. B. Dobson (1889–1976) was an English scientist who first detected the warm layer at high altitudes, and attributed it to the absorption of UV radiation by ozone. Ground-level measurements of ozone are made with a Dobson spectrophotometer (Figure 10.1). One Dobson unit (DU) refers

Figure 10.1 Dobson spectrophotometer (courtesy of NOAA).

to a layer of gas that would be 10-μm thick at standard temperature and pressure, or one 'milli-atmosphere-centimetre'. It is equal to 2.69×10^{16} ozone molecules per square centimetre of surface area. While ozone column amounts vary with both time of year and latitude, a typical value – at least outside the ozone hole region – is ~300 DU.

10.2 Catalytic Destruction

Based on the relevant reaction rates, and oxygen density profile, it is relatively straightforward to compute the expected profile of ozone in the atmosphere. When the predicted profile was compared with an extensive range of observations, it was found to systematically overestimate the data, with the total ozone column being smaller than the predictions by a factor of three to four. This clearly implies that there must be additional loss mechanisms at work. A number of such mechanisms have been identified, some involving naturally occurring chemical species, some not. The mechanism by which all the species involved remove ozone from the stratosphere is a catalytic cycle that may be written in generic form as

$$X + O_3 \rightarrow XO + O_2 \tag{10.6}$$

$$XO + O \rightarrow X + O_2 \tag{10.7}$$

The net effect of this cycle can clearly be seen to be

$$O_3 + O \rightarrow 2O_2 \tag{10.8}$$

X represents the catalyst, and XO an intermediate product. In the absence of another reaction for X or XO, this cycle can continue indefinitely, removing both O_3 and O (our 'good guys') from the stratosphere.

Why would a species undergo such a cycle? The answer is that X must have an odd valence (i.e. odd number of electrons) – either one or three – so that XO will also be odd. X and XO flip back and forth in the hope of finding another molecule with an odd number of electrons – hoping the grass might be greener – but rarely succeed.

So here is the second instalment in our story of odds and evens. Ozone and atomic oxygen have an even number of electrons (even valence), while the destructive catalytic species all have an odd number of electrons, and odd valence. So, this time we must reverse Orwell: 'even electrons good; odd electrons bad'!

10.2.1 Hydroxyl Radical

The first candidate suggested for the catalytic species was our old friend the hydroxyl radical, OH. It has (at least) three sources, all of which require atomic oxygen in its excited, singlet-D state (Chapter 2), which is more common in the mesosphere and upper stratosphere. It is formed by the photolysis of ozone:

$$O_3 + hf \rightarrow O(^1D) + O_2 \quad \lambda < 310 \text{ nm} \tag{10.9}$$

This more reactive species can create OH through the following reactions:

$$O(^1D) + H_2O \rightarrow 2OH \tag{10.10}$$

$$O(^1D) + CH_4 \rightarrow CH_3 + OH \tag{10.11}$$

$$O(^1D) + H_2 \rightarrow H + OH \tag{10.12}$$

Like any atom or molecule in an excited state, $O(^1D)$ may also be collisionally de-excited, converting the excitation energy to kinetic energy, and thus contributing to a rise in temperature. Both water vapour and methane mainly enter the stratosphere in the tropics. For methane this is straightforward, but for water vapour the cold tropical tropopause results in very low partial pressures. However, it is believed that some water crosses the tropopause via the anvil outflow of towering thunderstorms, which are prevalent in the monsoon circulation.

Below ~40 km OH from these sources acts as the catalyst, X, via the reaction cycle of reactions (10.6) and (10.7). Below 30 km, atomic oxygens are too scarce for reaction (10.7) to make a meaningful contribution. Instead, the following reaction becomes important:

$$HO_2 + O_3 \rightarrow OH + 2O_2 \tag{10.13}$$

The net effect of reactions (10.6) (with X = OH) and (10.13) is that two molecules of ozone become three of oxygen.

10.2.2 Nitrogen Oxides

Nitrogen is trivalent, while oxygen is divalent. Nitrogen forms a series of oxides – NO (nitric oxide), NO_2 (nitrogen dioxide), N_2O (nitrous oxide) – the first two of which contain an unpaired electron. In reality they are free radicals, and hence quite reactive, as we saw in Chapter 2. By contrast, nitrous oxide is stable and 'semi-permanent', and is well-mixed to at least the lower stratosphere. There it may react with excited oxygen atoms to form nitric oxide:

$$O(^1D) + N_2O \rightarrow 2NO \tag{10.14}$$

although it may also be photodissociated in the upper stratosphere. Nitric oxide is now another candidate for the catalytic species, X, in reactions (10.6) and (10.7).

A major source of atmospheric NO is incomplete combustion, and this is an important component of urban pollution, or smog (Section 2.3). Nitric oxide generated at or near the ground is far too reactive to make it to the stratosphere. However, around 1970 a new potential source of stratospheric NO, and hence a new threat to the ozone layer, was identified: high-altitude (supersonic) aircraft, such as the Concorde and the mooted US equivalent. This mode of transport failed to achieve commercial success, so this threat has largely dissipated (for now). However, current commercial aircraft flying polar (intercontinental) routes may actually spend time in the stratosphere due to the low altitude of the polar tropopause.

10.2.3 Halogens

At much the same time as the threat of NO from high-altitude aircraft was raising concerns, another man-made threat was being discussed: the halogens, chlorine and bromine, both natural (natural sources of Cl and Br in the stratosphere are methyl chloride (CH_3Cl) and methyl bromide (CH_3Br)) and, of more concern, those from industrially manufactured chlorofluorocarbons (CFCs), which contain carbon, chlorine, and fluorine (as the name implies), but – and this is the key to their inertness – no hydrogen. The most common are $CFCl_3$ (CFC-11) and CF_2Cl_2 (CFC-12). They were first produced in 1928, and marketed under the trade name Freon as non-toxic, non-flammable refrigerants. (Previous refrigerants had often proven deadly when leaks occurred.) They were also used as propellants in (aerosol) spray cans. In 1973 it was discovered that, as a result of their inertness, they had spread globally, and were expected to have a residence time of a hundred years or more in the troposphere.

So, what is their fate? What are the sinks? Once they rise to the mid-stratosphere (above ~25 km), they are able to absorb UV radiation in the wavelength interval 190–230 nm and photodissociate: for example

$$CFCl_3 + hf \rightarrow CFCL_2 + Cl \tag{10.15}$$

$$CF_2Cl_2 + hf \rightarrow CF_2Cl + Cl \tag{10.16}$$

The chlorine atom released can then serve as the catalyst, X, in reactions (10.6) and (10.7). By 1990, roughly 85% of stratospheric chlorine was of anthropogenic origin. It also happens that CFCs are strong absorbers in the infrared window region, making them powerful greenhouse gases.

In addition to the standard catalytic cycle, the following coupled cycle is also a powerful mechanism (ozone is far more abundant than atomic oxygen):

$$Br + O_3 \rightarrow BrO + O_2 \tag{10.17}$$

$$Cl + O_3 \rightarrow ClO + O_2 \tag{10.18}$$

$$BrO + ClO \rightarrow BrCl + O_2 \tag{10.19}$$

$$BrCl + hf\,(vis) \rightarrow Br + Cl \tag{10.20}$$

Again, the net effect is that two molecules of ozone are converted to three molecules of oxygen.

10.3 Biological Effects of UV Radiation

The UV spectral region is often subdivided into the UV-A, UV-B, and UV-C, based on the transition of ozone absorption from very strong to almost zero. While some authors use a slightly different breakdown, we will define UV-A as 320–400 nm, where the atmosphere is largely transparent; UV-B as the 290–320 nm transition region; and UV-C below 290 nm, where the atmosphere can be said to be fully opaque (Figure 10.2).

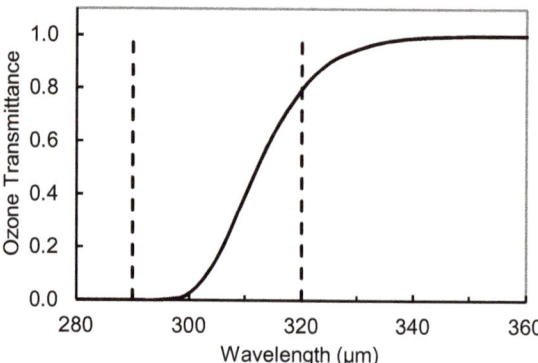

Figure 10.2 Ozone transmittance in the UV-B and UV-A. *Source: Physics of Radiation and Climate*, figure 12.9, Box and Box, CRC Press, 2016, reproduced by permission of Taylor & Francis Group.

Ultraviolet radiation with wavelengths of less than ~315 nm (with photon energies above ~4 eV) is capable of breaking bonds in key biochemical molecules, including DNA. UV-B and (especially) UV-C is thus deadly to most forms of life on land, and even in the surface layers of the ocean. If we were to lose our ozone layer for any reason it would not just be human skin that would suffer, but our food supply as well.

While we will focus on the harmful effects of UV radiation, we need to start with the major beneficial effect. UV-B exposure induces the production of vitamin D in the skin. The majority of positive health effects of UV are related to this vitamin. It has regulatory roles in calcium metabolism (vital for bone growth and the maintenance of bone density), immunity, cell proliferation, insulin secretion, and blood pressure.

Overexposure to UV-B radiation can cause sunburn (erythema) and some forms of skin cancer. However, the deadliest form, malignant melanoma, is mostly caused by indirect DNA damage (free radicals and oxidative stress). In humans, prolonged exposure to solar UV may result in acute and chronic health effects on the skin, eyes, and immune system. Moreover, UV-C can cause adverse effects that can be mutagenic or carcinogenic. Ultraviolet photons harm the DNA molecules of living organisms in different ways. In one common damage event, adjacent thymine bases bond with each other, instead of across the 'ladder'. This thymine dimer makes a bulge, and the distorted DNA molecule does not function properly.

In the past, UV-A was considered far less harmful, but today it is known that it can contribute to skin cancer via indirect DNA damage. It penetrates deeply, but does not cause sunburn. UV-A can generate highly reactive chemical intermediates, such as hydroxyl and oxygen radicals, which in turn can damage DNA. Accordingly, the DNA damage caused to skin by UV-A consists mostly of single-strand breaks in DNA, while the damage caused by UV-B includes direct formation of dimers, and double-strand DNA breakage. UV-A is immunosuppressive for the entire body, and is mutagenic for basal cell keratinocytes in skin. In 2011, the International Agency for Research on Cancer of the World Health Organization classified all categories and wavelengths of ultraviolet radiation as a Group 1 carcinogen. This is the highest-level designation for carcinogens and means 'there is enough evidence to conclude that it can cause cancer in humans'.

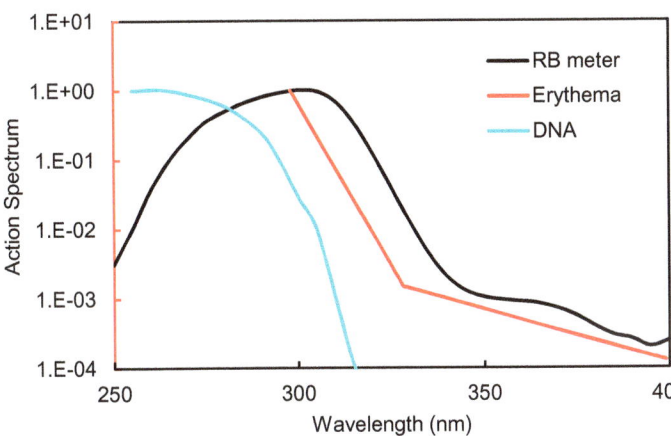

Figure 10.3 Action spectra for DNA and erythema, plus the response function of the Robertson–Berger meter. *Source: Physics of Radiation and Climate*, figure 12.10, Box and Box, CRC Press, 2016, reproduced by permission of Taylor & Francis Group.

High intensities of UV-B are hazardous to the eyes, and exposure can cause welder's flash (photokeratitis) and may lead to cataracts. UV is absorbed by molecules known as chromophores, which are present in the eye cells and tissues. If too much UV is absorbed, eye structures such as the cornea, the lens, and the retina can be damaged.

For all of these conditions, and others not included here, the biological response varies with wavelength. The relative response, or sensitivity, is known as the **action spectrum**, $A(\lambda)$, which is normalised to have a maximum value of 1.0. We may then define the **dose rate** by combining (weighting) the action spectrum with the solar flux at the surface. Mathematically (for those who have done calculus) this is by an integration:

$$D = \int A(\lambda)F(\lambda)d\lambda \qquad (10.21)$$

where $F(\lambda)$ is the surface UV flux. Figure 10.3 shows action spectra for the DNA molecule and **erythema** (sunburn), along with the response function of the Robertson–Berger (R-B) meter, which is designed to mimic the response of Caucasian skin. Erythemal dose is often quoted in terms of the **UV index**, which is a scaled version of the dose rate with wavelength limits of 250 nm and 400 nm, and a multiplying factor of 40 m^2 W^{-1}, designed to produce results in a range up to ~15.

The annual effective UV dose for erythema varies with latitude, primarily because solar elevation directly affects surface flux. (There is also some latitudinal variation in the average annual ozone column; however, outside an area of ozone depletion this has a relatively small effect on annual dose.) The dose varies by about 4% with each degree of latitude, so that Caucasians living closer to the equator have increasing risk. Australia has the world's highest skin cancer rate, with the state of Queensland highest overall. Annual dose will also be affected by changes in total ozone column which, as we have seen, is a topic of potential concern. The percentage increase in UV dose for a given percentage decrease in ozone is known as the radiation amplification factor. This is not a constant, as the ratio depends non-linearly on ozone depletion. However, for depletions of less than 5–10% the amplification factor may be taken as approximately 1.0, or a little higher.

10.4 Antarctic Ozone Hole

The ozone layer is clearly vital to all terrestrial life on this planet. Unfortunately, over the years a number of threats to this vital shield have been identified, including a number of man-made chemical species. These threats, and the responses to them, are sufficiently serious to deserve special attention. The first of these threats has already been alluded to: nitric oxide (NO) from high-altitude aircraft.

10.4.1 History

In 1973 Rowland and Molina, then at the University of California, Irvine, began studying the impacts of CFCs in the Earth's atmosphere. They discovered that CFC molecules were stable enough to remain in the atmosphere until they drifted up into the middle of the stratosphere, above most of the protective ozone shield, where they would finally (after an average of 50+ years for two common CFCs) be broken down by UV radiation, releasing a free chlorine atom. They then proposed that these chlorine atoms might be expected to cause the breakdown of large amounts of ozone in the stratosphere, via the catalytic process outlined above. Their argument was based on an analogy to contemporary work by Crutzen and Johnston, which had shown that nitric oxide could catalyse the destruction of ozone. In 1995, Paul Crutzen, Mario Molina, and Sherwood Rowland were awarded the Nobel Prize in Chemistry for their work on the role of both CFCs and nitrogen gases in stratospheric ozone depletion.

Then, in 1985, British Antarctic Survey scientists Farman, Gardiner, and Shanklin published results of abnormally low ozone concentrations above Halley Bay (76 °S). After carefully rechecking their data, they reported that, since 1977, they detected a 30% decrease in total column ozone in October (austral springtime), and speculated that this was connected to increased levels of CFCs in the atmosphere. The depletion was later confirmed by a re-analysis of satellite data, and dedicated field campaigns, including balloon flights, which showed depletion through the entire ozone layer.

The impact of these studies, the metaphor 'ozone hole', and the colourful visual representation in time lapse animation proved shocking enough for negotiators in Montreal to take the issue seriously. Since then, the Antarctic ozone hole has grown to occupy an area of around 25 million km^2, similar to that of North America. (An ozone hole is defined as an area with an ozone column of less than 220 Dobson units.) It now appears to have stabilised, as a result of the Montreal Protocol, and is likely to steadily recover over the next half century.

Figure 10.4 shows the situation on 4 October 2004, based on satellite data. (An animation from 1979 to 2019 can be found on NASA's Earth Observatory.)

10.4.2 Heterogeneous Chemistry

A number of key questions now arise. Why Antarctica? Why spring? The answers to these questions involve some fascinating science. Chlorine is certainly the primary culprit, but the chemistry is more complex than previously postulated. The catalytic reactions (10.6) and (10.7), with Cl as X, do not work well in the lower stratosphere, for two reasons. Firstly, the

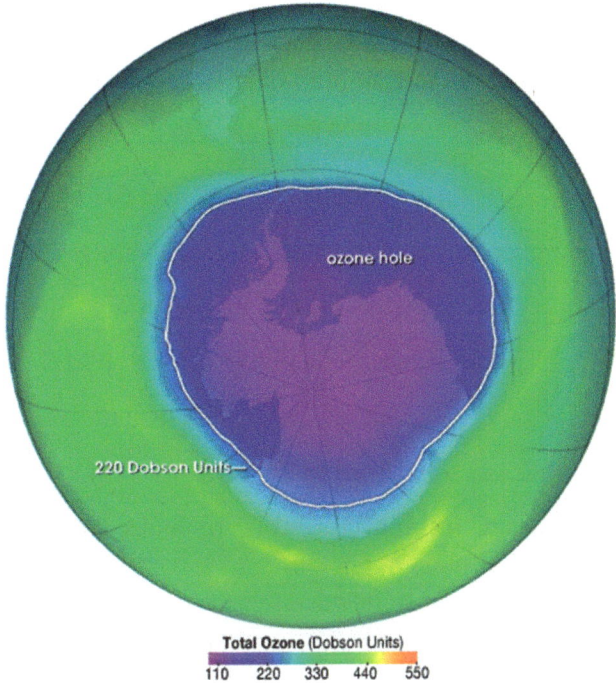

Figure 10.4 Ozone data for 4 October 2004 (courtesy NASA Ozone Watch).

concentration of oxygen atoms is low. The second reason is that much of the Cl and ClO in the stratosphere is quickly tied up in 'reservoir species' such as

$$ClO + NO_2 + M \rightarrow ClONO_2 + M \tag{10.22}$$

$$Cl + CH_4 \rightarrow HCl + CH_3 \tag{10.23}$$

These reservoirs are generally quite stable. However, on the surface of the ice crystals that constitute polar stratospheric clouds (PSCs), which form during the cold austral winter, the following heterogeneous reactions become significant:

$$ClONO_2(g) + HCl(s) \rightarrow Cl_2(g) + HNO_3(s) \tag{10.24}$$

$$ClONO_2(g) + H_2O(s) \rightarrow HOCl(g) + HNO_3(s) \tag{10.25}$$

$$HOCl(g) + HCl(s) \rightarrow Cl_2(g) + H_2O(s) \tag{10.26}$$

Here the parenthetical g denotes species in the gas phase, while s denotes species in the solid phase, in or on ice particles. Note that some of the ice particles will settle out of the stratosphere, taking $HNO_3(s)$ with them, and reducing the reservoir species $ClONO_2$: a process known as denitrification.

This set of five reactions takes place during the Antarctic winter, converting reservoir species into 'potentially' reactive species. Denitrification also contributes to this priming, and so the stage is set for the return of the Sun. The other change that occurs during the months of darkness is that atomic oxygen is converted to ozone, or otherwise removed, while ozone cannot be photolysed, which would be the source of fresh oxygen atoms.

In spring, the Sun returns and the following photolysis reactions are triggered:

$$Cl_2 + hf \rightarrow 2Cl \tag{10.27}$$

$$HOCl + hf \rightarrow OH + Cl \tag{10.28}$$

Ozone above Antarctica is then efficiently destroyed by the reactions

$$Cl + O_3 \rightarrow ClO + O_2 \text{ (twice)} \tag{10.29}$$

$$ClO + ClO + M \rightarrow (ClO)_2 + M \tag{10.30}$$

$$(ClO)_2 + hf \rightarrow Cl + ClOO \tag{10.31}$$

$$ClOO + M \rightarrow Cl + O_2 + M \tag{10.32}$$

The net effect is, again, the conversion of two molecules of ozone to three of oxygen, accompanied by the absorption of photons.

The following points should be noted:

- Two Cl atoms are regenerated for every two destroyed, so this is a catalytic cycle.
- Atomic oxygen is not required: this is important, as it is in short supply.
- The Cl atom in the ClO originates from CFCs via reactions (10.20) and (10.21). Normally, this would be quickly tied up as $ClONO_2$ via reaction (10.22), but denitrification has reduced the NO_2 available.
- Thanks to PSCs, reactive chlorine is released via reactions (10.24)–(10.26), which then release their Cl and ClO when the Sun rises.
- The dimer $(ClO)_2$ is formed by reaction (10.30) only at the low temperatures that are found in the lower stratosphere over the winter pole.

10.4.3 Antarctic Winter

We have answered the second of our questions, why spring? Now we must address the geographic question of why the ozone hole appears over Antarctica, but not at lower latitudes, or in the Arctic. The key to the ozone hole is the heterogeneous chemistry that occurs on ice crystals, or PSCs, which only occur at quite low temperatures. So, the question we need to address is why do temperatures drop so low over Antarctica in winter, and not elsewhere?

If you were to place yourself several thousand kilometres above the South Pole in July and look down (you'd need your own light source, of course), you would be looking at a very 'symmetric' picture. You would see a circle of ice, Antarctica and its sea ice, out to around latitude 65° or even 60°. Beyond that you would see the Southern Ocean, stretching on towards ~40° latitude, interrupted only slightly by the finger of southern South America. This is the symmetric geography of the Southern Hemisphere, which is ocean-dominated, but with a continent at its pole.

A direct consequence of all this symmetry is a symmetric wind field known as the polar vortex, which prevents warmer, lower-latitude air from interacting with air above Antarctica. The Southern Ocean is also able to circulate west-to-east in a circular flow, the circumpolar current. In the absence of sunlight this air radiates to space and becomes extremely cold, falling below $-80°C$. This allows PSCs to form (see Section 4.4.3).

Contrast all this with the view you would get above the North Pole in January. You would be presented with a very different picture, especially in the latitude belt of 50–65°. This belt includes the land masses of Eurasia and North America, which would have significant snow cover at this time of year, with temperatures dropping to $-20°C$ and lower in many parts of the continental interiors. But this band also covers the northern Pacific Ocean, and the Gulf Stream in the Atlantic, which remain above ~10°C all year. This far more asymmetric situation is unable to sustain a symmetric polar vortex type of wind flow – a wave pattern forms instead – so that warmer, lower-latitude air is more readily mixed in, keeping air temperatures above the Arctic higher. Occasionally, a vortex-type circulation is temporarily established, allowing polar stratospheric clouds to form, resulting in small, temporary ozone holes.

10.5 The Montreal Protocol

The Montreal Protocol on Substances that Deplete the Ozone Layer, a protocol to the **Vienna Convention** for the Protection of the Ozone Layer, is an international treaty designed to protect the ozone layer by phasing out the production of numerous substances believed to be responsible for ozone depletion, as discussed above. The treaty was opened for signature on 16 September 1987, and entered into force on 1 January 1989, followed by a first meeting in Helsinki in May 1989. During the 1990s it underwent seven revisions. It is believed that if the international agreement is adhered to, the ozone layer is expected to recover by 2050. Due to its widespread adoption and implementation, it has been hailed as an example of exceptional international co-operation, with Kofi Annan quoted as saying that 'perhaps the single most successful international agreement to date has been the Montreal Protocol'. The ozone treaties have been ratified by 197 states and the European Union, making them the most widely ratified treaties in UN history.

The treaty is structured around several groups of halogenated hydrocarbons that have been shown to play a role in ozone depletion. All of these ozone-depleting substances contain either chlorine or bromine. (Substances containing only fluorine do not harm the ozone layer as the C–F bond is too strong to be severed by the available UV radiation in the stratosphere.) For each group the treaty provides a timetable in which the production of those substances must be phased out and eventually eliminated.

The signatory states agreed to accept a series of stepped limits on CFC use and production, including:

1. From 1991 to 1992 the levels of consumption and production of the controlled substances in Group I of Annex A do not exceed 150% of the calculated levels of production and consumption of those substances in 1986.
2. From 1994 the calculated level of consumption and production of the controlled substances in Group I of Annex A does not exceed, annually, 25% of the calculated level of consumption and production in 1986;
3. From 1996 the calculated level of consumption and production of the controlled substances in Group I of Annex A does not exceed zero.

The substances in Group I of Annex A are:

- $CFCl_3$ (CFC-11);
- CF_2Cl_2 (CFC-12);
- $C_2F_3Cl_3$ (CFC-113);
- $C_2F_4Cl_2$ (CFC-114);
- C_2F_5Cl (CFC-115).

There was a slower phase-out of other substances (halons 1211, 1301, 2402; CFCs 13, 111, 112, etc.) and some chemicals were given individual attention (e.g. carbon tetrachloride, CCl_4). The phasing-out of the less-active HCFCs – transitional CFC replacements, used as refrigerants, etc. – only began in 1996 and will go on until a complete phasing-out is achieved by 2030. In terms of their ozone-depleting potential (ODP), in comparison to CFCs that have ODPs of 0.6–1.0, these HCFCs have lower ODPs, i.e. 0.01–0.5. There are a few exceptions for 'essential uses', where no acceptable substitutes have been found (for example, in the metered dose inhalers commonly used to treat asthma and other respiratory problems) or halon fire suppression systems used in submarines and aircraft (but not in general industry).

The provisions of the Protocol include the requirement that the Parties to the Protocol base their future decisions on the current scientific, environmental, technical, and economic information that is assessed through panels drawn from the worldwide expert communities. To provide that input to the decision-making process, advances in understanding on these topics were assessed in a series of reports entitled Scientific Assessment of Ozone Depletion. As of 16 September 2009, all countries in the United Nations and the EU have ratified the original Montreal Protocol, Timor-Leste being the last country to do so. Fewer countries have ratified each consecutive amendment. Only 167 countries have ratified the Beijing Amendment.

Chemical transport models now capture the essentials of the ozone hole, and we may use them to understand future depletion levels. The effective equivalent stratospheric chlorine (EESC) is a parameter that is used to quantify the combined effects of chlorine and bromine. Natural background levels are about 1 ppb. In 1985 the value was close to 2 ppb, and continued to rise until about 2000, due to transport from the troposphere, reaching about 4 ppb. It is now decreasing, and model results suggest that by 2060 it will again be close to 2 ppb. However, a recent paper by Western et al. shows the levels of five restricted CFCs actually rising, including CFC-115, which is an Annex I substance.

Summary

Ozone can be a pollutant in the lower troposphere, but in the stratosphere it is, quite literally, life or death. No terrestrial life could exist on our planet until an ozone shield could be established. This firstly required a rise in atmospheric oxygen, one of the interesting stories we will study in Section 16.2.

The creation and destruction of stratospheric ozone is an interesting photochemical process, as we have seen. While the basic reactions are quite simple, it turns out that Nature has decided to throw a couple of extra processes into the mix. (If you found the heterogeneous chemistry in Section 10.4.1 a struggle, you're not alone.)

The prediction and discovery of the ozone hole is one of the most interesting chapters in the history of science, and definitely worthy of a Nobel Prize. It is both a tribute to the scientists involved and a testament to the importance of basic science. Imagine what our world might be like today if the powers-that-be (funding agencies, promotion panels, etc.) decided to only support commercially oriented research?

FURTHER READING

The full details of stratospheric chemistry are well covered by

- *Theory of Planetary Atmospheres* by J. W. Chamberlain and D. M. Hunten (Academic Press, 1987).
- *Atmospheric Change, An Earth System Perspective* by T. E. Graedel and P. J. Crutzen (Freeman and Co., 1993).
- *Atmospheric Science for Environmental Scientists* by C. N. Hewitt (Wiley-Blackwell, 2020).

The effects of UV radiation are covered in some detail in

- *Biological Effects of Ultraviolet Radiation* by W. Harm (Cambridge University Press, 1980).

The original work on CFCs as a Cl source is

- Stratospheric sink for chlorofluoromethanes: Chlorine atom catalyzed destruction of ozone by M. J. Molina and F. S. Rowland, *Nature*, vol 249, 810–812, 1974.

The discovery of the ozone hole was

- Large losses in Antarctic ozone reveal seasonal ClO_x/NO_x interaction by J. C. Farman et al., *Nature*, vol 315, 207–208, 1985.

The recent paper just mentioned on the increase in certain CFCs is

- Global increase of ozone-depleting chlorofluorocarbons from 2010 to 2020 by L. K. Western et al., *Nature Geoscience*, vol 16, 303–313, 2023. doi-org/10.1038/s41561–023-01147-w.

REVIEW QUESTIONS

1. How is it that UV radiation both creates and destroys ozone?
2. Why were the reactions in Section 10.2 described as 'catalytic'?
3. Name one benefit of UV radiation.
4. Why do we get an ozone hole over Antarctic but rarely over the Arctic?
5. What are PSCs, and what is their role in the ozone hole?
6. What is the major group of chemicals being phased out under the Montreal Protocol?

EXERCISES

What information is available where you live on UV levels? Is it considered an important piece of information (at least in the summer months) by the local media?

11 Long-Wave Radiation Transfer

The inflow of solar radiant energy into the Earth System is relatively straightforward. A large fraction is absorbed by water vapour, in the lowest few kilometres of the troposphere, with most of the rest either being reflected or reaching the ground.

The outward flow of terrestrial/long-wave radiation is a different kettle of fish. The majority of the energy radiated by the surface is absorbed by gases in the atmosphere, again mostly in the troposphere. Much of this is then radiated back to the surface, warming it. This, of course, is the greenhouse effect. In Chapter 8 we introduced the basic concepts and then made use of simple models that, while ignoring some important physics, were still valuable in helping us visualise the flows of radiation which are central to these processes.

So now we must turn our attention to the central questions that we have so far glossed over. Why is it that some gases are radiatively active – i.e. greenhouse gases – while others are not? Exactly how do these gases absorb, and re-emit, long-wave radiation? This latter question is central to the whole field of global warming – also known as the enhanced greenhouse effect – as it is the key to being able to quantify the effects of any *increases* in these gases.

This chapter is *by far* the most technical in the book, and we know many of our readers will not fully grasp all of the details. Nevertheless, we urge you to give it a try. We have done our best to help guide you to an understanding of the most important information, so do feel free to skip the equations.

11.1 Atomic Spectroscopy

Why is it that oxygen, nitrogen, and argon, which make up 99.9% of our (dry) atmosphere, are not greenhouse gases, but carbon dioxide and water vapour are? The world would be vastly different if that were the case: in fact, we probably wouldn't be here to discuss it! The answer lies in the way atoms and molecules absorb and emit electromagnetic radiation. We hinted at this when talking about photochemistry in Chapter 2 when we introduced the photon concept: radiation is absorbed or emitted 'one photon at a time', by a single atom or molecule.

But that is only half the story; and by far the simpler half. The other half is all about atomic and molecular structure; the province of quantum physics. This is most definitely beyond the scope of this book! Nevertheless, we believe that the central concepts can be explained in sufficient detail, starting with a little history.

One quick point before we get into details. When discussing black-body radiation in Chapter 8, we stated that this was a puzzle that was beyond nineteenth-century physics, and finally solved by Planck and Einstein. There was a related problem: the spectra of atoms when heated in a suitable apparatus (e.g. a Bunsen burner). This showed a seemingly random set of bright lines – **spectral lines** – of emitted radiation, which varied from element to element without rhyme or reason. Solving this puzzle is what we are about to do.

11.1.1 Atomic Structure

This subsection is included to introduce you to some of the discoveries, and the scientists who made them, that opened up the atomic world.

The atom and its structure were discovered in the closing years of the nineteenth century. In 1897, **J. J. Thomson** discovered the electron, a negatively charged particle whose very large charge-to-mass ratio clearly indicated that it was very much lighter than a hydrogen atom: it must, therefore, be a 'part' of an atom (1906 Nobel Prize in Physics). About the same time, Henri Becquerel and the Curies discovered radioactivity (1903 Nobel Prize). **Marie Curie** later received the 1911 Chemistry Prize for discovering the radioactive elements radium and polonium: she is the only person to win a Nobel Prize in two sciences. This work was followed up by many scientists, most notably by (Lord) **Rutherford** (1908 Nobel Prize in Chemistry). These discoveries clearly showed that atoms were real, and that they must have some form of structure. So how were atoms composed?

Thomson suggested that they consisted of a blob of positively charged 'stuff', containing almost all the atom's mass, with the required number of electrons embedded in it to make it overall neutral: known for obvious reasons as the **Plum Pudding Model**. Lorentz placed these electrons on 'springs', with appropriately chosen spring constants (elasticity) so that, when energetically excited, they might oscillate at the appropriate frequency to emit the line spectra we just mentioned. Clever; but wrong.

However, when a team in Rutherford's lab fired a beam of α-particles through a thin gold foil, they found some actually 'bounced back'. In the Plum Pudding Model, the electrical forces could not have been strong enough for this to happen; they were too 'diffuse'. Rutherford concluded that all the positive charge (and the mass) must be concentrated in a tiny fraction of the atom's volume – the **nucleus** – leading to potentially much stronger electrical forces if an α-particle happened to approach it head-on. So atoms are mostly empty space, consisting of a tiny, positively charged nucleus, with virtually all the atom's mass and positive electrical charge, with the electrons orbiting somewhere off in a distant cloud.

But this immediately raised its own challenge. If the electrons were in some form of orbital motion, then they must be accelerating, even if only changing direction while undergoing uniform circular motion. But Maxwell's electromagnetism equations make it clear that such an accelerating electric charge must emit radiation, and spiral into the nucleus in a fraction of a second. Atoms were 'illegal'!

11.1.2 Particles and Waves

So atoms cannot exist, at least according to nineteenth-century physics. Yet atoms exist. Clearly we have a conundrum we need to solve. This subsection is also history, but the conclusions from it are a key to the rest of the chapter.

In 1913 **Niels Bohr** took the first step by 'legislating' the problem away with a new model (1922 Nobel Prize), based on four postulates, the first three of which are

1. The energies of electrons are **quantised**: electrons can only be in certain orbitals, or energy levels, or 'states', with certain allowed energies. *When in such states they don't radiate energy.* Perhaps the best translation of the word quantised is 'discrete'.
2. An electron can make a 'quantum jump' from one state (energy level) to another by emitting or absorbing a single photon with the appropriate energy: the difference between the two energy levels. (It is **never** 'in between', not even for a nanosecond.)
3. While in an allowed orbit, classical mechanics applies (e.g. the Coulomb force between the electrons and the nucleus).

He then added a fourth, ad-hoc postulate, which allowed him to calculate the energy levels of a hydrogen atom, and show that the photons emitted/absorbed in the resulting transitions between levels corresponded to the observed hydrogen spectrum. This worked well, and was clearly the germ of an idea, but couldn't be extended to any atoms heavier than hydrogen: almost the entire Periodic Table! One key problem was how to handle the mutual repulsion of the electrons in such atoms. So, one small step forward. The Great War then intervened, before new thinking arrived.

Einstein's photon postulate effectively claimed that electromagnetic radiation was both a wave and a (stream of) particles: certainly quite profound. What about the reverse: could a particle such as an electron also act like a wave? Even more profound. In 1924 **de Broglie** made the bold suggestion that the answer was, indeed, yes, including a formula for the connection between a particle's momentum and its wavelength (1929 Nobel Prize). While this idea was initially greeted with a degree of scepticism, it is the key to the quantum revolution, from the theory we are about to sketch out, to the electron microscope, silicon chip, and so much more.

In the standard version of quantum mechanics we teach to undergraduates, the application starts with the solution of a wave equation (due to **Schrödinger**: 1933 Nobel Prize) for the electrons in an atom, or whatever system you are investigating. This is trivial for hydrogen, with just one electron (and the results are consistent with the Bohr model, but richer), but more of a challenge when electron–electron repulsion must be included. However, that is really only a computational challenge, not a conceptual challenge.

The solution of this equation is known as the **wave function** (for obvious reasons), and contains all the information on the system – atom, molecule, crystal, etc. – that Nature can provide, *ever*. From this we can extract such information as the energy levels and the shape of the electron orbitals. The spectrum of an atom is determined by these various energy levels, or more specifically the differences between them, as Bohr had postulated. The shape of the orbitals is important in chemistry, as it relates to the covalent bonds many atoms form to produce molecules, including the ones in the atmosphere we have been discussing. For example, it explains why the carbon dioxide molecule is linear, while the water molecule is bent.

11.1.3 Atomic Spectra

So, what are the implications of all this for the question we asked above? That is to say, why are some gases radiatively active, and others not? Central to the greenhouse effect, as we have seen, is the absorption and re-emission of radiation in the spectral range from about 4 μm to about 100 μm and longer. Based on Einstein's formula quoted in Chapter 8,

$$E = 1.24/\lambda \ (\lambda \text{ in } \mu m, E \text{ in eV})$$

we see that this corresponds to energies in the range 0.31–0.0124 eV. Let's just say 'smaller than 0.3 eV'. So now we must turn our focus to this energy range.

Argon exists in the atmosphere as an atom, and its energy levels are widely spaced. This means that the only photons it can absorb are of quite high energy, and this places them in the UV part of the spectrum: hence, argon is not radiatively active in the region of interest, and so is not a GHG. We must now turn our attention to molecules.

11.2 Molecular Physics

11.2.1 Rotation and Vibration

With the major exception of argon, virtually every chemical species in the atmosphere is molecular, not atomic. Now, as we briefly noted when discussing the Kinetic Theory of Gases (Section 5.4), molecules have additional 'degrees of freedom'. They can rotate, and they can vibrate, about their centre of mass, and these motions are also quantised (i.e. have discrete energy levels). Note that for a diatomic molecule these motions are very simple. The two atoms can oscillate back and forth, stretching and compressing their bond length, and they can rotate about the two axes at right angles to that bond. Triatomic molecules have more options, and the results can be more complex, even 'messy', depending on their shape (bent vs. linear).

Rotational energy levels have the lowest energies, and most molecules are rotating at room temperature. Kinetic Theory gave us a connection between energy and temperature; from that we find that room temperature corresponds to an energy of ~0.025 eV: this is generally higher than their lowest rotational energy. This is the reason why, in Section 5.4, we said that the nitrogen and oxygen molecules, which dominate our atmosphere, have five degrees of freedom (three translational and two rotational), not the three which applies to atoms.

Vibrational energies are, in general, a good deal higher, so that almost all molecules should normally be in their 'ground state', i.e. not vibrating. However, just as some gas molecules will have twice the translational kinetic energy of the average (remember the Maxwell–Boltzmann distribution function), so a few molecules may be 'thermally agitated' to vibrate, although mostly only in their so-called first excited state.

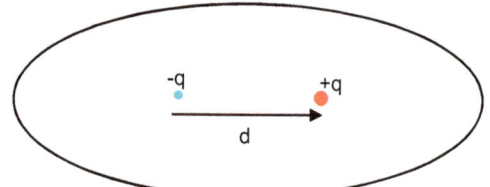

Figure 11.1 Schematic illustration of a dipole moment.

11.2.2 Dipole Moment

There is one technical point we need to introduce.

Like atoms, molecules are electrically neutral, having equal numbers of positively charged protons and negatively charged electrons. However, in some molecules, the 'average' position of all the positive charges does not correspond to the 'average' position of all the negative charges. (If you are familiar with the concept of centre of mass, then 'centre of charge' is very similar.) In this case the molecule is said to be polarised; to possess an electric dipole moment.

This is illustrated schematically in Figure 11.1, for a hypothetical distribution of charges. The average positions of the positive and negative charges are indicated by $+q$ and $-q$, and the distance between them is d. The dipole moment is defined as the product, qd. Note that this 'separation' of charges creates an electric field, not unlike the magnetic field created by a bar magnet.

One of the best examples is the water molecule, where the greedy oxygen atom grabs more than its fair share of the electrons that are bonding the hydrogen atoms to it. This leads to strong electrical attractions between the oxygen atom in one water molecule and a hydrogen atom in another. This is a key part of the ice crystal structure, and why it is so strong.

The dipole moment is central to the absorption and emission of radiation, because it is this that the electric field in the (electromagnetic) radiation 'grabs hold of' (interacts with). If the molecule does not have a dipole moment, then even if it is vibrating and/or rotating it cannot interact with the electromagnetic wave, and so cannot absorb or emit radiation. (Such a molecule can still increase or decrease its rotational energy via collisions with other molecules, swapping some translational kinetic energy for rotational energy.) The nitrogen and oxygen molecules are what is known as homonuclear: they consist of two atoms (i.e. nuclei) that are identical: thus, the molecule is totally symmetric. Such a molecule can never possess a dipole moment, as they will always retain this symmetry, even when rotating or vibrating. *It is for this fundamental reason that nitrogen and oxygen are not 'radiatively active', i.e. greenhouse gases.*

11.2.3 Absorption Bands

So, how do molecules 'behave', rotationally and vibrationally speaking? More specifically, what are the quantum rules? As already noted above, the rotation of triatomic molecules like CO_2 and H_2O can be a lot messier than for a simple diatomic molecule, depending, for example, on whether it is linear (e.g. CO_2) or bent (e.g. H_2O). We will focus on the simplest case, which is enough to get the key ideas across. Rotational energies of linear molecules are given by the simple formula

$$E_J = BJ(J+1) \tag{11.1}$$

where B is known as the rotational constant of the molecule (basically its mass and bond length) and $J = 1, 2, 3, \ldots$ is the label for the particular energy level (it is an example of a quantum number; they're usually just numerical). There is also a 'selection rule' that says that J can only change by one unit at a time, so that the allowable transitions are given by

$$\Delta E = E_{J+1} - E_J = 2B(J+1) \tag{11.2}$$

So, the rotational spectrum of such a molecule would appear as a series of linearly spaced lines for the different values of J. (Of course, when we say 'appear' we must remember that these spectral lines are well down into the infrared, and are certainly not visible to the human eye.)

Vibrational energies are invariably a good deal higher than rotational energies, so that, at room temperature, most molecules are not vibrating. However, infrared photons may have enough energy to 'promote' a molecule up to its lowest vibrational level. Of course, the photon will need just the right amount of energy to do this, suggesting that vibrational spectra of most molecules should be just a single line.

However, a photon with a little more than the right amount of energy might be able to promote a molecule to vibrate *and* increase the level of rotational energy of the molecule by one step, using the excess. Similarly, a photon with a little less than the needed amount of energy might still be able to promote the molecule by grabbing some of its rotational energy to complete the job. So molecular spectra are often composed of a series – or band – of lines involving a one-step change in vibrational energy, combined with a whole range of rotational changes: known, for obvious reasons, as a rotation–vibration band.

11.2.4 Spectral Line Shapes

We need to take yet another technical excursion.

If quantum mechanics can compute atomic and molecular energy levels with absolute precision, then the energies of the photons emitted or absorbed in a transition should be precisely defined. So only those photons with *exactly* the right energy should be absorbed or emitted. However, to use slightly strained language, the number of photons with precisely a particular energy is precisely zero. If that really were the case, the greenhouse gases would only be able to absorb a negligible amount of the energy emitted by the Earth's surface, leading to a negligible greenhouse effect. Clearly this is not the case.

Spectral lines have a finite spectral width, and there are three factors that contribute to this. Firstly, there is the so-called **natural line width**. Heisenberg, one of the founding fathers of quantum mechanics (1932 Nobel Prize), enunciated two **Uncertainty Principles**. The better known of the two tells us that we can never know both the position and momentum of an electron (for example) with absolute precision. The lesser known relates an uncertainty in energy to an uncertainty in time, which we may express via

$$\Delta E \, \Delta t \sim h \tag{11.3}$$

where h is Planck's constant. (One 'application' of this result is to say that we can violate the fundamental Conservation of Energy Law by an amount ΔE, but only for an amount of time Δt; this is only relevant at the sub-atomic level.)

Now excited states of atoms and molecules have a finite lifetime, which is typically of the order of 10^{-8} seconds. Heisenberg's Uncertainty Principle takes this as the Δt. From that we may calculate ΔE, which we interpret as the uncertainty in the energy of the excited state, and from that the range of photon energies which might be absorbed or emitted. This turns out to be about 10^{-5} nm: very small, but finite. So now a finite range of photon energies may interact with – i.e. be absorbed or emitted by – an atom or molecule.

On its own, the natural line width would not produce a noticeable greenhouse effect. Fortunately (for life on Earth), there are two further mechanisms that contribute to additional broadening of spectral lines. The first of these is known as **pressure broadening**, or collision broadening. We saw in Chapter 5 that molecules collide with one another at a prodigious rate. In the process, the electrons in one atom or molecule will exert electrical forces on the electrons in the other, and this will have the effect of slightly altering energy levels, and hence the range of photon energies which might be absorbed or emitted: in other words, broadening spectral lines.

Both the natural line shape/profile, and the pressure-broadened line shape, follow the so-called **Lorentz profile** (the guy who put J. J. Thomson's electrons on springs), given by

$$\varphi(f) = \frac{1}{\pi} \frac{\alpha}{(f - f_0)^2 + \alpha^2} \tag{11.4}$$

Here f is the radiation frequency, f_0 is the 'line centre' frequency, and α is the 'line width'. Note that the total area 'under the curve' is equal to 1.0: that is, it is normalised. Figure 11.2 shows the shape of a pressure-broadened line, for several degrees of broadening. As a line broadens, you will see that it also 'flattens': the total area under the curve remaining constant.

The rate of molecular collisions depends on two factors. Firstly, the collision rate will be proportional to density, i.e. the number of molecules to be collided with. Secondly, collision rates must increase with temperature, as the average molecular speed is related to temperature, as we saw in Chapter 5. Density and temperature are connected to pressure, of course, via the ideal gas equation, so that when we combine these various bits and pieces it turns out that spectral line widths are proportional to pressure; hence the name, pressure broadening. Because

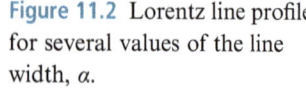

Figure 11.2 Lorentz line profile for several values of the line width, α.

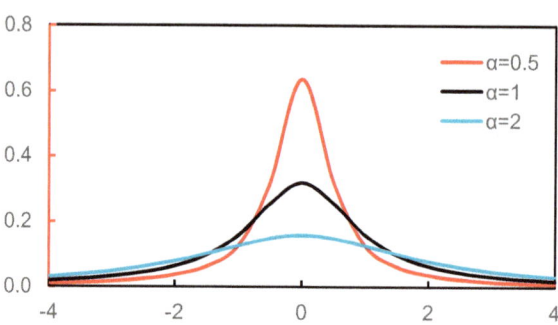

pressure decreases with altitude, so does line width, one of many challenges when doing detailed calculations.

The third mechanism that broadens spectral lines is the Doppler effect: the effect which causes a train whistle to sound higher in pitch as a train approaches, and then lower in pitch as it moves away from you. Molecules, as we know, are moving with a range of speeds, given by the Maxwell–Boltzmann distribution. The Doppler effect then spreads out spectral lines, with the same type of distribution. **Doppler broadening** is mainly important in the stratosphere, where pressure is low and pressure broadening quite small.

11.3 Greenhouse Gas Spectroscopy

Now that we understand at least the basics of the physics of how a gas absorbs and emits electromagnetic radiation, we can look at the specifics for the various gases which contribute to the greenhouse effect.

11.3.1 Carbon Dioxide

The carbon dioxide molecule is linear and symmetrical (O=C=O), with a bond length of 116 pm. Its symmetry means that it has no permanent dipole moment, and no (pure) rotation band. Its fundamental vibration modes are shown in Figure 11.3. Of these, the so-called v_1 mode also involves no change in dipole moment, and so is 'inactive'.

Carbon dioxide has a strong rotation–vibration band centred at 15.0 μm, based on the two v_2 modes (which are degenerate). This band is very important in climate, and is also vital in satellite remote sensing for the collection of weather data, as will be discussed in the next chapter. There are also strong bands at 4.3, 2.7, and 2.0 μm, the first of which is climatically important, and two bands at 9.6 and 10.6 μm, which are used in CO_2 lasers.

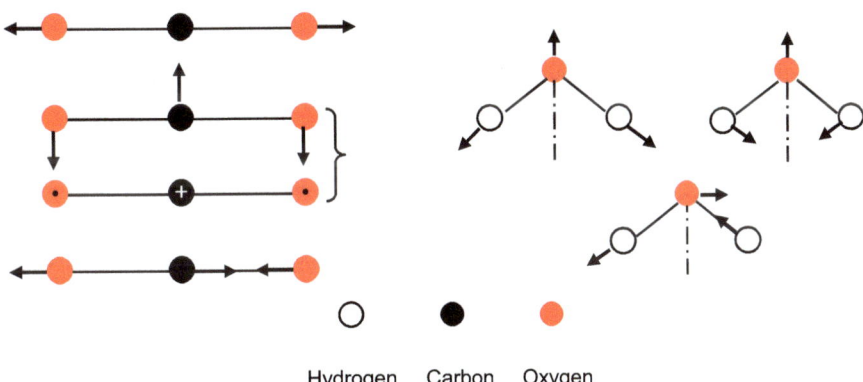

Hydrogen   Carbon   Oxygen

Figure 11.3 Vibrational modes of CO_2 (left) and H_2O (right). *Source: Physics of Radiation and Climate,* figure 9.2, Box and Box, CRC Press, 2016, reproduced by permission of Taylor & Francis Group.

11.3.2 Water Vapour

The water molecule is what is known as an asymmetric top molecule with a bond length of 95.8 pm, and an apical angle of 104.5°. As a result, the three components of its moment of inertia (i.e. how it rotates about its three axes) are different, giving rise to a complex/messy spectrum. As already noted, it has a large dipole moment (6.16×10^{-30} C m) producing strong rotation bands. Its three fundamental vibration modes are also shown in Figure 11.3.

The main climatically important bands of water vapour are the very strong rotation band, from roughly 11.0 μm (effectively extending into the radio range), and the 6.3 μm v_2 band. There are also bands at 2.7 μm, as well as a number of bands in the near infrared and the visible, which were briefly noted in Chapter 9.

11.3.3 Methane

The methane molecule is what is known as a 'spherical top' molecule, with a bond length of 109.3 pm, and no permanent dipole moment (hence no pure rotational spectrum). It has four vibration modes, with the one at 7.6 μm being important for climate.

11.3.4 Nitrous Oxide

The nitrous oxide molecule is linear, and asymmetric, with the configuration NNO, and a small permanent dipole moment. The N–N bond length is 112.6 pm; the N–O bond length is 118.6 pm. It has vibration bands centred at 4.5, 7.8, and 17.0 μm. The first two contribute to the greenhouse effect, while the third is 'swamped' by the carbon dioxide and water bands.

11.3.5 Ozone

The ozone molecule is bent at an apical angle of 116.8° and has a bond length of 127.8 pm. Two of its vibration modes combine to produce the important 9.6-μm band. It also has bands at 14.27 μm, which is masked by carbon dioxide and water, and 4.75 μm, on the edge of the thermal emission spectral range. Its absorption in the visible, and its roles in the stratosphere, were covered in the previous two chapters.

11.3.6 Window Region

Figure 11.4 provides a broad-brush picture of the spectral absorption by the gases we have just discussed.

A very important feature of these data is the region between about 8.0 and 12.0 μm, known as the **window region**. The vast majority of the long-wave radiation that escapes directly to space is in this spectral region. The one distinct absorption feature is the relatively narrow 9.6-μm band of ozone. In addition, there is a continuum absorption, primarily due to water vapour, although there is some uncertainty as to the exact mechanism.

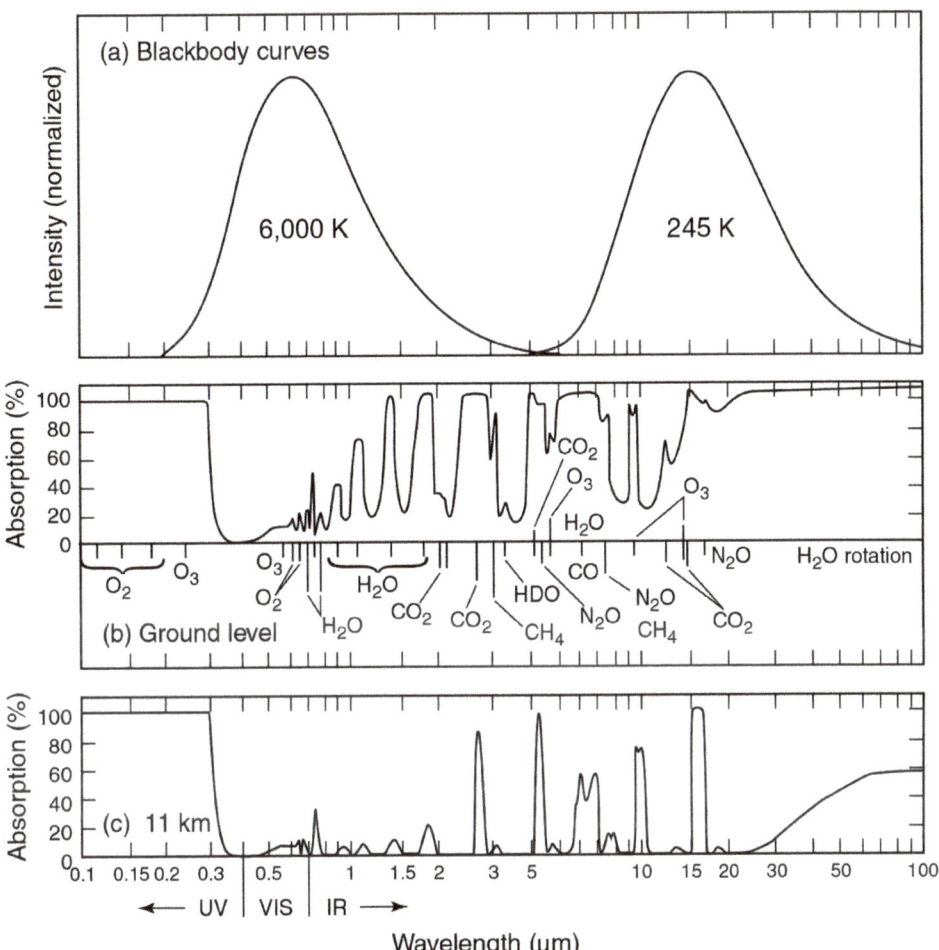

Figure 11.4 Atmospheric transmission of radiation across the spectrum. Panel (a) shows the normalised Planck spectra for temperatures of 6,000 K and 245 K (c.f. Figure 8.3). Panel (b) shows the absorption fraction for a full atmospheric column, along with the gases which absorb in each spectral region. Panel (c) shows the fractional absorption from 11 km – let's say the tropopause – to the top of the atmosphere. *Source*: After *Atmospheric Radiation, Theoretical Basis*. 2nd Edition, Goody and Yung, Oxford University Press, 1989.

There is one other contributor to absorption in this region that we need to discuss: the man-made chemicals known as chlorofluorocarbons (CFCs), such as $CFCl_3$ and CF_2Cl_2, plus a suite of related compounds. Although all are present in the atmosphere in minute amounts, the fact that they absorb in an otherwise largely transparent spectral region is a cause for concern. While some of these have been phased out because of their impact on the ozone layer (as we saw in the previous chapter), many of their replacements also absorb in this region. We should point out that a key property of these replacements is that they have much shorter lifetimes than the CFCs (due to a C–H bond which is susceptible to attack by OH), which means that their long-term impacts are much reduced.

11.4 Infrared Transfer

Now that we understand the basic ideas of atomic and (especially) molecular structure, the various energy levels and transitions, and the corresponding photon energies which may be absorbed and emitted, including the line shapes, we can address the practical implications for the outward flow of long-wave radiation through the greenhouse gases that are found in the Earth's atmosphere. This is a very daunting but very necessary challenge. Some mathematics has been included for those readers with the appropriate background: and bravery. For others, we hope our explanations will at least give you a 'feel' for what is going on.

The question being asked is simple enough: given a certain amount of radiation emitted by the Earth's surface, what is its fate? That is, how much is absorbed, and how much makes it to space? Remember that the emission follows Planck's law, and is thus spectrally 'smooth', while the absorption by the greenhouse gases is composed of the various lines and bands we have just discussed. Thus, radiation of some wavelengths/frequencies will be strongly absorbed, while at other wavelengths/frequencies that radiation will have an easier passage. We need to compute the upward flow, or flux, of all this radiation, in order to determine the so-called **transmittance** (the fraction of the energy that makes its way to space) and its complement, the **absorptance** (the fraction absorbed in the atmosphere). That means we will have to integrate across the full long-wave spectral range from about 4.0 μm, effectively to infinity. (In practice we can probably stop at ~1,000 μm.)

11.4.1 Line-by-Line Calculation

The most accurate way to do this is to take baby steps in wavelength, look at the relevant data on absorption at that wavelength, follow the radiation flow from surface to the TOA (top of the atmosphere), and add that result to the progressive total. How many such steps would we need to get a result in which we could have (close to) absolute confidence? Well, around the world there are a number of compilations of spectral line data – spectral atlases – containing over one million lines! So, this is an indication of the work that needs to be done, and is done, as a benchmark against which to compare simpler, more practical, formulations.

11.4.2 Practical Calculations

Clearly, such line-by-line calculations have an important role to play in climate science, but they are totally non-cost-effective in practice, and also less than informative. So, we will now turn our focus to the sort of models which are used in day-to-day investigations, which, while not as accurate, are capable of being implemented in a climate model with sufficiently reliable accuracy. Because they are fundamentally semi-analytic, they also provide the insight that line-by-line output can't. These models start by breaking the full long-wave spectral range down into a series of bands – hence the name, band models – and use calculus to integrate across a set of spectral lines within that band. The resulting mathematics may be characterised as anywhere from 'such fun' (MB) to 'bloody horrible' (GB).

We will, of course, avoid the details. However, we can get a feel by looking at the very simplest model: just a single Lorentzian spectral line. That is, we assume initially that, in the spectral region of interest, there is just a single line, with the shape given by Equation (11.4). We'll step you through the basics (and try to keep it as general as possible to begin with); the pieces should be easy enough to understand, and that will help you appreciate the result.

We start with the basics: **Beer's law** for the attenuation of a light beam as it passes through a scattering and/or absorbing medium from the surface to TOA;

$$I(f) = I_0(f) \, exp \, [-\tau(f)]$$

where τ is the optical thickness, and we have explicitly indicated that it is likely to vary with frequency, f. We'll ignore scattering, and just focus on absorption. Now the absorption by a given gas along such a path will depend on the total amount of the absorbing gas, u, and how strongly it absorbs at a particular frequency, $k(f)$ – the latter depends on both the 'line strength', S (which can be found from the detailed quantum mechanics), and the line profile, φ – so giving

$$\tau(f) = uk(f) = uS\varphi(f)$$

Now we define the monochromatic transmittance (or transmission function) as the fraction of the emitted radiation that escapes to space:

$$T(f;u) = \frac{I(f)}{I_0(f)} = exp \, [-uS\varphi(f)] \qquad (11.5)$$

and we have explicitly indicated that it depends on both radiation frequency, and the amount of absorbing gas. Figure 11.5 illustrates the physics of this, for the case of a single Lorentz line, for five values of the product, uS: 1 (top), 3, 7, 15 and 30 (bottom). We see that for the largest value (30), the line centre is **saturated**, and T is (effectively) zero.

The next step is to average (i.e. integrate) this over the appropriate spectral range, to obtain the so-called **band transmittance**:

$$T(u) = \frac{1}{\Delta f} \int exp \, [-uS\varphi(f)] \, df \qquad (11.6)$$

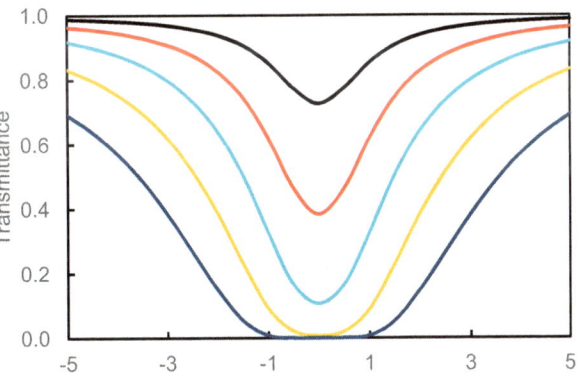

Figure 11.5 Spectral transmittance in the vicinity of a single Lorentz spectral line, for five values of uS.

It turns out to be both more informative, and mathematically easier, to focus on the complement of this quantity:

$$W(u) = \int \{1 - exp\,[-uS\varphi(f)]\}\,df \tag{11.7}$$

W is known as the '**equivalent width**', and can be regarded as a measure of how much of the band is 'blocked' by the absorbing gas (hence the name). The variation of W, as a function of the amount of absorbing gas, u, is known as the "**curve of growth**".

The formulation we have just arrived at is quite general: we have said little about the number or type of absorption lines in the band. The heavy lifting comes next (but not in this book).

11.4.3 Band Models

Any realistic attempt to compute the flow of long-wave radiation through the atmosphere must use bands of sufficient spectral width that they contain many spectral lines. (Otherwise, we might just as well be doing a line-by-line calculation.) Two different models have been used in an effort to characterise a set of lines in ways that are both 'reasonably' realistic and amenable to some form of analytical mathematics.

When we looked at the absorption by linear molecules such as CO_2, we noted that it should produce a set of equally spaced lines; Equation (11.2). This is the basis of the Regular Band model, which assumes a set of identical lines with a uniform spacing – a bit like the teeth of a comb. By contrast, because of the differing moments of inertia about its three axes, the water vapour spectrum is a total mess. This has been modelled via the Random Band model, which uses a probability distribution for the variation of line strengths. Both models have parameters that can be adjusted after comparing their results with those of benchmark standards. The results can be found in some of the references at the end of this chapter.

11.4.4 Limiting Cases

We can gain important insights by looking at two extreme cases: the weak and strong limits. We start with the situation where either the amount of gas, u, is small, and/or the line strength, S, is weak, or the product uS is sufficiently small. That means the argument in the exponential function in Equation (11.7) is small enough that we may turn to its Taylor expansion, namely

$$exp(x) = 1 + x + ..$$

Using this in Equation (11.7) we find

$$W(u) = \int uS\varphi(f)df = uS \tag{11.8}$$

by virtue of the normalisation of the line profile function. This result tells us that the amount of radiation absorbed is directly proportional to the product of the amount of gas and the line strength. That is, all the photons get a chance to 'see', and be absorbed by, all the molecules. This result is independent of the line width. Both of the band models just discussed give the same limit, as should be expected.

At the other extreme we consider the situation where the product uS is large. This time the mathematics is messy: in the case of a single spectral line we may derive the result

$$W = 2\sqrt{uS\alpha} \tag{11.9}$$

This time W only increases as the square root of the gas amount, u. This is because all photons with frequencies near the line centre are fully absorbed – the line is saturated (Figure 11.4) – and absorption is only taking place in the line wings. When we move beyond a single line to the band models mentioned above, we find that their strong limits are different (and too opaque to include here), as the line wings overlap with one another, but in different ways.

11.5 Radiative Forcing

If we lived in a stable climate – one with stable greenhouse gas concentrations – all of this would be of little more than academic interest. A handful of scientists could perform a small number of sufficiently accurate calculations in order to demonstrate – mainly to themselves – that they really did understand how our climate works. For example, it is just such calculations which we use to decide how much radiation each of the greenhouse gases intercepts (traps), as noted in Chapter 8: their contribution to the greenhouse effect. The rest of us would not have the slightest interest in the results: and why should we?

But we don't live in a stable climate, because the concentrations of key greenhouse gases are changing: increasing. This is the reason for the public's interest – and, we trust, of yours. So, one of the central questions at the heart of climate change science is 'how much extra long-wave radiation is being trapped by the increased greenhouse gases?' The answer to that question is directly tied to the curve of growth. We will now discuss the key takeaways from the previous section, and leave the quantification, based on the latest science, for Chapter 17.

The atmospheric concentration of carbon dioxide is such that its key absorption bands are well and truly saturated. That means we need to look at the strong limit, based not on the single line band model, but a model more suited to its spectrum. It turns out that the increase in absorption is proportional to the logarithm of the increase in CO_2 concentration. This is just as well. We know that the pre-industrial concentration of 280 ppm was enough to produce a greenhouse effect of ~33°C. Thank an appropriate deity that the 50% increase in the past 200 years has not produced a proportionate increase in the temperature of our world.

The concentrations of both methane and nitrous oxide are such that they are now in the square root region of the curve of growth – Equation (11.9) – so their concentration increases are having a somewhat larger proportionate impact on climate, making them – potentially – more dangerous (on a per mole basis). Finally, the various man-made chemical species such as the CFCs are of such low concentrations that they are still in the linear range.

Summary

The material in this chapter is some of the most challenging in the entire book, but it is also some of the most important. With an absolute minimum of mathematics, we have taken you on a journey through the quantum physics that governs the atomic world, to the spectra of atoms, then on to the rotation and vibration properties of molecules. You should have at least acquired a rough idea of the rotation/vibration spectra of molecules. And, also, why nitrogen and oxygen are not greenhouse gases.

From there we took you on a photon's journey 'up through the atmosphere', and gave you the key ideas on whether it will be absorbed (and re-emitted; perhaps many times), or escape to space unscathed. These are the sorts of considerations, and computations, that are crucial to any climate, and climate change, modelling. Finally, we were able to present you with the answer to one of the most crucial questions there is: how much more radiation will be absorbed as greenhouse gases increase?

It is unlikely that all this information will be fully assimilated just yet: it will almost certainly be very new and very different. We strongly suggest you re-read this chapter when you get to the climate change chapters (Part V), starting with Chapter 17.

FURTHER READING

Atomic and molecular structure is a branch of quantum mechanics, and is covered in many texts. We note the following (each with a different emphasis):

- *Quantum Physics of Atoms, Molecules, Solids, Nuclei, and Particles* by R. Eisberg and R. Resnick (Wiley, 1974).
- *Quantum physics* by S. Gasiorowicz (Wiley, 1974).
- *Radiative Processes in Astrophysics* by G. B. Rybicki and A. P. Lightman (Wiley, 1979).
- *Molecular Spectra and Molecular Structure* by G. Herzberg (Van Nostrand, 1950).
- *Physics of Radiation and Climate* by M. A. Box and G. P. Box (CRC Press, 2016).
- *Atmospheric Science for Environmental Scientists* by C. N. Hewitt (Wiley-Blackwell, 2020).

The radiative transfer material in this chapter is covered extensively in specialist texts, such as those listed at the end of Chapter 8. One additional reference worth checking out is

- *Environmental Physics* by E. Boeker and R. van Grondelle (John Wiley & Sons, 1995).

REVIEW QUESTIONS

1. What is the essence of Bohr's quantum hypothesis?
2. What are the two extra 'degrees of freedom' that a molecule has but an atom does not?
3. Explain why nitrogen and oxygen are not greenhouse gases.
4. What do we mean by a molecular rotation–vibration band? Why does it arise?
5. What are the key factors that broaden a spectral line?
6. What does it mean when we say that a spectral line is 'saturated'?

7. Explain how a saturated spectral line can still absorb some radiation.
8. Why are the CFCs important greenhouse gases, despite their tiny concentrations in the atmosphere?

EXERCISES

1. Write a small computer program (or just hand calculate) to reproduce Figure 11.2, using Equation (11.4). We chose $\alpha = 0.5$, 1.0, and 2.0, with $f - f_0$ ranging from -4 to $+4$. Try some different values as well as those.
2. Write a small computer program (or just hand calculate) to reproduce Figure 11.5, using Equations (11.4) and (11.5). We chose $\alpha = 1.0$ with $f - f_0$ ranging from -5 to $+5$. Try some different values as well as those.

12   Remote Sensing of a Dynamic Environment

Some of the most important data on the state of the environment, and how it may be changing, are supplied by satellites, which have the enormous advantage of global or near-global coverage. They supply some of the most important information – data – that is used in quantifying environmental change, as will become apparent in later chapters.

What sort of measurements do satellites make, and how are we able to obtain meaningful, *quantitative* information about the atmosphere, or the Earth's surface, from a sensor more than 500 km above? Those are the questions we address in this chapter.

One key to this question is found in the previous chapter. There we saw that molecules have discrete absorption lines/bands: chemists use these as unique signatures in their laboratories to identify an unknown substance. With a suitable space-based instrument we can do much the same, using either thermal emission, or absorption. The details of these processes may depend on the temperature, so this information may also be inferred.

By contrast, aerosol particles (mainly) scatter light, with relatively little wavelength variation. The first challenge here is to separate the light scattered by molecules from the light scattered by particles. This may be reasonably straightforward, or it may be quite challenging, depending on the circumstances.

12.1   Remote Sensing

'Remote sensing' refers to a broad range of techniques whereby we learn about something of interest by observing, and, whenever possible, *quantifying*, electromagnetic radiation that has come from it, but without us actually making 'physical contact' with it. Vision is a prime example – one we take for granted! Photography is another example; as is all of astronomy.

The first piece of technology the photographer usually adds to their basic camera is a telephoto lens, to gain a closer look at a subject of interest. The next technological advance might be to add some filters, in order to bring out certain desirable features of a scene we're interested in. Polarised filters reduce glare – unwanted light – and thus improve the signal-to-noise ratio. Spectral filters which selectively transmit certain wavelength bands, while rejecting others, might be used for similar purposes. Our eyes, of course, have no option but to accept all incoming light (polarised sunglasses help, of course), while modern technology enables us to choose only the light that contains certain relevant **information**.

Our eyes can only 'see' light within the visible wavelength range – 400–700 nm (with some variation from person to person). By using a detector that is designed to be sensitive to other wavelength ranges, our generic camera may now see light across a much wider spectral range. You have probably seen infrared images, taken using a detector which is sensitive to wavelengths on the edge of the thermal range (around 4–5 μm), where the Planck function is strongly temperature-dependent.

Atmospheric remote sensing makes use of these and other techniques to learn about the 'thermodynamic state' and composition of the Earth's atmosphere, as well as its surface. In this chapter we will give our readers a quick overview of what is, in reality, a very large and fascinating branch of science. Over the years we have been part of a number of research projects that have involved the analysis of remotely sensed data, starting with the two years we spent working for NASA. We have also established many connections, both professional and personal, in this field. This will, naturally, be one key influence on the choices we make.

12.1.1 Practical Considerations

A full treatment of remote sensing, and especially space-based remote sensing, would require an adequate discussion of all of the following topics:

- The underlying radiation transfer physics, which covers how the radiation measured at a detector has interacted with the atmosphere (and/or the surface).
- The nature of the detection systems employed: the technical properties of individual detector elements (pixels); their combination into linear or rectangular arrays to allow for simultaneous detection; the optical system which images the scene; etc.
- Instrument calibration, and how this can be maintained/verified once in orbit.
- The platform on which the detector/optics are mounted, and which powers it (and often other instruments).
- Data storage and transmission to ground receiving stations.
- Satellite orbital characteristics.
- Geolocation: where on the ground, or in the atmosphere, the instrument was looking when it recorded a data element. This is particularly important when we wish to combine data from multiple instruments, especially when mounted on different platforms (e.g. the A-Train; below), to obtain a fuller understanding of the environment at a given location.
- The mathematical algorithms that are used to convert the data into information about the composition or state of the atmosphere.

In this chapter we will focus almost entirely on just the first of these items, and that only lightly. The mathematical algorithms are also a very important (and very different) challenge, and one with which we have had some involvement. We will only be able to sketch some key points.

12.2 Ground-Based Remote Sensing

Although it is space-based instrumentation that provides the vast majority of the remotely sensed data we use in weather and climate studies today, some important measurements are made by ground-based instruments which employ similar physical principles. Many of these are deployed to help provide a cross-check on satellite measurements. Although their spatial coverage is tiny compared with that of a satellite, they have the one key advantage that they can be easily serviced and recalibrated; something all but impossible in space.

12.2.1 Radiometry

A radiometer on the ground can measure sunlight that has been attenuated as it passes through the gases and aerosols above it. (We assume that all measurements are for a cloud-free period.) We may express this process mathematically via Beer's law:

$$I(\lambda) = I_0(\lambda)\, exp\, [-\tau(\lambda)sec\theta]$$

Here, $I(\lambda)$ is the intensity measured by the radiometer, for wavelength (i.e. channel) λ; $I_0(\lambda)$ is the intensity at the top of the atmosphere; θ is the solar zenith angle; and $\tau(\lambda)$ is the optical thickness (at wavelength λ). The 'air mass factor', $sec\theta$, accounts for the extended pathlength of the solar beam when the Sun is not overhead. (The $sec\theta$ factor assumes a flat atmosphere, and is valid for zenith angles up to $\approx 75°$. If the Sun is closer to the horizon, 'spherical corrections' are needed.)

The optical thickness will, in general, have several factors (Section 9.3.3): molecular scattering, aerosol scattering and absorption, and possibly molecular absorption. With a ground-based instrument it is generally easier to separate out the different contributions, as molecular attenuation (away from absorbing gases) is easy to remove. Such instrumentation is used to characterise ozone amounts, and also aerosol loadings and size distributions.

The standard way of analysing such measurement programs is to make a series of careful measurements throughout a day, or at least half a day, which implies a range of solar zenith angles, θ. Now we take the logarithm of the equation above, converting it to read

$$ln\, [I(\lambda; sec\theta)] = ln\, [I_0(\lambda)] - sec\theta\; \tau(\lambda)$$

A graph of the log of the measurements vs. $sec\theta$ should give a straight line, with a negative slope which is the optical thickness at that wavelength – assuming the atmosphere is 'stable' during the measurement period. Such a graph is known as a Langley plot. The intercept tells you what the instrument would have measured if placed above the atmosphere. If that value is considered sufficiently accurate, then the instrument is said to be calibrated, and may be used directly in the above equation.

Calibration is often performed at a high-altitude observatory, where atmospheric variability (e.g. aerosols) is less likely to be an issue. Examples are the Mauna Loa observatory in Hawaii (3,397 m above sea level), and the Izaña observatory on Tenerife (2,390 m above sea level).

Figure 12.1 Cimel radiometers at the calibration facility on Tenerife. *Source*: Photo courtesy of Cimel Electronique, Paris, France.

Calibration may also be performed against a reference (or 'master') instrument, with 'absolute' calibration sometimes being performed in a specialist laboratory.

A Dobson spectrophotometer (Figure 10.1) makes a pair of measurements in the UVA and UVB spectral region (300–340 nm), one of which is significantly attenuated by ozone, the other less so. Their ratio can then be converted to the 'total column' amount of ozone. We have already noted that ozone column amounts are 'measured' in Dobson units.

Aerosol optical thickness (or column loading) can be obtained from a series of measurements in the UVA, visible, and near IR. We can also 'invert' such data sets to obtain the aerosol size distribution: the variation of particle number with radius. The simplest analysis looks at how rapidly optical depth decreases with wavelength: if it decreases rapidly, the size distribution is dominated by small particles; if it decreases slowly, the distribution contains more large particles, as noted in Section 4.7. A number of sophisticated algorithms have been developed that provide more detailed information.

NASA has a number of space-based instruments, which we shall look at below, that are designed to obtain data on aerosols. For this reason, they have deployed the Cimel CE318 Sun Photometer (Figure 12.1), with measurements channels at 380, 440, 500, 670, 870, and

1,020 nm, across their Aerosol Robotic Network (AERONET) to provide 'ground truth' data to anchor these satellite data sets: check out the website.

12.2.2 Lidar

Lidar (light detection and ranging), or laser radar, can be used to infer vertical profiles; mainly of scattering by aerosols, but also of absorption by various gases. By firing a short (vertical) pulse of light, and measuring the strength of the backscattered signal as a function of time (which 'translates' to round-trip distance; radar uses similar principles), we obtain a profile of scattering 'strength' as a function of altitude: for example, aerosol profile. The vertical resolution depends on the digitisation rate. (To help guide your thinking here, we point out that a more useful value for the speed of light is 'one foot per nanosecond'.)

Several more sophisticated approaches are available, if we have the technology. DIAL (differential absorption lidar) involves the use of two nearby wavelengths, one absorbed by a (pollutant) gas, the other not. This can provide information on the amount of the gas along the path, assuming we have a suitable reflector at the other end. We may even learn about the profile of the pollutant along the path, although this is more of a challenge.

In a Raman setup, returned radiation is detected at a different wavelength from that emitted; in most cases a longer wavelength. This means that the photon has (effectively) lost some of its energy: what has happened to it? It has been transferred to a molecule, leaving it in a higher rotational state, as we discussed in the previous chapter. This technique can provide profile information for a particular gas.

Figure 12.2 shows the lidar system at the Institute of Physics, Belgrade, Serbia. It uses an Nd: YAG laser, which has a fundamental wavelength of 1,064 nm, plus additional emissions at 532 and 355 nm (i.e. double and triple frequencies). It sends out 5 ns pulses, with energies of 105, 45, and 65 mJ, at a rate of 20 Hz, allowing a 7.5 m vertical resolution. The receiver uses a 250-mm diameter Cassegrain telescope, with a field of view ranging from 0.5 to 3 milliradians. It also detects Raman-shifted backscattered radiation at 387 nm.

The system operates as part of EARLINET, the European Aerosol Lidar Network, which was established to supply a comprehensive, quantitative database of the horizontal and vertical distribution of aerosols across Europe, and use these data for studies related to the impact of aerosols on a variety of environmental problems.

12.2.3 Airborne Observation

NASA has versions of a number of its satellite sensors in a form that can be easily mounted on an aircraft, such as the ER-2 High-Altitude Airborne Science Aircraft. It has a typical cruising altitude of 65,000 feet (~20 km), which places it above 95% of the Earth's atmosphere. It can carry a payload of 2,600 lb, which can be rapidly installed and removed, as required. That payload may include sensors for testing prior to deployment, or already in space, and has the advantage of recent calibration, along with full control of the flight path.

This capability makes it a very powerful contributor to large-scale field campaigns to study aerosols, ozone depletion, etc. During the winter of 1999/2000, an ER-2 participated in the

Figure 12.2 Lidar system of the Institute of Physics Belgrade, Pregrevica 118, 11080 Belgrade, Serbia.

SAGE III Ozone Loss and Validation Experiment (SOLVE), based in Kiruna, Sweden. This was the largest field campaign to measure ozone in the Arctic stratosphere. A few years earlier it was involved in a similar campaign, from Fairbanks, Alaska. NASA also flies a much larger Douglas DC-8.

12.3 Orbits and Observations

There are two common types of orbits used for satellites whose task it is to observe/study our environment and gather important information such as meteorological data. Geostationary satellites (orbital radius of 35,800 km; period 24 hours) look down at the Earth, and continuously observe the same ~40% of the globe. Such orbits are used for some weather observations and communication. Low Earth Orbiting (LEO) satellites (altitude range 500–1,000 km; period ~95–100 minutes) are generally polar orbiting sun-synchronous, retrograde (i.e. inclination slightly west of north) orbits.

12.3.1 Geostationary Satellites

Geostationary satellites are used to provide real-time visible and infrared imagery of clouds and surface features. There are generally (at least) five in orbit. The United States operates three GOES (Geostationary Operational Environmental Satellites), at 60°W, 75°W, and 135°W; the

Europeans operate their two Meteosats at 0° and 3.5°; and the Japanese have one at 140°E. (Russia, India, and China are also involved.) All provide multispectral (i.e. 'colour') visible and infrared images, as well as data on temperature and moisture (with high temporal, but low spatial resolution).

The western Pacific region is served by Japan's Himawari-8 satellite. It has four channels in the visible and near IR, with a spatial resolution of 1 km (or better), and 12 channels between 1 and 14 μm with a spatial resolution of 2 km, and can image the entire region below in just 10 minutes. This enables meteorologists to obtain much better information on the centres of tropical cyclones as they develop over the ocean; thunderstorms as they develop; volcanic ash; fire and smoke; and also fog and low cloud. Researchers are finding many ways to use this high-quality information source to improve our understanding of the atmosphere.

12.3.1.1 Clouds

On a black and white image clouds show up as white, as they are very efficient at reflecting sunlight. However, for half of the time, a satellite is looking at the night side of our planet, so such photography is useless. So the next step was infrared imagery, with a particular focus on the atmospheric window region ~10 μm. This time the detector is measuring the emitted radiation from the surface and atmosphere, which strongly depends on temperature. If there are no clouds present, the satellite sees the surface, with a temperature ~285 K, but if the scene is cloud-covered the satellite will see the cloud, which will invariably have a lower temperature, depending on cloud-top altitude. Thus, a high/thick cloud will emit rather less radiation than the surface, and hence look (relatively) 'black' on such imagery. Because that is the reverse of the visible imagery, we are shown the photo-negative version.

Cloud imagery has been used to develop a climatology of the world's distribution of cloudiness, via the International Satellite Cloud Climatology Project (ISCCP), under the auspices of the World Climate Research Programme, and the WMO.

12.3.2 Polar-Orbiting Satellites

If the Earth were a perfect sphere, satellite orbits would remain 'in a fixed plane in space' as the Earth moves around the Sun during the year. This would mean a satellite would not observe the same surface features at the same *local* time, because the part of the Earth facing the Sun will change in an annual cycle. However, the Earth is an oblate spheroid. This causes the satellite's orbital plane to *precess*, at a rate which depends on both its inclination and altitude. By choosing the right combination (97° to 100°, for altitudes of 500 to 1,000 km) we can ensure that the orbit precesses exactly once per year: rocket science!

12.3.2.1 Limb Scanning/Occultation

Some space platforms/instruments are designed to study radiation coming from the atmospheric limb: whether it be attenuated or scattered sunlight, or thermal emission. Sunlight passing through the limb of the Earth's atmosphere will be attenuated, and may be detected

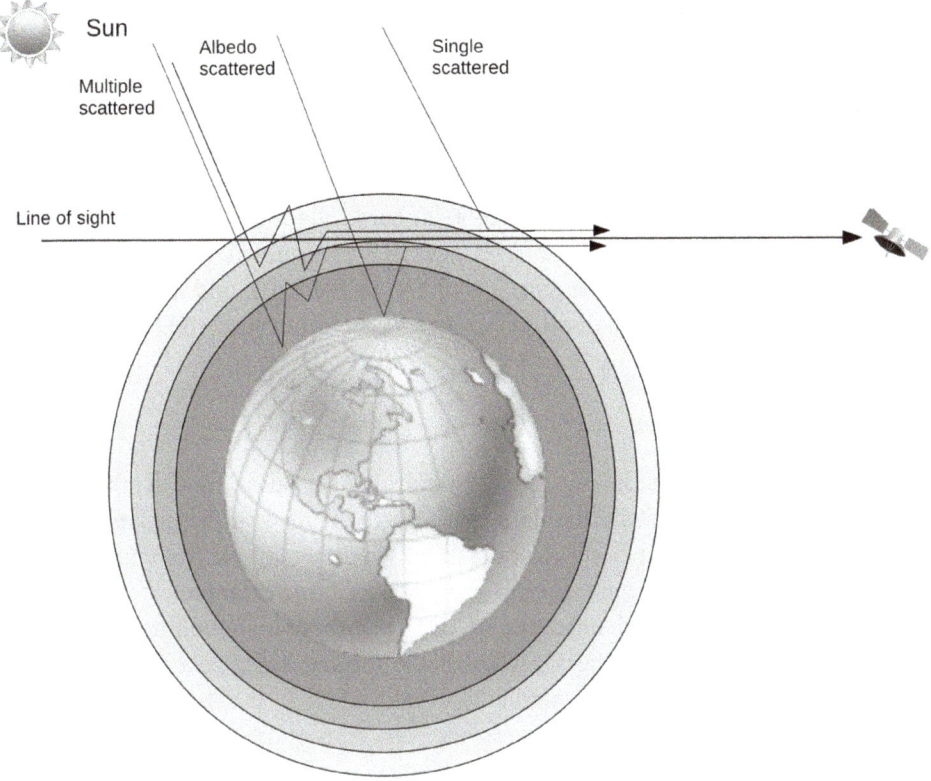

Sun

Multiple scattered

Albedo scattered

Single scattered

Line of sight

Figure 12.3 Limb scan geometry. *Source*: courtesy T. Trautmann, German Aerospace Center DLR.

on the other side, although the mathematical details are somewhat more complicated as Earth curvature is clearly a key factor in this process. Depending on its orbital details, a satellite may observe such radiation twice per orbit, providing information on atmospheric composition, at least above the cloud tops.

Such 'occultation' measurements are particularly valuable for obtaining vertical profile information on stratospheric constituents such as aerosols, ozone, and the gases that may damage the ozone layer. Calibration is not a problem as the sensor may measure the unattenuated solar beam when it is between the Earth and the Sun.

Figure 12.3 shows a generic depiction of the limb viewing modes.

12.3.2.2 Vegetation

Satellites are also used to study the Earth's surface properties. If we place some form of camera on a space platform we may 'photograph' the scene below. From such images we may extract valuable information on the surface itself: for example, primary productivity (including crop growth/health) and geology (including potential mineral deposits). We will not discuss the mineral resource issues (the province of the **Landsat** programme which began in 1972), and leave issues to do with military/security surveillance to our readers' imaginations.

While a spectrally broad image is sufficient to identify clouds, and to separate different surface types – land vs. ocean, forest vs. desert – more spectrally selective imagery may be valuable in teasing out subtle variations within those broad categories. Green vegetation strongly reflects sunlight in the near IR (from ~750 nm), compared with how much it reflects in the visible. By contrast, the reflectivity of bare soil (or unhealthy vegetation) does not vary so much over this spectral range. The AVHRR instrument (below) makes measurements of reflected light at the wavelengths $\lambda_1 = 0.63$ μm (i.e. red), and $\lambda_2 = 0.86$ μm (near IR). From these we define the Normalised Difference Vegetation Index (NDVI) by

$$NDVI = \frac{I(\lambda_2) - I(\lambda_1)}{I(\lambda_2) + I(\lambda_1)}$$

In order to make use of any satellite data, it must be 'ground truthed': that is, we need to compare a number of satellite data sets with the reality on the ground; a form of calibration. Once we are confident that we understand what's going on, we can use the satellite data to provide a valuable measure of crop growth/health, for example.

The NDVI is a good indicator of live (chlorophyll-containing) vegetation, which varies throughout the growing season, whether the surface is agricultural or natural, so that regular monitoring can be used to predict harvest yields. It has been said to 'capture the Earth breathing'. Vegetation monitoring is also very important for constraining the carbon budget, and for carbon accounting and carbon trading.

12.4 Satellite Observations for Weather Forecasting

Weather forecasts today typically go out for seven days (and beyond, although mainly for research purposes). To achieve their growing success, the models on which they are based (covered in Chapter 15) must be fully global, so all such models require input data from around the world: much of it satellite data.

12.4.1 Early History

In 1929, Robert Goddard launched a rocket-borne package which included a camera, a thermometer, and a barometer: this may be considered the founding step in meteorological remote sounding. (Both NASA's Goddard Space Flight Center (GSFC) outside Washington, DC, and Goddard Institute for Space Studies (GISS) in New York City, are named in his honour.) As is so often the case, 'military necessity' is the spur to technological advances which occasionally bring other, more useful applications, and so it was with rocketry during the Second World War, which led to the first high-altitude pictures of clouds in the late 1940s.

After the first satellites were launched in the late 1950s, mainly for national prestige and (potential) military reasons, atmospheric scientists argued for the inclusion of some form of camera, looking down, rather than towards the heavens. A simple black and white photograph was enough to show storms for the first time. Today these are a feature of the weather segment on every television news service.

The first weather satellite was the polar-orbiting Television InfraRed Observational Satellite-I (TIROS-I), launched on 1 April 1960, which was able to show storm systems. The first two Applications Technology Satellites, ATS-1 and ATS-3, were launched in 1966 and 1967. The idea of temperature profiling (below) was first raised in 1958 by Kaplan. A feasibility test was carried out on NIMBUS-III in 1969.

12.4.2 Vertical Profiling

Obtaining profile information (temperature, or water vapour concentration, as a function of altitude) is an interesting exercise. The instrument makes a series of measurements in an absorption band of a relevant gas (CO_2, H_2O) but at a series of wavelengths with a range of absorption strengths. To understand what's going on, it helps to 'think backwards'. (This is known as the adjoint form of the problem.) At wavelengths where emission/absorption is weak, the satellite can effectively 'see' deeper into the atmosphere, and so obtain 'information' from the lower atmosphere. Where absorption is strong, the satellite can only 'see' higher parts of the atmosphere.

The 'probability distribution' of where a detected photon might have been emitted in the atmosphere is known as the weighting function. This can also be interpreted as the source of the information which that particular measurement contains, or the region of the atmosphere whose temperature (or H_2O concentration) has been 'measured'. Figure 12.4a shows some example weighting functions. Sophisticated algorithms are then used to process this information, to produce a 'smooth' profile. Note that the weighting functions for limb scanning (Figure 12.4b) are much sharper than for nadir viewing.

Weather satellites provide two profiles. By making measurements in the 15-micron band of CO_2, a well-mixed gas of known concentration, the only variable which affects the signal is the temperature at the level of emission. Hence, these data provide information on the vertical

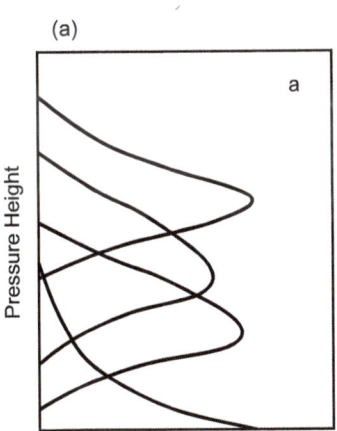
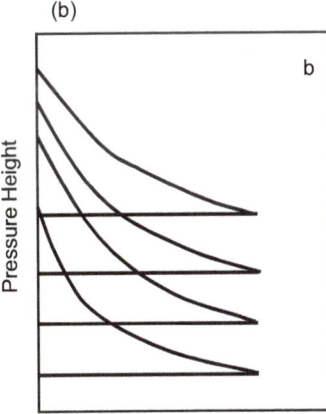

Figure 12.4 Schematic weighting functions for (a) nadir viewing, and (b) limb viewing. *Source: Physics of Radiation and Climate*, figure 14.3, Box and Box, CRC Press, 2016, reproduced by permission of Taylor & Francis Group.

profile of temperature. A second set of measurements is also made in a water vapour band. This time, having already obtained the temperature profile, the water vapour concentration is the unknown, so that these data provide that information. Today, US weather satellites are mostly operated by NOAA, the National Oceanic and Atmospheric Administration.

12.4.3 NOAA Platforms

TIROS-N, the first of a new generation, was launched on 13 October 1978. Its instrument package, TIROS Operational Vertical Sounder (TOVS), consisted of three instruments:

- High-Resolution Infrared Radiation Sounder (HIRS) – 16 channels, mainly in the 15-μm band of CO_2 – designed to obtain the vertical profiles of temperature and water vapour, plus surface and other 'correction' channels.
- Microwave Sounding Unit (four channels, 50–60 GHz) complements HIRS because (being a much longer wavelength) it can see through clouds.
- Stratospheric Sounding Unit (three channels).

One of the latest satellites is known as NOAA-N Prime (or NOAA-19), and considerable information, and an indication of its data, are available on the NOAA website. (NOAA-20 is now also in orbit.) We will give a brief description of its main instruments.

The AMSU (Advanced Microwave Sounding Unit) is a dual 20-channel scanning passive microwave radiometer. AMSU-A uses 15 channels to obtain vertical temperature profiles, as well as surface water and precipitation. The MHS (Microwave Humidity Sounder) is a new five-channel microwave radiometer for profiling water vapour, which flew for the first time on NOAA-18. Channels 3–14 sample the 50.3–89.0 GHz O_2 band to obtain the temperature profile up to ~50 km (the stratopause). Channels 1 and 2 (23.8–31.4 GHz) are in a window region and (along with channel 15) provide data on surface spectral emissivity, liquid water and total precipitable water, to 'correct' some of the information from the other channels (that is, to improve the analysis of the data).

The AVHRR/3 (Advanced Very-High-Resolution Radiometer/3) first carried on NOAA-15, is a six-channel cross-track scanning imaging radiometer operating in the visible, near IR, mid IR, and far IR regions. Its purpose is to study land-surface imagery, cloud imagery, sea-surface temperature, ice and snow cover, vegetation, the land/water boundary, volcanic ash, and aerosols. Its channels are spectrally broad by the standards of many 'modern' instruments. Channels 1 (600 nm), 2 (900 nm), and 3A (1.6 μm) measure reflected radiance, while channels 3B (3.5 μm), 4 (11 μm), and 5 (12 μm) have weighting functions that peak near the surface and thus are able to provide information on sea-/land-surface temperature.

The HIRS/4 is a 20-channel infrared sounder comprising one visible (0.69 μm), 12 long-wave IR (from the window to the 15-μm CO_2 band) and seven short-wave IR (3.75–6.5 μm) channels. Its purpose is to study temperature profiles, total column ozone, cloud-top height, surface albedo, water vapour profile, and sea- and land-surface temperature. HIRS/4 and AMSU are complementary in many respects.

12.5 Birth of the Satellite Era

Satellite observation is one of the most important tools we use to study our changing planetary environment, and especially the changing composition of its atmosphere (and surface). This presents a separate set of challenges from weather observation: if we are to see small **changes**, we need good **calibration/precision**. The key here is **repeatability.** We would also like to sample as much of the globe as practical, although repeating an identical measurement every day is not as important as it is for weather forecasting. The story here is all about finding ingenious ways to observe different aspects of the Earth's environment.

First question: what would we like to observe in order to detect the causes, or the effects, of environmental change?

- Radiation budget: e.g. more incoming solar radiation than outgoing long-wave radiation (however, the difference is very small, and clouds are very inhomogeneous).
- Changes at the surface: sea/land ice; changes to forests and other ecosystems; land-use changes; sea-level rise.
- Changes in gas concentrations: greenhouse and other climate forcing gases; gases affecting the ozone layer.
- Changes in aerosol loadings and properties. (Aerosols scatter light, but so do molecules and the land surface: separating these components can be challenging.)

Satellites can measure both scattered/reflected sunlight, and thermal emission, from both the atmosphere and the land or ocean surface. They are generally placed in quasi-sun-synchronous orbits in order to provide (near) global coverage, with a fixed **repeat cycle**, so that any environmental changes can be reliably detected. Among the many compromise decisions that need to be made are **resolution**: spatial and spectral. (We can only collect a certain amount of data per second). High spectral resolution usually has to be traded off against surface coverage: spatial resolution and/or swath width.

12.5.1 Nimbus-7

Although satellites have been used to monitor changes in our atmosphere since the early 1960s, the big breakthrough came with the launch of Nimbus-7, on 24 October 1978 from Vandenberg Air Force Base in California: altitude 950 km; inclination 99.3°. At least for climate observation purposes, the 'satellite era' is said to date from 1979, following the launch of Nimbus-7 and TIROS-N. Nimbus-7 carried eight ground-breaking instruments: many of the instruments in use today can trace their origin to those on this mission (Figure 12.5).

ERB (Earth Radiation Budget) was designed to measure the Earth's radiation budget across the full spectral range, from 0.2 to 50.0 μm (and beyond). This was derived from 22 channels, with differing spectral ranges and fields of view. A special feature was its ability to provide data on the angular distribution of the Earth–atmosphere reflectance. This radiometer was the first one stable enough to detect short-term and long-term solar irradiance variability.

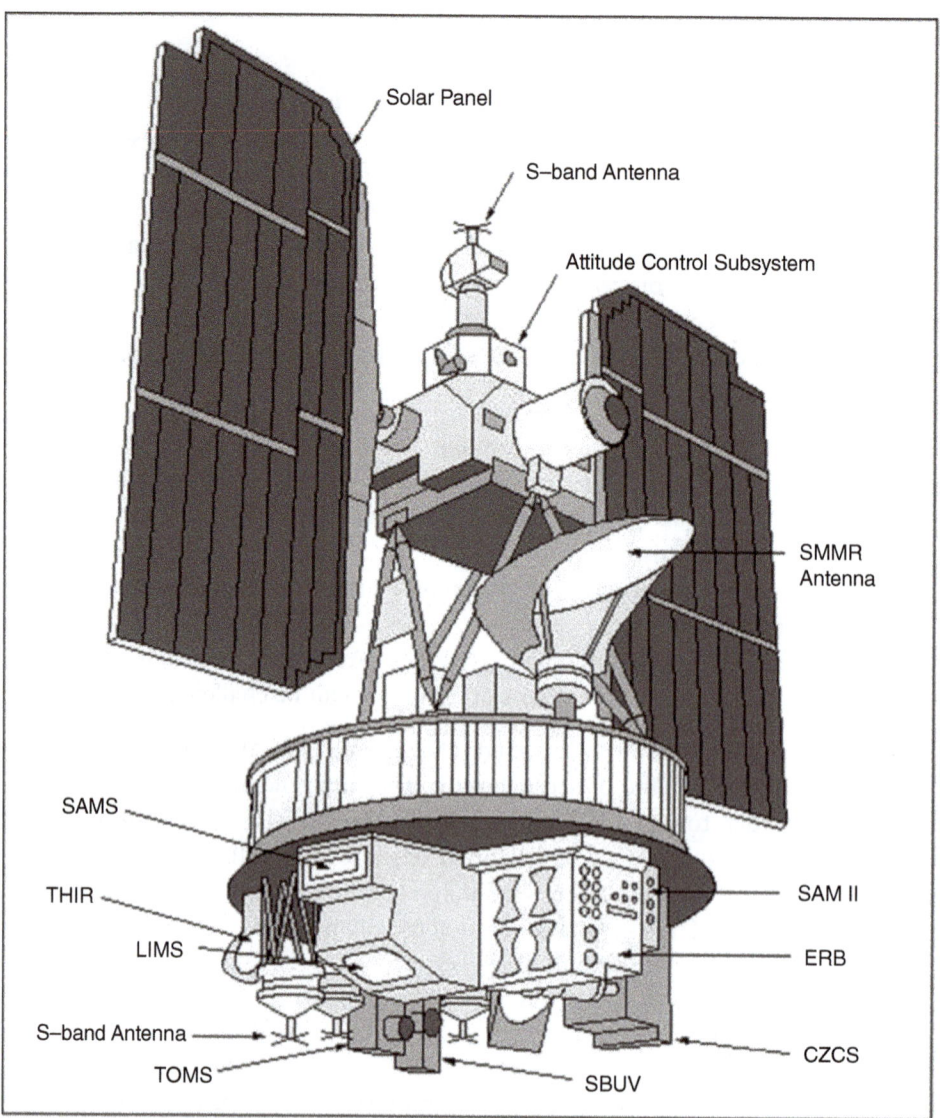

Figure 12.5 Nimbus-7, showing the arrangement of its instruments. *Abbreviations*: CZCV, Coastal Zone Color Scanner; ERB, Earth Radiation Budget; LIMS, Limb Infrared Monitor of the Stratosphere; SAM II, Stratospheric Aerosol Measurement II; SAMS, Stratosphere and Mesosphere Sounder; SBUV, Solar Backscatter UV; SMMR, Scanning Multichannel Microwave Radiometer; THIR, Temperature/Humidity Infrared Radiometer; TOMS, Total Ozone Mapping Spectrometer. *Source*: NASA.

LIMS (Limb Infrared Monitor of the Stratosphere) was a limb-scanning instrument with six channels: five narrow band channels at 6.25, 6.75, 9.65, 11.35, and 15.25 μm; and one broad channel from 13.3 to 17.2 μm. It was designed to obtain profile information on both temperature and the concentrations of the trace gases O_3, H_2O, NO_2, and HNO_3. **SAMS**

(Stratospheric and Mesospheric Sounder) was designed to measure profiles and concentrations of H_2O, CH_4, CO, and NO. A total of nine channels were used, between 4.1 and 15 μm, and 25–100 μm.

The group we worked with at NASA Langley Research Center was responsible for *SAM II*. Its 'big brother' SAGE flew on its own satellite (below).

12.5.1.1 Total Ozone Mapping Spectrometer

TOMS has been our primary source of ozone data since its first incarnation on Nimbus-7. It then flew on the Russian Meteor-3 (launched in 1994), and on Earth Probe (TOMS-EP; launched in 1996). It has been superseded by OMI (ozone measuring instrument) on the Aura satellite, and GOME (Global Ozone Monitoring Experiment) on the European Space Agency's ERS-2 satellite.

It measures reflected sunlight in six channels (wavelengths) in the near UV, which have different absorption strengths of ozone. Of these it is mainly the 317.5 and 331.2 nm channels that are used under normal conditions, and the 331.2 and 360 nm channels for high solar zenith angle conditions. By ratioing appropriately, TOMS can determine the total amount of ozone in the atmospheric column. Some channels are also sensitive to sulphur dioxide.

Unfortunately, it did initially miss the ozone hole, but since then has given us all the vital data to keep track of it. The reason for this can be traced to the processing algorithm, not the hardware: it had simply been assumed that such low values could only be due to contamination in the field of view. However, as all the raw data had been retained, a simple coding change was all that was needed to show the history of Antarctic ozone depletion.

12.5.1.2 Ocean Productivity

Primary productivity in the ocean is a very important quantity to monitor, for a number of reasons. Firstly, it represents a very important element of the global carbon budget: the conversion of inorganic carbon to organic carbon, and hence its (potential) removal as a greenhouse agent. Secondly, it is the first step in the marine food chain, and thus a vital indicator of the ocean's biological health.

Chlorophyll-a, the most important photosynthetic pigment, is a valuable indicator of the ocean's primary production. Such pigments selectively absorb blue light, so that the ratio of the 'water leaving radiance' at 443 and 550 nm has been used as a measure. Water masses where the reflectance is dominated by photosynthetic pigment absorption are referred to as Case I waters: in Case II waters the reflectance is determined by suspended particulates and dissolved organic matter (also known as 'yellow stuff' or gelbstoff).

The Coastal Zone Color Scanner (*CZCS*) was originally flown on Nimbus-7 as a test mission. It proved sufficiently successful that many follow-up missions have included a suitable pair of channels to allow for the extraction of some measure of ocean chlorophyll concentration and primary production. Follow-on missions have included the Sea-viewing Wide Field-of-view Sensor (SeaWiFS), and the MODIS sensor on both NASA's Terra and Aqua satellites, to be discussed below.

12.5.2 SAGE, ERBS

12.5.2.1 SAGE and SAGE-II

SAGE (Stratospheric Aerosol and Gas Experiment) uses the solar occultation technique. Because it can only make two measurements per orbit (sunrise and sunset), it requires an oblique orbit to provide adequate global coverage. (SAM-II on Nimbus-7 similarly uses occultation, and because of its orbit it sees only the polar regions.) For this reason, it flew on its own bus/platform/satellite, which was launched on 18 February 1979, from Wallops Flight Center, Virginia. It was followed a few years later by SAGE-II, which flew on the ERBS platform (below). SAGE-II had seven channels (384.6–1,019.7 nm) designed to obtain stratospheric profiles of O_3, NO_2 (an ozone threat), H_2O, and aerosol. Later incarnations were SAGE III Meteop-3M, and SAGE III ISS (on the International Space Station).

SAGE-II and SAM-II do a great job of detecting volcanic aerosol, including mapping Mt Pinatubo, as we saw in Figure 4.1. SAM II was also responsible for identifying the polar stratospheric clouds, which are important to the formation of the Antarctic ozone hole.

12.5.2.2 Earth Radiation Budget Satellite

Understanding the Earth's radiation/energy budget, and especially the role of clouds, has been a long-standing challenge for science. Space observation is, clearly, the only realistic approach to obtain the global data required. ERBE is a broad project using instruments on a number of NOAA's polar-orbiting satellites. To complement this, ERBS (the ERB Satellite), was launched from the space shuttle *Challenger* in October 1984 with a 57° inclination. SAGE-II also flew on the ERBS platform. The current incarnation is CERES (below).

Data are presented in the form of short-wave (reflected solar) and long-wave (emitted thermal) radiation; split into 'all-sky', and 'cloud-free'. Many colourful monthly data sets have been published, often gracing the covers of scientific journals. The short-wave cloud-free data show surface albedo, which is dominated by ice and snow, but also show the reflectivity of deserts. Similarly, the long-wave cloud-free data provide a good measure of surface temperature. The all-sky data are dominated by cloud-top reflection and emission, of course.

12.6 Environmental Observation in the 1990s

12.6.1 Upper Atmosphere Research Satellite

UARS was launched on 12 September 1991, from the Space Shuttle *Discovery* (STS-48) into a 575-km orbit with an inclination angle of 57°, and finally deactivated on 14 December 2005. It carried 10 instruments (six of which were still operational on deactivation) covering chemical studies, dynamics, and energy inputs. There is considerable overlap in their target species, but that is the nature of a proof-of-concept mission.

CLAES (Cryogenic Limb Array Etalon Spectrometer) measured atmospheric infrared limb emission in the spectral range 3.5–12.7 μm. This information was used to detect the vertical profiles of members of the N and Cl families, O_3, H_2O, and CH_4. *ISAMS* (Improved

Stratospheric and Mesospheric Sounder) was an infrared radiometer measuring thermal emission from the Earth's limb. These data were used to determine the temperature profile above the tropopause; the distribution of nitrogen oxides; as well as H_2O, CH_4, and CO.

MLS (Microwave Limb Sounder) detected thermal emission in the microwave region. The 63 GHz radiometer was used to measure temperature; the 183 GHz radiometer measured water vapour and ozone; and the 205 GHz radiometer was used to measure ClO, O_3, SO_2, HNO_3, and H_2O. It was the first instrument to provide a global data set on ClO, a key species in ozone destruction, as we saw in Chapter 10. *HALOE* (HALogen Occultation Experiment) was a solar occultation instrument measuring sunlight in the spectral range 2.43–10.25 µm. These data were used to provide vertical profiles of O_3, HCl, HF, NO, NO_2, CH_4, and H_2O.

HRDI (High Resolution Doppler Imager) and *WINDII* (wind imaging interferometer) both used Doppler-shift techniques to obtain upper air wind fields.

SUSIM (Solar Ultraviolet Spectral Irradiance Monitor) measured solar UV emissions from 120 to 400 nm, with a resolution of 0.1 nm. High-energy photons have the potential to cause instrument degradation. To understand these effects, two instruments were used, one almost continuously, and the other just occasionally to verify the calibration of the first. *SOLSTICE* (Solar Stellar Irradiance Comparison Experiment) was similar to *SUSIM*, measuring the solar spectrum from 115 nm to 430 nm, with a 0.12-nm resolution. *ACRIM2* (Active Cavity Radiometer Irradiance Monitor II) was designed to measure total solar irradiance.

12.6.2 ADEOS

The Japanese Space Agency, NASDA (subsequently JAXA), orbited two environmental monitoring spacecraft from its Tanegashima Launch facility: ADEOS I and ADEOS II (ADvanced Earth Observing Satellite). ADEOS I was launched on 17 August 1996 into an 800-km sun-synchronous orbit with an inclination of 98.6°. It failed on 30 June 1997, due to a solar panel problem. ADEOS II was launched on 14 December 2002 into a similar orbit. It also failed, on 23 October 2003, possibly due to debris impact.

ADEOS I carried eight instruments: AVNIR, OCTS, IMG, ILAS, RIS, NSCAT, TOMS, and POLDER. ADEOS II carried five primary instruments: AMSR, GLI, ILAS-II, POLDER, and SeaWinds. We will discuss some of these below. As with other missions, much of the power of ADEOS came from combining data from several sensors. Some of the 'primary research goals' of the ADEOS mission were to:

- determine an increased number of physical and optical properties of tropospheric aerosols;
- improve climate-relevant descriptions of the physical and radiative properties of clouds;
- document cloud–aerosol–radiation interactions to assess impacts on the radiation budget;
- document the variability of ocean primary production to improve our understanding of the role of the ocean in the carbon cycle;
- improve the physical characterisation of the reflectances of vegetated surfaces to derive indices and parameters to model the dynamics of the continental biosphere.

AVNIR (Advanced Visible and Near-Infrared Radiometer) had four bands (blue, green, red, near IR) to detect chlorophyll and sediment in waters, and to study plant growth; and a panchromatic band (520–690 nm) to identify land surfaces. Its focus was mostly around Japan. *OCTS* (Ocean Colour and Temperature Scanner) had a similar mission to *AVNIR*.

IMG (Interferometric Monitor for Greenhouse gases) was a Michelson interferometer able to operate in three modes: nadir viewing, limb emission, and limb extinction. It had three bands: SW1 (3.3–4.3 μm); SW2 (4.0–5.0 μm); and LW (5.0–14.0 μm) for sensing CO, CO_2, CH_4, H_2O, O_3, N_2O, and HNO_3. *ILAS* (Improved Limb Atmospheric Spectrometer) was a solar occultation instrument to measure the vertical profiles of CH_4, H_2O, O_3, N_2O, NO_2, HNO_3, and N_2O_5, as well as the two key CFCs, $CFCl_3$ and CF_2Cl_2.

RIS (Retroreflector In Space) was, as its name implies, little more than a corner cube from which a laser beam could be reflected. By using a CO_2 laser, it was able to sense many of the chemical species already listed for IMG and ILAS.

NSCAT (NASA SCATterometer) was designed to provide wind-speed data over at least 90% of the ice-free ocean every two days. Its microwave radar used backscatter from the ocean surface. As wind speeds increase, wave motion increases, and the backscatter increases. By using a geophysical model, signal strength can be related to wind speed.

TOMS (Total Ozone Mapping Spectrometer) is an instrument we have already discussed: it was included to cover a data gap.

GLI (GLobal Imager) was a 36-channel multipurpose instrument covering the visible to mid-infrared spectral range (0.375–12.5 μm). Its performance is similar in many ways to MODIS, which will be discussed below.

12.6.2.1 POLDER

POLarization and Direction of the Earth's Reflectances (POLDER) was an innovative CNES (French) instrument. POLDER used a wide-field-of-view camera to image the Earth scene into 244×274 elementary pixels in a CCD matrix, with a swath width of 2,400 km. Each pixel (at nadir) images an area of ~6×7 km^2. It aimed to measure ocean colour, vegetation, and aerosols.

A rotating filter carried spectral and polarising filters for three of its nine channels. It is the polarisation component that is POLDER's key innovation. The reflected signal will contain components from surface reflection, molecular scattering, and aerosol scattering, as well as multiply scattered contributions. Surface reflection is essentially non-polarised, and this dominates the signal for longer wavelengths. Rayleigh (molecular) scattering is highly polarised, in a known way, and this is a significant contribution at shorter wavelengths. The aerosol scattering contribution is somewhere in between. The combination of intensity and polarisation makes it possible to separate the contributions. The combination of measurements at the three polarised wavelength channels (plus the 765-nm channel) provided unprecedented information on aerosol properties.

12.6.3 Sea Level

One of the most important questions in climate science/change is sea-level rise, and the role of melting ice. (The contribution from thermal expansion is a 'relatively simple' physics exercise.) What role might remote sensing play?

Sea level can be monitored by radar. Microwaves have the great advantage that they can see through clouds, which are very common at high latitudes where ice – both floating sea ice and land-based ice sheets/glaciers – predominates. Useful data may be obtained from instruments such as the Advanced Microwave Scanning Radiometer (AMSR) on ADEOS II, and GCOM-W1 (below).

Topography Experiment (TOPEX)/Poseidon was a joint venture between CNES and NASA, launched in 1992. Its primary instrument was the dual-frequency NASA Radar Altimeter, which sent pulses at 13.6 and 5.3 GHz towards the surface, and measured the characteristics of the echo. These data were then combined with data from a microwave radiometer, allowing it to measure ocean surface topography. It enabled scientists to forecast the 1997–1998 El Niño. The mission ended in January 2006. Among its other achievements it:

- measured sea levels with an unprecedented accuracy of better than 5 cm;
- Monitored the effects of currents on global climate change;
- provided global data to validate models of ocean circulation;
- mapped year-to-year changes in heat stored in the upper ocean;
- produced the most accurate global maps of tides ever;
- improved our knowledge of the Earth's gravity field.

12.7 Environmental Observations: Twenty-First Century

NASA spent many years planning the next generation of (polar-orbiting) satellites and payloads, under the broad title of Earth Science Enterprise/Mission to Planet Earth. The EOS (Earth Observing System) science plan goes back to the mid-1990s under NASA's Office of Earth Science.

ICESat (ice, cloud, and land elevation satellite) was the benchmark mission for measuring ice sheet mass balance, cloud and aerosol heights, as well as land topography and vegetation characteristics. From 2003 to 2009, the ICESat mission provided multiyear elevation data needed to determine ice sheet mass balance as well as cloud property information, especially for stratospheric clouds common over polar areas. It also provided topography and vegetation data around the globe, in addition to the polar-specific coverage over the Greenland and Antarctic Ice Sheets.

12.7.1 Terra (EOS-AM)

NASA now has a suite of spacecraft in sun-synchronous orbits making outstanding contributions to our knowledge of the atmosphere, oceans and land surface. **Terra** was launched from Vandenberg Air Force Base in December 1999, in a 713-km orbit with a 16-day repeat cycle and a descending node of 10:40 a.m. It carries five instruments.

MODIS is an extremely versatile instrument, flying on NASA's Terra and Aqua platforms. It has 36 spectral channels (0.415–14.235 μm) measuring both reflected sunlight and emitted terrestrial radiation. It is designed to provide information on 'the climatology and dynamics of atmospheric properties, the impact of human activity on the regional and global environment, and the interaction and subsequent impact of man on terrestrial and marine biota'. It provides data on land/cloud/aerosol properties; ocean color and productivity; chemistry; temperature; water vapour; and much more.

MISR measures reflected sunlight in the bands 446, 558, 672, and 866 nm. It has nine cameras, one in the nadir, and the others viewing at 26.1°, 45.6°, 60.0°, and 70.5° both forward and aft. This helps to separate reflection from the ground and the atmosphere: aerosols are its main target (characterised by the 'underwear model': small/medium/large; clean/dirty).

CERES (Clouds and the Earth's Radiant Energy System) continues measurements of the Earth's radiation budget using two broadband scanning radiometers.

MOPITT (measurement of pollution in the troposphere) is a Canadian instrument designed to measure CH_4 and CO, in order to determine their sources and sinks.

ASTER is a Japanese instrument that provides very-high-resolution images (15 m in certain channels) of the land surface, water, ice, and clouds.

12.7.2 The A-Train

The Afternoon Constellation (or A-Train) is a suite of platforms carrying a broad range of instruments that complement each other. They fly in formation (ascending node ~1:30 pm) in order to observe the same footprint. A total of seven platforms have comprised the system, but one had to be switched off ('de-orbited'). We will give brief accounts of their instruments.

GCOM-W1 (Global Change Observation Mission – Water) leads the constellation. Its goal is to provide long-term observations of changes in global water circulation, using a six-band instrument (AMSR2) to observe water vapour, precipitation, water levels on land, and snow depth; data which may be used to improve weather forecasts.

AQUA (EOS-PM) is similar to Terra, and contains both the MODIS and CERES instruments. It also carries AIRS: the Atmospheric Infrared Sounder, which collects data on the infrared energy emitted from the Earth's surface and atmosphere. Its instrument suite consists of a hyperspectral instrument with 2,378 infrared channels, and four visible/near infrared channels, along with the AMSU-A instrument with 15 microwave channels. Its data provide 3D measurements of water vapour, plus a host of other information.

CALIPSO (Cloud-Aerosol Lidar and Infrared Pathfinder Satellite Observation) carries the Cloud Aerosol Lidar with Orthogonal Polarisation (CALIOP) to measure the vertical profiles of clouds (including ice/water phases) and aerosols.

CLOUDSAT complements CALIPSO's cloud studies with a Cloud Profiling Radar (94 GHz), which is not affected by water vapour. It can detect multiple layers of cloud in order to supply better data for use in both weather and climate models.

PARASOL (Polarisation and Anisotropy of Reflectances for Atmospheric Sciences coupled with Observations from Lidar) carries a POLDER-like radiometer that measures the polarisation of reflected sunlight to study aerosols. It was de-orbited in December 2013.

AURA travels 15 minutes behind the rest, as it is a limb-scanning instrument, focused on stratospheric ozone, air quality (tracing pollutants back to source), and climate change. The nadir-viewing OMI, an evolution of TOMS, provides ozone and NO_2 data.

OCO-2 (Orbiting Carbon Observatory) makes global measurements of CO_2.

12.7.3 The European Space Agency

The Gravity Recovery and Climate Experiment (GRACE) was a joint partnership between NASA and the German Space Agency (DLR), launched in March 2002. Its mission is designed to accurately map variations in Earth's gravity field. GRACE consists of two identical spacecraft that fly about 220 km apart in a polar orbit. GRACE maps Earth's gravity field by making accurate measurements of the distance between the two satellites, using GPS and a microwave ranging system. The results from this mission are yielding crucial information about the distribution and flow of mass within Earth and its surroundings.

The gravity variations studied by GRACE include: changes due to surface and deep currents in the ocean; runoff and groundwater storage on land masses; exchanges between ice sheets or glaciers and the ocean; and variations of mass within Earth. GRACE results are making a huge contribution to global climate change studies. The GRACE mission ended in 2011, and has been replaced by GRACE-FO (GRACE Follow-On), launched in May 2018.

12.7.3.1 ERS and ENVISAT

ESA, a consortium of 20 countries, has its own programme. It launched two small platforms (ERS-1 and ERS-2) from its base in French Guiana for 10- to 15-year missions. This was followed by **ENVISAT**, launched in 2002 (~800 km; 98.4°). Unfortunately contact was lost in 2012. It carried a suite of nine instruments, including a radar altimeter to measure sea level.

Its primary instrument was *SCIAMACHY* (the Scanning Imaging Absorption Spectrometer for Atmospheric Cartography), which viewed in both nadir and limb mode. (Spectral coverage 240–2,380 nm – UV-NIR – resolution 0.2–0.5 nm.) It was able to detect O_3, O_2, $(O_2)_2$, H_2CO, SO_2, BrO, OClO, ClO, NO, NO_2, NO_3, H_2O, CO, CO_2, CH_4, N_2O, aerosols, and clouds.

12.7.3.2 Copernicus/Sentinel Programme

ESA is currently developing and deploying a series of next-generation Earth Observation missions, on behalf of the joint ESA/European Commission initiative GMES (Global Monitoring for Environment and Security), to meet the operational needs of the Copernicus Programme. The goal of the Sentinel Programme is to replace the older Earth Observation missions that have reached retirement or are currently nearing the end of their operational lifespan. This will ensure a continuity of data so that there are no gaps in ongoing studies. Each mission will focus on a different aspect of Earth observation – Atmospheric, Oceanic, and Land monitoring – and the data will be of use in many applications.

Sentinel-1 is composed of two polar-orbiting satellites operating day and night, and will perform radar imaging, enabling them to acquire Land and Ocean monitoring imagery regardless of the weather. The first Sentinel-1 satellite was launched in April 2014. Sentinel-2 is land monitoring, and the mission will be composed of two polar-orbiting satellites providing high-resolution optical imagery. Vegetation, soil, and coastal areas are among objectives. The first Sentinel-2 satellite was launched in June 2015. Sentinel-3's primary objective is marine observation, and it will study sea-surface topography, sea- and land-surface temperature, ocean and land colour. Composed of three satellites, the mission's primary instrument is a radar altimeter, but the polar-orbiting satellites will carry multiple instruments, including optical imagers.

Sentinel-4 and -5 are dedicated to air quality monitoring. The mission aims to provide continuous monitoring of the Earth's atmospheric composition, with high temporal resolution over Europe (Sentinel-4) and lower temporal resolution over the rest of the world (Sentinel-5). A precursor mission, Sentinel-5P, aims to fill in the data gap and provide data continuity between the retirement of the Envisat satellite and NASA's Aura (OMI) mission and the launch of Sentinel-5. It was successfully launched on 13 October 2017 from the Plesetsk Cosmodrome in Russia.

The main objective of the Sentinel-5P mission is to perform atmospheric measurements, with high spatiotemporal resolution, relating to air quality, climate forcing, ozone, and UV radiation. For example, it is providing global data on emissions of methane from pipelines and coal mines. Focusing on specific aspects of our changing atmosphere of relevance to our book, we will briefly list the following Sentinel objectives:

- Greenhouse gases. The OLCI instrument on Sentinel-3, together with Sentinel-5P and Sentinel-5, will support monitoring of greenhouse gases, changes in the ozone layer, and aerosols.

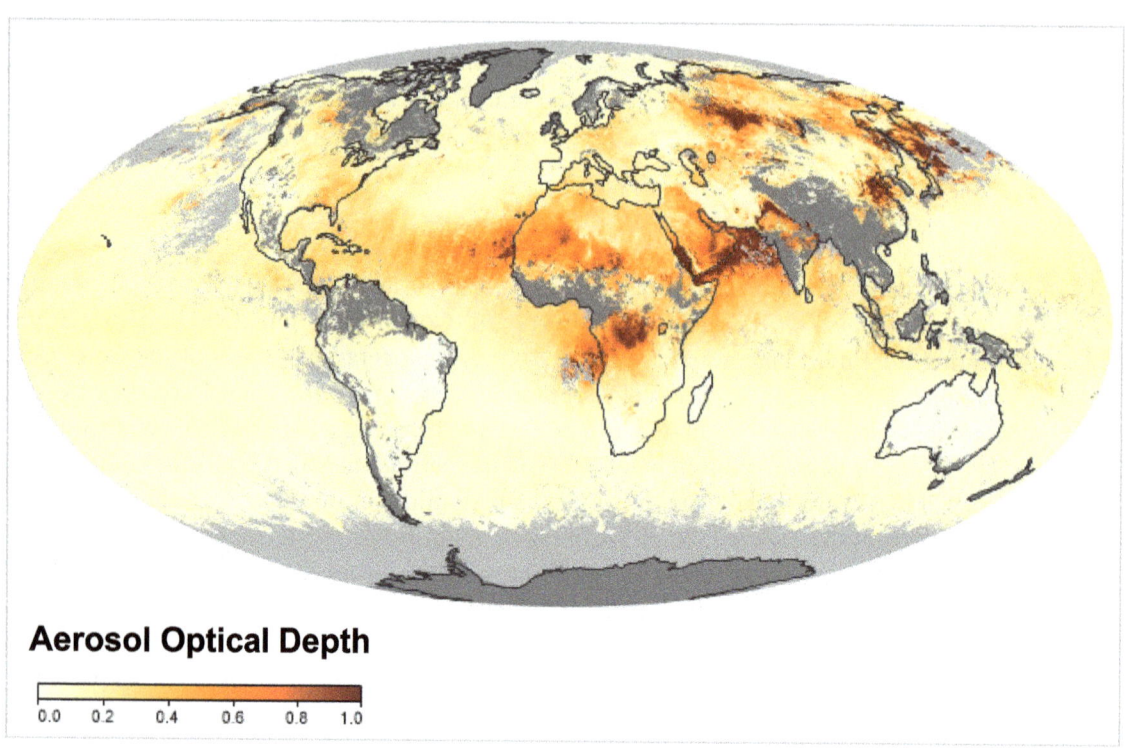

Figure 12.6 Aerosol optical depth for July 2012. *Source*: NASA Earth Observatory.

- Reactive gases. These are gases which, when exposed to sufficient heat and light, chemically react and absorb ozone. These include OClO, BrO, as well as NO_2.
- Ozone and UV. Measurement of changes in the ozone layer is now operationally possible in near-real-time, allowing ongoing updates of the effects of pollution, for which the atmosphere missions Sentinel-5P, Sentinel-5, and Sentinel-4 provide essential data.
- Aerosols. Satellites will help to monitor potentially harmful levels in the atmosphere, and distinguish between natural and human-made aerosols, to prevent further pollution.

Looking to the future, six high-priority missions (Copernicus Expansion Missions) are currently being studied, both to address EU policy and gaps in Copernicus user needs, and to expand the current capabilities of the Copernicus space components.

12.8   NASA's Earth Observatory

NASA has collected much of the key data from its satellite instruments (primarily MODIS), and made them available in the form of global maps of monthly data. A total of 16 data sets are presented, mostly with time running from early 2000: Snow Cover, Net Radiation, Cloud Fraction, Rainfall, Water Vapour, Aerosol Optical Depth, Fire, Land Surface Temperature

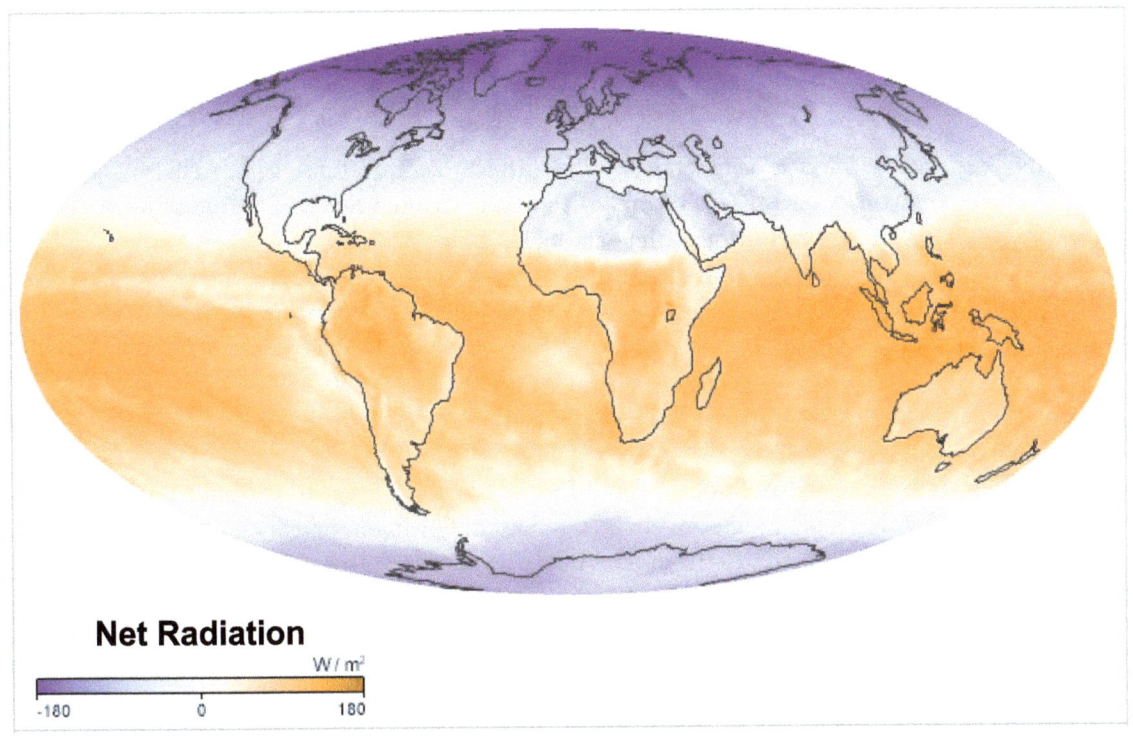

Net Radiation

W / m²

-180   0   180

Figure 12.7 Net radiation for October 2010. *Source*: NASA Earth Observatory.

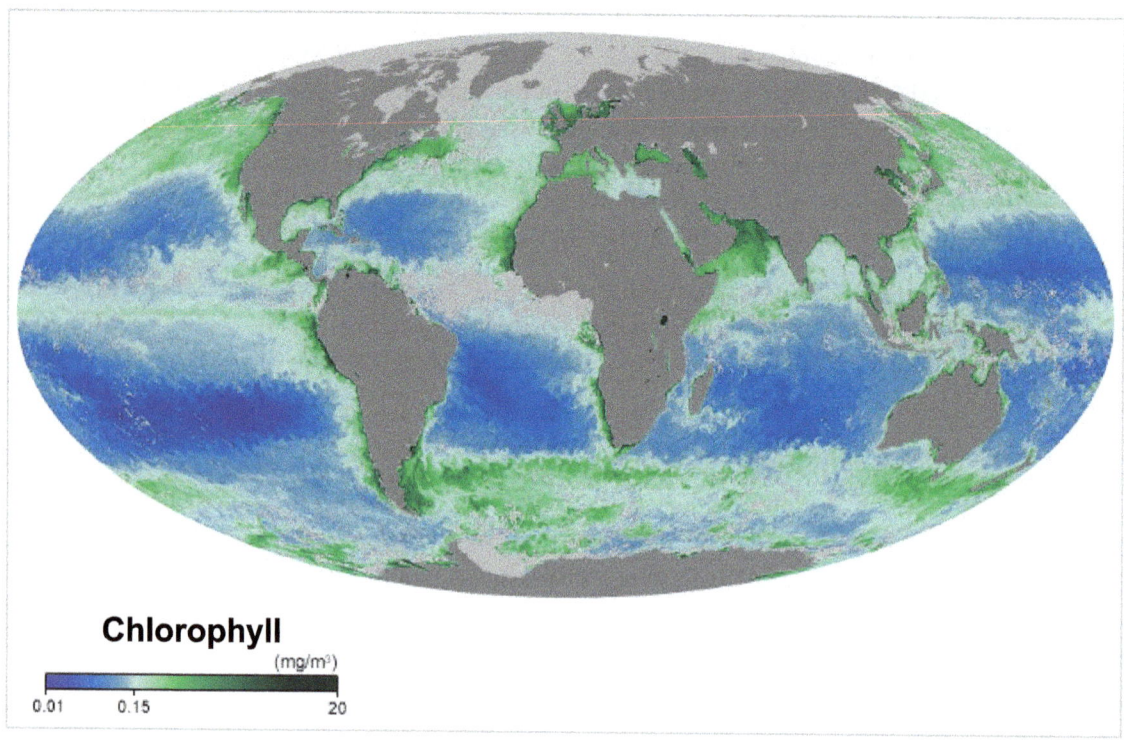

Chlorophyll
(mg/m³)

0.01   0.15   20

Figure 12.8 Chlorophyl for February 2014. *Source*: NASA Earth Observatory.

Anomaly, Land Surface Temperature, Sea-Surface Temperature, Chlorophyl, Vegetation, Carbon Monoxide, Aerosol Size, Primary Productivity, Sea Surface Temperature Anomaly. You can click on and run through the data like a movie.

You can also compare data sets. When you so choose, the website has a paragraph or two telling you how the two data sets may be related: how one may influence the other (telling you what to look for). For example, CO (from MOPITT) and fires (from MODIS), and aerosol optical thickness (also MODIS). We'll now present a selection of data snapshots from this library, and invite readers to explore this excellent resource for themselves.

Figure 12.6 shows aerosol optical depth for the month of July 2012 (from MODIS/Terra).

Figure 12.7 shows net radiation – that is, radiation flowing in minus radiation flowing out – for the month of October 2010 (from CERES). The scale runs from -180 W m^{-2} to 180 W m^{-2}. Seasonal and latitudinal effects are obvious.

Figure 12.8 shows chlorophyll for the month of February 2014 (from MODIS/Aqua), in units of mg/m^3. We see that productivity is mostly found in the colder waters and continental margins where nutrients tend to be abundant.

Figure 12.9 shows the vegetation index for the same month (from MODIS/Terra).

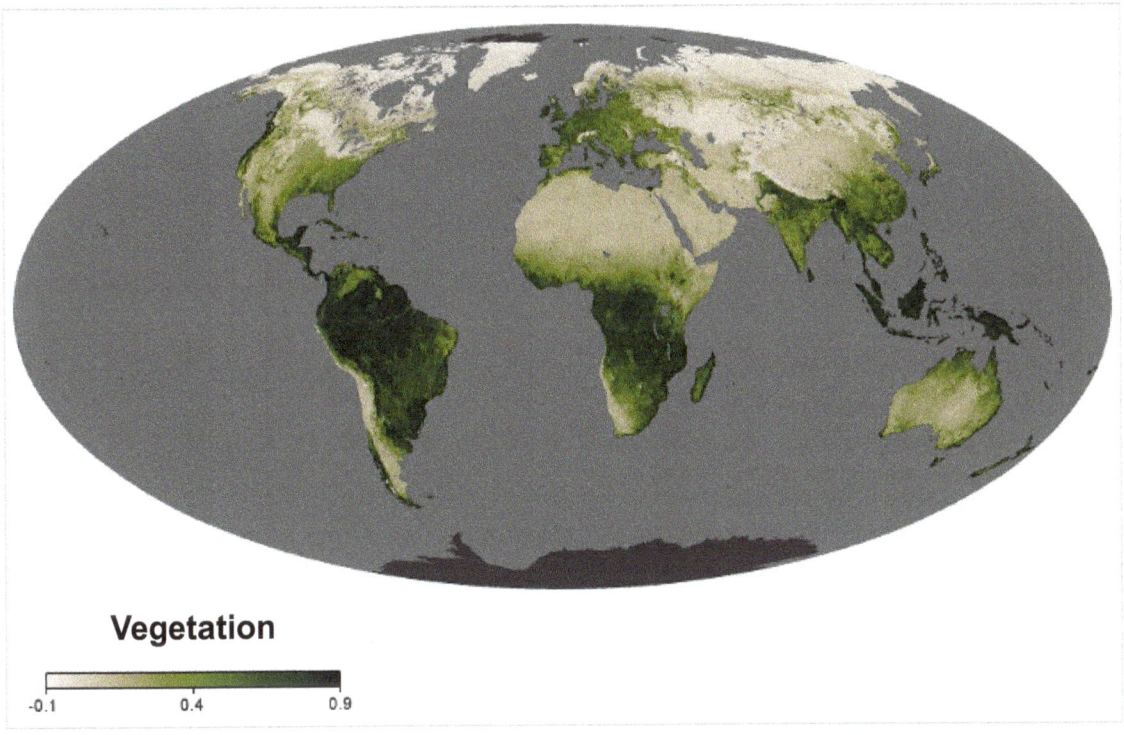

Vegetation

-0.1   0.4   0.9

Figure 12.9 Vegetation for February 2014. *Source*: NASA Earth Observatory.

Acknowledgement

We wish to thank NASA for all the information/data/figures available on its websites, and especially its Earth Observatory. We also thank them for the wonderful science undertaken to benefit us all.

FURTHER READING

Remote sensing, and especially satellite observation, is a source of vast amounts of useful data on our environment and any threats it may face (provided we know how and where to look). Some of the material covered in Sections 12.5, 12.6, and 12.7 will help to underpin our investigations of climate change in Part V of this book.

We have only been able to scratch the surface of this fascinating subject; should you like to learn more, there are many places to look.

- *Earthwatch, The Climate from Space* by J. Harries (Ellis Horwood, 1990): although dated, it provides an excellent overview of many applications and missions.
- *SCHIAMACHY – Exploring the Changing Earth's Atmosphere* edited by M. Gottwald and H. Bovensmann (Springer, 2011).

- *An Introduction to Atmospheric Radiation* by K. N. Liou (Academic Press, 2002) provides a thorough introduction to the theoretical radiative transfer aspects of remote sensing.
- *SCIAMACHY The Need for Atmospheric Research from Space* by J. P. Burrows, A. P. H. Goede, C. Muller, and H. Bovensmann (Springer, 2010).
- *Atmospheric Aerosols: Characteristics and Radiative Effects* by S. Ramachandran (CRC Press, 2017).
- *Introduction to Satellite Remote Sensing* by W. Emery and A. Camps (Elsevier, 2017).

Two delightful 'coffee table books' looking at what we have learned from satellite remote sensing are:

- *Atlas of Satellite Observations Related to Global Change* edited by R. J. Gurney et al. (Cambridge University Press, 1993).
- *Our Changing Planet, The View From Space* edited by M. D. King et al. (Cambridge University Press, 2007).

Finally, the list of websites is endless, starting with:

- http://eospso.gsfc.nasa.gov/
- http://earthobservatory.nasa.gov/global-maps
- http://science.nasa.gov/earth-science
- www.esa.int
- www.noaa.gov
- https://aeronet.gsfc.nasa.gov
- www.bom.gov.au/australia/satellites/shtml

REVIEW QUESTIONS

1. Why are geostationary satellites placed at such a high altitude?
2. Why are polar-orbiting satellites usually placed in mildly retrograde orbits?
3. What is the key to detecting healthy vegetation?
4. What are the *two* steps involved in obtaining a water vapour profile from satellite observations?
5. Which satellites were used to learn about snow and ice?
6. What are the advantages of having a constellation of satellites, such as the A-Train, flying in close formation?

EXERCISES

1. Explain the limb scan weighting functions in Figure 12.4b.
2. Use your imagination to explore one of the websites and prepare a report.

Part IV

The Climate System

In the first three parts of this book, we have developed the three branches of science – chemistry, physics, radiant energy – that are central to understanding the climate of planet Earth. In doing so, our primary focus has been the atmosphere, with occasional brief excursions into the oceans. It is now time to take a broader look, and start by asking some important questions.

What, exactly, is climate? Is it just a function/property of the atmosphere? What information – both 'data', and scientific understanding – might be needed to better understand climate? And the big one, which follows from the previous: what do we need in order to 'predict' climate, especially in a time of change? We have placed the key word here in quotes, as we wish to use it in as precise, or as vague, a manner as circumstances dictate.

Firstly, what is climate? Superficially, at least, climate is the distillation of weather statistics. It is average values of temperature, rainfall, and the rest of what we experience from day to day. But it is, of course, far more than that. While averages are a simple place to start, climate is at least as much about variability as averages: today that usually means deciles. How often is the temperature for a particular month significantly above, or below, the average? Does the rainfall come slowly and steadily, day by day, or might we get a month's worth of rain in a couple of days? This is one we know from experience can often be the case.

But then we might probe even deeper. Even after doing all the statistical analysis, is one year significantly different from another? And why? A drought may last for a number of years, so that all the averages from that period will be statistically different from the long-term average. After that, we might find the average is 'restored' as a result of a disastrous flood. These and similar questions/considerations are all a part of what we understand as climate.

So now we can turn to the other questions: understanding and predicting climate. Certainly, we need to interrogate our weather data more thoroughly, but that is just the start. If we are to 'predict', especially in a time of apparent change, we must first understand 'why'. So, does that simply mean understanding our atmosphere better, or do we need to dig deeper?

Our atmosphere is a relatively simple system, all things considered. This is one of the reasons we are able to predict next weekend's weather, with steadily growing accuracy. Good observations, and great computer power, are also essential, but understanding comes first. Now, our atmosphere is not a 'self-contained' system: it has boundaries: the oceans, the cryosphere, and the land. For a seven-day weather forecast, these will mostly remain close enough to fixed, and can usually be treated as such. But climate 'evolves' month by month, year by year, decade by decade. None of these boundaries will remain unchanged over such time frames. We need to understand how potential changes will impact on the atmosphere, on appropriate timescales.

So now we can understand the brief comments made in Chapter 1. The climate system is, in fact, a set of interlocking subsystems. The final goal of this book is, of course, to understand how, and why, our climate is changing, and whether or not it will continue to change into the future: from years, to decades, and even to centuries. That is the focus of Part V. However, we cannot hope to achieve that if we do not fully understand the climate system, including all of its component subsystems. This is the goal of Part IV.

In Chapter 13 we will start with the obvious candidate for next subsystem, the ocean. Being a fluid, it has many properties in common with the atmosphere, and obeys the same physics. However, its far greater density (by at least a factor of a thousand) has obvious consequences, starting with its 'response time'. The oceans are also confined by the continents, of course, which control their circulation.

The other two boundary subsystems are the cryosphere and the land surface. These are both more complex systems, and we will only touch on them briefly in Chapter 14. The main thrust of that chapter is how different elements of the system might interact, and, more significantly, feed back on one another. Once we have a reasonable understanding of the various pieces, and how they interact, we are in a position to build a model of the system; the subject of Chapter 15. In both of these chapters we will introduce you to simplified models, and invite you to explore.

With all of this under our belts, we are ready to tackle the primary focus of the book: our changing climate. This will be covered in the four chapters that comprise Part V. But before we tackle that issue, there are some useful insights and perspectives to be gained by looking back in time, before human activity had any consequences. This is the subject of Chapter 16.

13   The Ocean's Role in Climate

Oceans cover two-thirds of the Earth's surface, and as such are the major lower boundary of the atmosphere. They influence the atmosphere in two major and several minor ways. They are, of course, the overwhelming source of water vapour to the atmosphere, which also includes a latent heat flux. They are also a source of sensible heat to the atmosphere, and this is largely dictated by sea-surface temperature. The minor influences are the fluxes, in both directions, of various chemical species, such as salts, DMS, and CO_2.

The oceans circulate, just as the atmosphere does, transporting heat, with predictable latitudinal and seasonal variations. So we'll start by looking at the key properties of sea water, and the basic ocean circulation patterns; both the surface currents, and the deep ocean circulation.

Sea-surface temperatures do exhibit subtle variations, and it is for this reason that the oceans are the major source of the interannual variability of regional climate. While the influences of these oceanic 'modes of variability' are mostly regional, some global-scale impacts are also known. The best known – and best understood – of these is ENSO: El Niño and the Southern Oscillation, which has climatic impacts around the Pacific and beyond. We also examine a number of the lesser-known modes which are now recognised as regionally important.

13.1   The Ocean

13.1.1 Properties of Sea Water

Sea water is not 'pure' water, for reasons we have already explored in Chapters 3 and 4. In Chapter 3 we looked in some detail at the role and impacts of carbon on ocean chemistry. In Table 4.1 we looked at the ionic composition of dissolved substances, primarily salts. While the chemistry of sea water is important, it is the physical properties that we need to note here.

The **salinity** of sea water varies between ~31 g/kg and ~38 g/kg (also expressed as ~31‰ to ~38‰), for reasons which depend on surface evaporation, precipitation, and river inflow. As a result, its density is higher than pure water, typically in the range 1,010–1,030 kg/m^3. Under the high pressures in the ocean depths, it can reach 1,050 kg/m^3. The equation of state for sea water connects density to both temperature and salinity: there is no simple expression.

13.1.2 Vertical Structure

The vertical temperature structure of the ocean may be thought of as comprising three layers. At the surface, the action of the wind causes wave motion, which effectively stirs the surface layers to a depth of the order of the wavelength. This **mixed layer** is essentially uniform in structure, extending to a depth of 50–100 metres. This is also the approximate distance to which sunlight can penetrate. The lowest layer, the **abyssal layer**, making up over 95% of the ocean, is largely separated from surface influences. In between is a transition zone, down to ~200 metres, the **thermocline region**. The boundaries are not sharp, and vary both in space and time.

Surface temperatures vary considerably with latitude, as expected. In the tropical western Pacific, from about 150°E to the coast of South-East Asia, and between latitudes of about 20°N and 15°S, temperatures regularly exceed 28°C all year round. This region is known as the **Tropical Warm Pool**, and along with the islands and peninsulas in the region is also known as the **Maritime Continent.** At the other extreme, the Southern Ocean approaches 0°C near the Antarctic Ice Sheet. Abyssal temperatures decrease slowly to typically 3–4°C. Summer temperatures in the mixed layer are typically 5°C warmer than winter temperatures for the extratropical oceans.

Salinity also varies with depth, although far less than temperature. In the mixed layer it has comparatively large values, due to evaporation. It reduces through the thermocline, before slowly rising through the deeper layers. The slow decrease in temperature and slow increase in salinity at great depth are both a consequence of the fact that the density of water increases with both of these trends: the densest water is, of course, the deepest. In polar regions the top two layers all but disappear, and cold water extends to the surface. When ice forms it is nearly fresh, having left most of its salt content behind. The resulting water is thus saltier, and denser, encouraging it to sink. Not surprisingly, most of the so-called **deep water formation** occurs in polar regions.

Water parcels that are no longer in contact with the surface tend to conserve both temperature and salinity as they move throughout the ocean depths. Such **water masses** may thus be tracked back to the regions in which they formed. They can also be dated by measuring their ^{14}C and ^{3}H (tritium) concentrations, as both experienced a spike due to atmospheric testing of nuclear 'devices'. Trace species such as the CFCs, with known production histories, may also be used. Such substances are collectively referred to as **tracers**.

13.2 Ocean Circulation

13.2.1 Surface Currents

The ocean circulates, as does the atmosphere, although the driving forces are somewhat different. The ocean surface layer responds primarily to the wind field, with the prevailing winds supplying the energy which drives surface currents. This wind-driven circulation is mainly confined to the upper few hundred metres, particularly in warmer regions where vertical stratification is strong. In the circumpolar region where stratification is much weaker, this circulation can extend to the ocean floor. For regions where surface topography leads to convergence (divergence), downwelling (upwelling) must result.

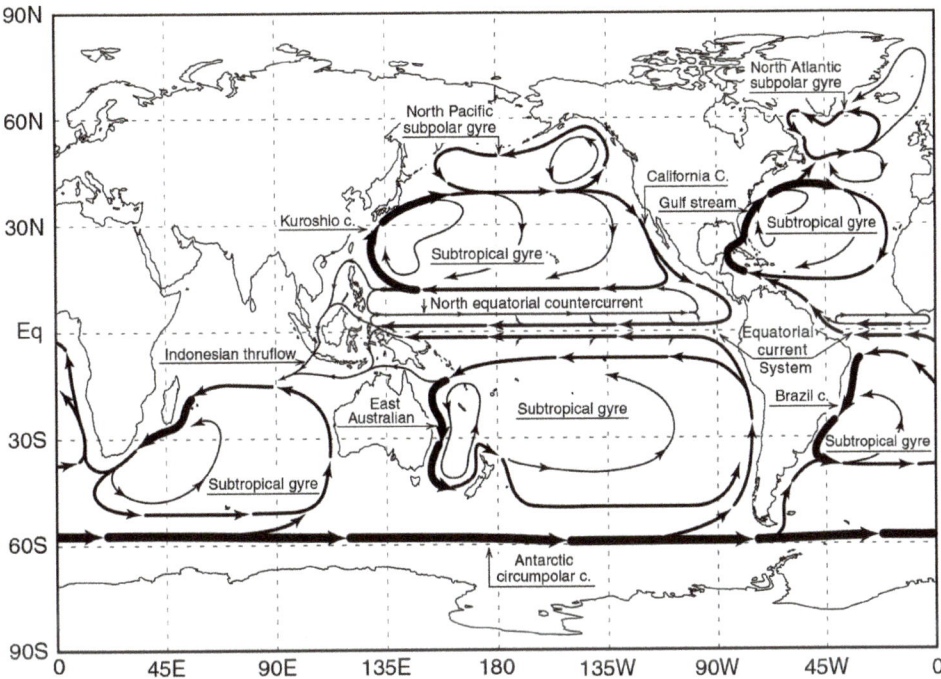

Figure 13.1 Major ocean surface currents. *Source*: *Climate Change and Climate Modelling*, J. D. Neelin, 2011. Copyright Cambridge University Press, with permission.

The main features are the **subtropical gyres** in the North and South Pacific, North and South Atlantic, and Indian Oceans: clockwise in the Northern Hemisphere, anticlockwise in the Southern Hemisphere (see Figure 13.1). The trade winds along the equatorial belt produce an east-to-west surface flow, while at high latitudes the prevailing westerlies produce a west-to-east flow. These are then partially steered by the Coriolis force, causing deflection; to the right in the Northern Hemisphere, and to the left in the Southern Hemisphere. The south-to-north and north-to-south flows that complete the gyres are more tightly focused on the western side of these ocean basins, known as **western boundary currents**.

Surface flows also connect the different Oceans. The **Antarctic Circumpolar Current** flows all the way around Antarctica, connecting the world's three major oceans. The **Indonesian Thruflow** is an important connector from the tropical Pacific to the Indian Ocean. The **Agulhas Current** flows southwards off the east coast of Africa, and partially into the Atlantic Ocean.

13.2.2 Deep Ocean Circulation

Water movement in the deep ocean is quite slow, and follows a complex, but well-defined pattern. This is a very large-scale circulation, which involves flows through the depths of all major ocean basins: see Figure 13.2. It is, as it obviously must be, a continuous loop – hence the

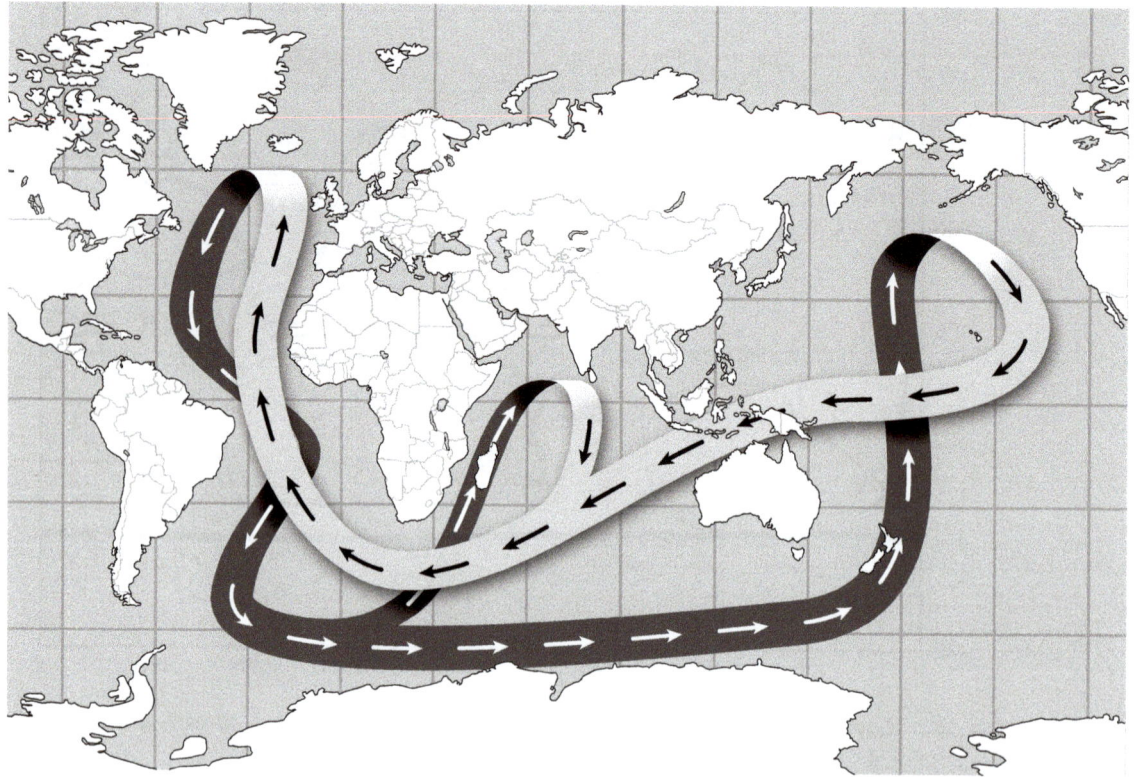

Figure 13.2 The thermohaline circulation. *Source*: Joe LeMonnier, from a sketch by Tom Page.

nickname **conveyor belt** – but it also necessarily requires a 'driver' to maintain it against dissipation by friction and other effects.

So, we will start our 'voyage' at the traditional key location. Waters in the North Atlantic, in the vicinity of Iceland (the Greenland and Norwegian Seas), are cooled for a number of reasons. Firstly, there is, of course, more outgoing long-wave energy than incoming short-wave radiation. Westerly winds blowing from North America, across the relative warmth of the Gulf Stream, produce significant evaporation, increasing salinity. These winds also pick up significant amounts of heat, giving western Europe its relatively mild climate, while cooling the water. This combination of low temperature plus high salinity produces unusually dense water, which is thus able to sink to the ocean floor: hence the more formal title **thermohaline circulation** ('thermo' meaning heat; 'haline', or 'saline', meaning salt). It is now more formally known as the **Atlantic Meridional Overturning Circulation** (AMOC): think of the reasons for each of these four words.

The water mass produced when the Gulf Stream sinks is known as **North Atlantic Deep Water** (NADW), and may be detected by its high salinity, low nutrient content, and the presence of soluble tracers. It then moves slowly down the Atlantic Ocean. At about 40°S it joins a somewhat faster-moving deep current that encircles Antarctica. The Circumpolar

Current is the ocean's principal mixing region, blending the NADW emerging from the Atlantic Ocean (~50%) with water from the Pacific and Indian Oceans at the Antarctic margin. The NADW loses its identity in half a circuit of Antarctica.

Just as important as the formation of NADW is the formation of **Antarctic Bottom Water** (AABW). When sea ice forms in the winter months, salt is expelled into the surrounding waters, again enhancing their density. These are the densest waters in the oceans. Two hundred and fifty trillion tonnes of cold, salty water a year cascade down underwater canyons, eventually moving into the Pacific and Indian Oceans. This forces warm water up, pushing particles left by dead marine organisms back to the surface: a 'biological pump' which underpins critical parts of the ocean's ecosystems. It is estimated that nutrients exported from the Southern Ocean support about three-quarters of global phytoplankton production.

Much of this combined water mass then flows from the south-west Pacific to the north central Pacific through the Samoan Passage, west of Samoa. Here the bathymetry is such that the flow is confined, and has to pick up speed. This leads to significantly increased turbulence: 1,000 to 10,000 times greater than normal for abyssal flows, and including breaking waves. It splits around Hawaii and rejoins to the north, upwelling and flowing back to the South Pacific at mid-depths. This is the start of the return flow. Much of it passes through the Indonesian Thruflow, and around the tip of Africa via the Agulhas Current. (Waters in the North Pacific and Indian Oceans are considerably less dense than the North Atlantic water – the former comparatively fresh, and the latter comparatively warm. As a result, there is a general upwelling in both of these basins.) It then flows northward, along the African coast, before crossing the central Atlantic to join with (or create) the Gulf Stream to complete the loop.

The flow in the deep ocean is extremely slow, and essentially adiabatic. As mentioned above, this allows scientists to track its flow by examining the concentrations of certain tracers, such as man-made chemicals whose atmospheric history we know, or radioisotopes from nuclear testing. For example, water starting out just below the mixed layer in the mid-Pacific can be followed to the Indian Ocean, and from there to the Atlantic and back to the Pacific, 2 or 3 km below its starting point, after a period of several decades.

13.2.2.1 Evolution

The present form of the conveyor belt appears to have commenced following the closing of the Panamanian sea passage between the two American continents. Geological evidence suggests a gradual closing from ~13 to 2 million years ago (Mya), and that closure was sufficient by 4.6 Mya to produce a marked change in large-scale ocean circulation. This intensified the Gulf Stream, allowing the transport of warm water to high latitudes, and intensifying NADW formation. This led to an increase of atmospheric moisture at high latitudes, helping to trigger the growth of the Northern Hemisphere ice sheets.

The Gulf Stream transports heat to the North Atlantic, which is a major moderating influence on the climate of much of Europe. Thus, the flow of the conveyor belt has very significant implications for regional climate. Because of the central role the entire system plays in heat transport, modelling this component of the ocean is vital if we are to understand both the present climate and any potential changes, past or future. Modelling involves balancing the budgets of water, salt, and heat. For a steady-state heat balance, the input of cold abyssal water (NADW plus AABW) must be balanced by the downward convection of relatively warm water (e.g. from the Mediterranean) and downward diffusion of heat across the thermocline.

In recent years, a number of potential threats to the AMOC have been identified. A significant melting of the Greenland ice cap could add sufficient fresh water to the North Atlantic Ocean, inhibiting the formation of NADW. We will look at this in some detail in later chapters. More recently, a threat to AABW formation has been identified. Using a combination of the analysis of available data, coupled with oceanographic modelling, a recent study found that increasing meltwater around Antarctica was freshening the water, and weakening the main driver of the oceanic overturning. They suggest that this system could largely break down by mid-century, with multiple consequences, but especially to ocean productivity.

13.3   ENSO

We have seen that the thermohaline circulation is a major transporter of heat, with very significant effects on the climate of Europe. The elevated **sea-surface temperatures** (SSTs) have an obvious effect on the winds that blow across the ocean surface and onto the continent, and also, of course, on evaporation and rainfall. In fact, SSTs have a major influence on regional climates across the globe, and their variability is a major factor in interannual climatic variability. What causes sea-surface temperature to vary? Seasonal effects are obvious, of course, but we now understand that there are causes of variability that make one year's winter rainfall (for example) differ significantly from another. This is **interannual** variability.

In the mid-latitudes we understand that our weather is dominated by synoptic-scale disturbances such as high- and low-pressure cells, fronts, etc. These tend to pass across us in a week, more or less, and we certainly understand that weather is variable on such timescales. The ocean also experiences 'disturbances'. However, because of factors such as its much greater density, the timescales of these are generally measured in months to years, and even decades.

The best known of these '**modes of variability**' is ENSO – **El Niño and the Southern Oscillation** – an oscillatory phenomenon that takes place across the central Pacific, with a return period which varies from ~2 to ~7 years. This phenomenon has been known about for at least a century, although it has only been during the past few decades that scientists have been able establish the key details of its development, and the full range of its regional impacts.

13.3.1   History

In the 1800s Peruvian sailors and fisherman noted a coastal current that appears after Christmas in some years. For this reason, they named it El Niño, the boy, but by implication the Christ Child. This warmer water contained fewer nutrients than the cold water that usually upwelled from the deep, leading to smaller catches. In years in which this phenomenon started earlier and lasted longer, the effects were much worse.

In the 1920s Sir Gilbert Walker was working in India, on behalf of the British administration, on the problem of the monsoons – vital to the subcontinent's food supply – looking for correlations that might explain the variability from one year to another. He found a negative correlation between mean sea-level pressure in the western Pacific Ocean (Darwin, Australia) and eastern Pacific Ocean (Tahiti), which he named the **Southern Oscillation**. He later showed that this irregular oscillation is associated with changes in rainfall and winds.

The **Southern Oscillation Index** (SOI) is a monthly statistic defined by

$$SOI = 10\frac{P_{diff} - P_{diffav}}{SD(P_{diff})}$$

where P_{diff} is the average Tahiti MSLP – the average Darwin MSLP, P_{diffav} is the long-term average MSLP difference for the month, and $SD(P_{diff})$ is the long-term standard deviation of the difference for the month.

Let us explain what this means. The SOI is a 'statistically weighted' difference in the monthly mean sea-level pressures (MSLPs) at Tahiti and Darwin. Basically, it asks how the monthly pressure difference between Tahiti and Darwin differs from the normal range. The 'scale factor' of 10 (which is sometimes omitted) is for convenience, with the SOI ranging between about –35 and +35, and (usually) quoted to the nearest integer. Sustained negative values below –7 often indicate an El Niño, while sustained positive values above +7 indicate a La Niña.

Walker's work received little follow-up until the 1960s, when Jacob Bjerknes at UCLA looked at a number of atmospheric and oceanic parameters and hypothesised an ocean–atmosphere coupling that is key to the development of El Niño: the **Bjerknes hypothesis**. In 1975 Klaus Wyrtki (University of Hawaii) noticed that a rise in sea level in the western Pacific tended to precede an El Niño, but he focused entirely on the ocean.

In 1982–1983, the biggest El Niño of the century (to that time) caught the experts by surprise. As a result, the Tropical Ocean–Global Atmosphere (TOGA) program was established in 1985 by the World Meteorological Organization's World Climate Research Program, and the International Council of Scientific Unions, involving a comprehensive range of oceanographic and satellite observations. By the end of TOGA in 1995, prediction schemes had moved from the research community to the national weather services.

13.3.2 Phenomenology

We now understand the importance of the coupling between the ocean and the atmosphere, and as a result we refer to the phenomenon as ENSO: El Niño–Southern Oscillation. ENSO is essentially an oscillatory phenomenon. Under 'normal conditions', there is a warm pool of water in the western central Pacific: the Tropical Warm Pool. The prevailing trade winds blow from east to west, as we saw in Chapter 7, pushing this warm pool further west. (One consequence is that sea level is about 50 cm higher in the western Pacific than the eastern Pacific.) This allows cold, nutrient-rich water to rise off the coast of South America.

In so-called El Niño years, these trade winds weaken, and may even reverse. Hence the eastern central Pacific becomes abnormally warm, and the nutrient-rich deep water is unable to rise, leading to a reduced Peruvian fish catch. In La Niña years (the girl) the opposite happens,

Figure 13.3 Walker circulation for neutral, El Niño, and La Niña conditions. *Source: Physics of Radiation and Climate,* figure 6.9, Box and Box, CRC Press, 2016, reproduced by permission of Taylor & Francis Group.

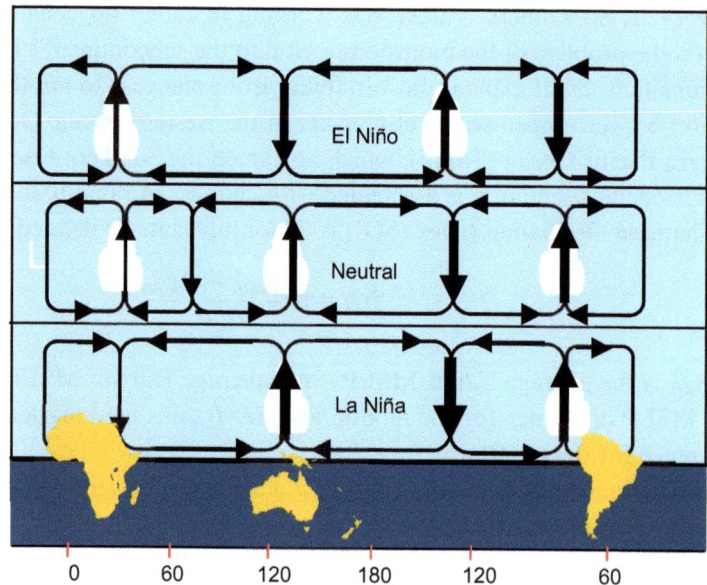

and the trade winds actually strengthen, pushing the warm pool further west, and making the waters in the eastern tropical Pacific colder than usual. This reversal of the trade winds is a central feature of the Southern Oscillation.

Atmospheric scientists now study sea-surface temperature anomalies in the central Pacific using satellite and other data. For convenience, they have divided this region into a number of boxes, as follows. NINO1+2: 0–10°S, 90°W–80°W; NINO3: 5°N–5°S, 150°W–90°W; NINO4: 5°N–5°S, 160°E–150°W; and NINO3.4: 5°N–5°S, 170°E–120°W. NINO3.4, which overlaps 3 and 4, is the most often cited. Generally, a high NINO3.4 index corresponds to a low SOI and vice versa, although the SOI contains more high-frequency variability, because the atmosphere varies on shorter timescales than the ocean.

If we examine in 3D the pressure/wind patterns (observations) in the south and central Pacific region, the first feature we would find is the Hadley cell, where the air rises near the equator – along the ITCZ – and descends at around 30°S latitude. If we then 'subtract' this circulation from our data, we find a secondary feature: the so-called Walker circulation, which is illustrated in Figure 13.3. Another way to think about these two components is that the Hadley circulation gives the north–south circulation after averaging over longitude, while the Walker circulation is the east–west circulation after averaging over latitude.

Under normal conditions, the warm water in the western tropical Pacific produces low pressure to Australia's north – and hence a (relatively) low MSLP at Darwin – and rising air, while the cold water off the coast of Peru produces high pressure and sinking air – and hence a (relatively) high MSLP at Tahiti. These two arms are connected by the easterly trade winds at the surface, and westerlies aloft: a typical thermal circulation pattern. These are the conditions that contribute much of the input to P_{diffav}. During an El Niño, the central Pacific is particularly warm, with low pressure there and rising air. The circulation pattern now has two descending branches, one in the eastern Pacific and one in the far western Pacific. This leads

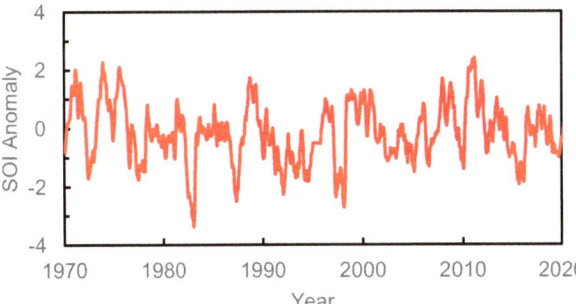

Figure 13.4 Time series of the Southern Oscillation Index. *Source*: NOAA.

to a higher MSLP at Darwin, and hence a negative SOI. During a La Niña, the increase in rising air over northern Australia is associated with lower MSLP at Darwin, and hence a positive SOI.

The National Oceanic and Atmospheric Administration (NOAA) has an enormous database of atmospheric and oceanic data. Figure 13.4, based on those data, shows the SOI from 1970 through to 2020. To reduce the complexity (high-frequency variability), we have employed a three-month rolling averaging approach.

13.4 Climatic Impacts of ENSO

13.4.1 Global Impacts

The warming of the central Pacific during an El Niño event is largely the result of a redistribution of heat within the upper layers of the tropical Pacific Ocean. Because of this enhancement in SSTs, there is a transfer of heat from the ocean to the atmosphere, via increased fluxes of both latent and sensible heat. As a result of this, one signature of a strong El Niño event is an increase in global average temperature. This, of course, is not climate change but climate variability. A large La Niña event has a contrasting impact, with a slight lowering of global average temperatures.

13.4.2 Regional Impacts

The Walker circulation and its perturbations are central to many of the climatic effects of El Niño and La Niña, with couplings to the large-scale atmospheric motions. One example is the flow of the jet streams over North America during winter. We quote from the NOAA/NWS website:

During La Niña, a variable Pacific jet stream in association with a polar jet stream shifted further south favours below normal precipitation across the southern US, with below normal temperatures across the northern US. During El Niño, a strong and amplified Pacific jet stream extending across the southern US in association with a polar jet stream shifted further north into Canada favours above normal precipitation across the southern US, and above normal temperatures over the northern US.

This is illustrated in Figure 13.5 from their website.

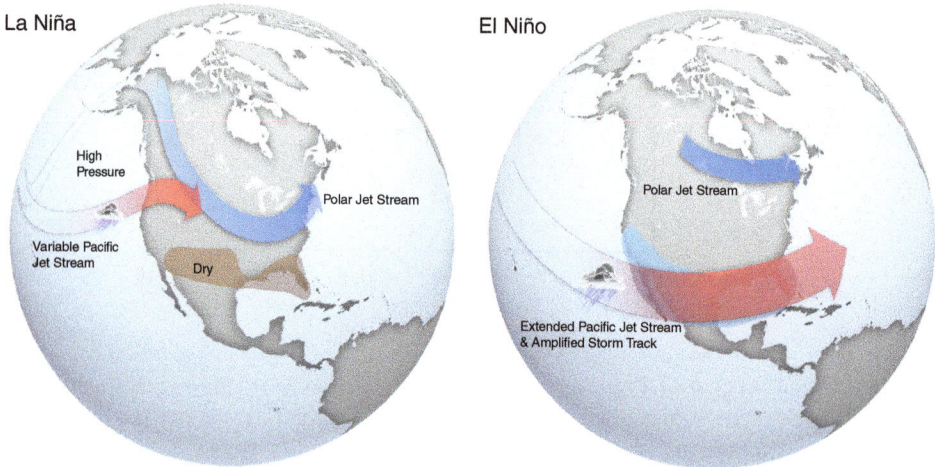

Figure 13.5 Typical late autumn through early spring upper-level Jet Stream positions associated with moderate to strong La Niña (left) and El Niño (right) events. *Source*: NOAA/NWS website.

The Oklahoma (and surrounding states) 'dust bowl' of the mid- to late 1930s is an event etched in history and literature (John Steinbeck, Woody Guthrie). The start of the drought, 1933–1935, occurred during a strong La Niña. Others followed in 1936 and 1939–1940. The secondary cause was poor land management practices: clearing the ground for agriculture made it susceptible to wind erosion under low rainfall conditions. (A mistake, or accident of timing, which is far from unique.)

Recent research has shown another interesting impact of ENSO. During a La Niña, beaches on the east coast of Australia tend to be denuded of sand, due to wave action, while beaches in the Americas tend to be replenished. During an El Niño the opposite happens.

13.4.3 Case Study: 2015–2016 Extreme El Niño

This case study is based on some of the material in box 11.4 of AR6's chapter 11: Weather and Climate Extreme Events in a Changing Climate. We cover much of that chapter in our Chapter 19, but feel that this example belongs here.

An extreme El Niño has the potential to produce multiple extreme events in different parts of the world. As we will see in later chapters, global warming due to greenhouse gas increases can exacerbate extremes in many parts of the world, even under normal El Niño conditions. The 2015–2016 El Niño was one of three extreme El Niños since the 1980s, and the availability of satellite rainfall observations. By some measures it was the strongest El Niño in the past 145 years, and this exceptional warmth was *unlikely* (IPCC terminology: Section 1.5) to have occurred entirely naturally. Both the ENSO amplitude and the frequency of high-impact events are higher since 1950 than over the pre-industrial period (*medium confidence*), which suggests that we are likely to see more of the extremes associated with this event as the climate warms.

Table 13.1 Some events associated with the 2015-2016 extreme El Niño (*source*: see Acknowledgement at the end of the chapter).

Region	Period	Events
Indonesia	July 2015 to June 2016	Droughts, forest fires
N. Australia	Late 2015 to early 2016	High temperature and drought
Amazon	September 2015 to May 2016	Droughts, forest fires
South America north of 20°S	Austral spring and 2015–2016 summer	Droughts
Ethiopia	February–September 2015	Droughts
Southern Africa	November 2015 to April 2016	Droughts
Europe	Boreal 2015–2016 winter	Effects on circulation patterns
India	May 2016	High temperature
	December 2015	Extreme rainfall
China	June–July 2016	Extreme rainfall
W. North Pacific	Boreal summer 2015	13 Category 4–5 cyclones
E. North Pacific	Boreal summer 2015	Record number of cyclones
Global	2015–2016 El Niño	High CO_2 release from fires

Table 13.1 {box 11.4, table 1} lists some of the key events associated with this El Niño. Several regions were strongly affected by droughts, including Indonesia, Australia, the Amazon region, Ethiopia, southern Africa, and Europe. Indonesia also experienced forest fires, with pronounced impacts on the economy, ecology, and human health. The Amazon region experienced an extreme drought, impacting most of tropical South America. The incidence of forest fire increased by 36% compared to the previous 12 years, which also increased the CO concentration in the area. Tropical cyclone activity was notably high in the North Pacific, with the western region recording twice its average of Category 4 and 5 storms. Atmospheric CO_2 growth rate was particularly high in 2015, possibly related to the droughts. Overall, tropical forests were a source of carbon to the atmosphere during 2015–2016.

13.5 Other Modes of Variability

While ENSO is by far the best known, and so far the best understood, of the oceanic modes of variability, there are a number of others that have been more recently identified, and studied in increasing detail. In this section we will present a brief outline of several of those that are important in different parts of the world. The definitions and understandings of some of these modes do change over time, as better data and analysis become available, including using proxy data to look back in time. The IPCC now includes a discussion of some of these modes, and any perceived changes, in chapter 3: Human Influence on the Climate System

13.5.1 Indian Ocean Dipole

The Indian Ocean Dipole (IOD) is a coupled ocean–atmosphere phenomenon in the equatorial Indian Ocean that affects the climate of countries that surround the Indian Ocean

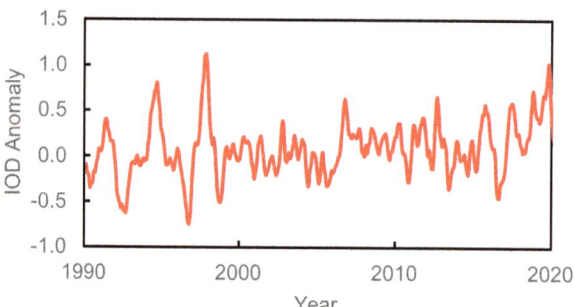

Figure 13.6 Dipole Mode Index of the Indian Ocean Dipole. *Source*: NOAA.

basin. The IOD is commonly measured by an index based on the difference between sea-surface temperature (SST) anomalies in the western (50°E to 70°E and 10°S to 10°N) and eastern (90°E to 110°E and 10°S to 0°S) Indian Ocean. The index is called the **Dipole Mode Index** (DMI).

A positive IOD period is characterised by cooler than normal water in the tropical eastern Indian Ocean and warmer than normal water in the tropical western Indian Ocean. A positive IOD SST pattern has been shown to be associated with a decrease in rainfall over parts of central and south-east Australia. Conversely, a negative IOD period is characterised by warmer than normal water in the tropical eastern Indian Ocean and cooler than normal water in the tropical western Indian Ocean. A negative IOD SST pattern has been shown to be associated with an increase in rainfall over parts of southern Australia. Figure 13.6, again using NOAA data, shows this index from 1990 to 2020, again using a rolling average.

The discovery of the IOD is a fascinating and illuminating story. ENSO is important for Japan's rice harvest, so that their meteorological agencies endeavour to provide forecast guidance to their farmers. In the 1990s one forecast turned out to be well astray, so the scientists set out to find an explanation. They concluded that the Southern Oscillation Index, a key element for forecasting El Niño, was 'contaminated'. Remember that it is based on the difference in pressure between Tahiti and Darwin. Now Tahiti is certainly in the Pacific, but Darwin lies mid-way between the Indian and Pacific Oceans. Thus, Indian Ocean variability must exert some influence on the pressure at Darwin, and hence on the SOI. Their careful analysis opened up a new area of investigation.

Although it was only formally identified in 1999, other evidence was there when you knew where to look, and had a reason to look. Data from fossil coral reefs suggest that the IOD has operated since the mid-Holocene, 6,500 years ago. An average of four, each positive/negative IOD events, occur during each 30-year period with each event lasting around six months. However, there have been 12 positive IODs since 1980 and no negative events from 1992 until a strong negative event in late 2010. The occurrence of consecutive positive IOD events is extremely rare, with only two such events recorded: 1913–1914 and the three consecutive events from 2006 to 2008 which preceded major bushfires in south-east Australia (Bureau of Meteorology website). The latter half of 2019 saw a large positive IOD, with devastating consequences for countries around the Indian Ocean, including Australia, Indonesia, South Sudan, Somalia, and Kenya. Some of these will be examined in Chapter 19.

13.5.2 Southern Annular Mode

The Southern Annular Mode (SAM), also known as the **Antarctic Oscillation** (AAO), describes the north–south movement of the westerly wind belt that circles Antarctica, dominating the weather patterns of the middle to higher latitudes of the Southern Hemisphere (e.g. the roaring forties). The changing position of the westerly wind belt influences the strength and position of cold fronts and mid-latitude storm systems, and is an important driver of rainfall variability in southern Australia.

In a positive SAM event, the belt of strong westerly winds contracts towards Antarctica. This results in weaker than normal westerly winds and higher pressures over southern Australia, restricting the penetration of cold fronts inland, and hence reducing rainfall. Conversely, a negative SAM event reflects an expansion of the belt of strong westerlies towards the equator. This shift results in more/stronger storms and low-pressure systems over southern Australia.

During autumn and winter, a positive SAM can mean cold fronts and storms are pushed farther south, and hence southern Australia generally misses out on rainfall. However, in spring and summer, a strong positive SAM can mean that southern Australia is influenced by the northern half of the high-pressure systems, and hence there are more easterly winds bringing moist air from the Tasman Sea, turning to rain as the winds hit the Great Dividing Range.

In recent years, a high positive SAM has dominated during the austral autumn–winter, and has been a significant contributor to the 'big dry' observed in southern Australia from 1997 to 2010. Some modelling studies have suggested that the emergence of the Antarctic ozone hole may be having an impact on the SAM in ways that reduce Australian rainfall.

One of the many questions the IPCC poses for itself is to seek out any changes in the various modes of variability. Its latest conclusion is that 'since the late nineteenth century, major modes of climate variability show no sustained trends, with the notable exception of the Southern Annular Mode, which has become systematically more positive (*high confidence*)'. This is not good news for agriculture in southern Australia.

Figure 13.7 shows the mode index data taken from the NOAA website. We have plotted a trend line to show this effect.

13.5.3 Pacific Ocean Modes

Pacific Decadal Variability (PDV) is the generic term for the modes of variability in the Pacific Ocean that can vary on decadal to inter-decadal timescales. It includes the Pacific Decadal Oscillation (PDO) – an anomalous sea-surface temperature pattern in the North Pacific – as

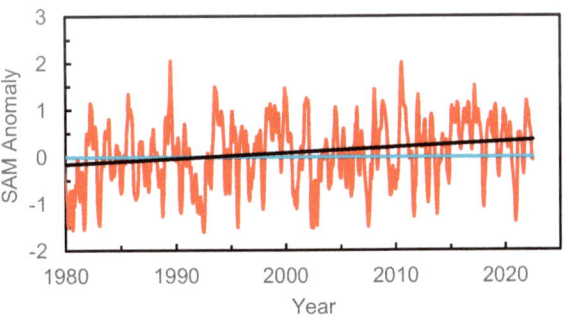

Figure 13.7 SAM Index, including trend line. *Source*: NOAA.

Figure 13.8 The PDO index. *Source*: NOAA.

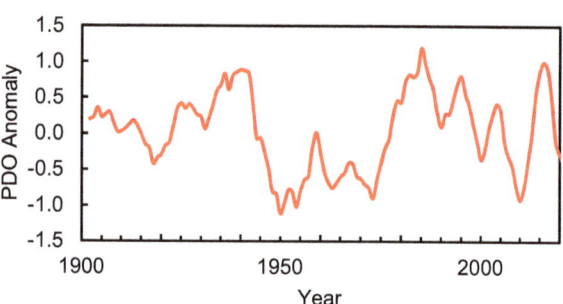

well as a broader structure associated with Pacific-wide SSTs termed the Interdecadal Pacific Oscillation (IPO).

The **Pacific Decadal Oscillation** is an anomaly in the SST distribution in the Pacific Ocean, north of 20°N. During the warm or positive phase, the western Pacific becomes cool, and part of the eastern Pacific warms: during a cool or negative phase, the opposite pattern is observed. It shifts phase on a timescale of 20–30 years. Tree-ring data have been used to reconstruct the PDO back to 1661. During its positive phase, temperatures in the north-west Pacific and Alaska tend to be above average, while Mexico and south-east USA are below average. Precipitation is above average in the Alaska coastal range and Mexico and south-west USA, while it is below average in Canada, eastern Siberia, Australia, and the Indian summer monsoon. In its negative phase these trends are reversed. Figure 13.8 presents the PDO index, again using data from NOAA, with a rolling average.

During the first decade or so of the present century, the rate of 'global warming' appeared to have slowed (or even stopped). While part of the explanation lies in the anomalously warm year of 1998, which can be traced to a very strong El Niño, a more important factor is the increased fraction of trapped heat entering the oceans, with one study pointing a finger at the PDO. During the positive phase from 1976 to 1998, much of this heat resulted in a warming of the surface layers. However, since the PDO switched to a negative phase in 1999, more than 30% appears to have penetrated below 700 m. This suggests that 'natural decadal variability modulates the rate of change of global surface temperatures'.

The **Interdecadal Pacific Oscillation** is a similar pattern of SST and MSLP anomalies, which affects both the North and South Pacific. It also has a cycle time of 15–30 years. When the IPO is in its positive phase, the tropical Pacific is warmer than average, while the northern and southern Pacific are cooler than average. This tends to produce more El Niños. When in its negative phase, the reverse occurs, and there tends to be more La Niñas. It was in a positive phase from 1913 to 1944; a negative phase from 1945 to 1976; a positive phase from 1977 to 1998; and a negative phase from 1999 to 2013. Figure 13.9 shows annually averaged and five-year averaged values of its index for the past 150 years.

13.5.4 Atlantic Ocean Modes

The **North Atlantic Oscillation** (NAO) is a reversal of pressure patterns over the Atlantic that affects the weather of Europe and the east coast of North America. In winter, if atmospheric pressure in the vicinity of the Icelandic low drops, and pressure in the region of the

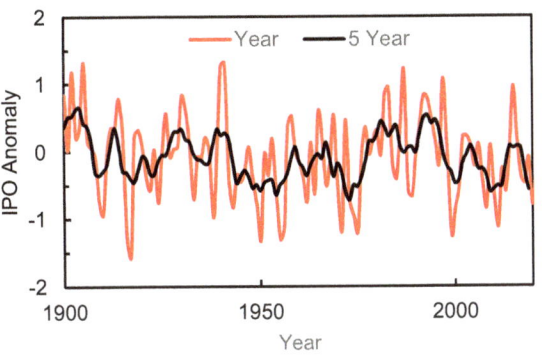

Figure 13.9 The IPO index (data from https://psl.noaa .gov/gcos_wgsp/Timeseries).

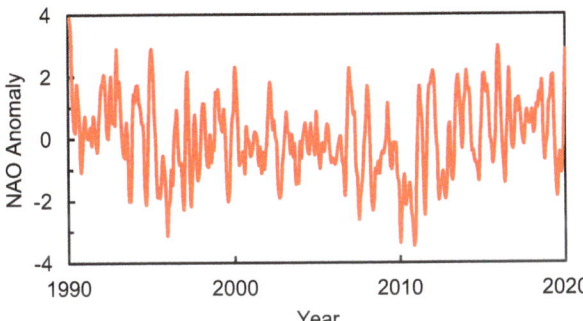

Figure 13.10 NAO anomaly. *Source*: NOAA.

Bermuda–Azores high rises, we are in the positive phase. This enhancement of the pressure gradient leads to a strengthening of the westerly winds, which direct strong storms on a more northerly track into northern Europe, where winters are normally wet but mild (due, of course, to the Gulf Stream). Winters in southern and central Europe are mild and relatively dry. Winters in the eastern United States tend to be wet and relatively mild, while northern Canada and Greenland are usually cold and dry.

The negative phase involves a rise in pressure around the Icelandic low, and a drop around the Bermuda–Azores high. This reduces the pressure gradient, and weakens the westerlies, leading to weaker winter storms. Southern Europe and the region around the Mediterranean Sea are wet, while northern Europe is usually cold and dry. Greenland and northern Canada experience milder winters.

The NAO is a relatively short-term phenomenon, but it may remain in one phase for several years. Over the past ~30 years there has been a trend to more of the positive phase. Figure 13.10 shows the NAO index, based on NOAA data, and a rolling average. These data still show a significant high-frequency structure, which is of minor significance.

Closely related to the NAO is the **Arctic Oscillation** (AO), involving changes in the pressure gradient between the Arctic and regions further south. This leads to changes in upper-level westerlies, with consequences similar to the NAO. These phenomena are sometimes also referred to as the Northern Annular Mode (NAM), although it remains unclear whether they are distinguishable, or just variations on a theme.

Atlantic Multidecadal Variability (AMV) refers to a climate mode representing basin-wide multidecadal fluctuations in surface temperatures in the North Atlantic. The Atlantic Zonal

Mode (AZM), also known as the Atlantic Equatorial Mode, or Atlantic El Niño, and the Atlantic Meridional Mode (AMM) are the two leading patterns of variability in the tropical Atlantic. Both are associated with changes in the strength and location of the ITCZ, and affect the African and American monsoon systems.

13.6   Case Study: Australian Weather Extremes

Roughly half of Australia – particularly the centre and west – is arid or semi-arid (as defined by a mean annual rainfall of 250 mm or less), and is the second driest continent (after Antarctica). Its interannual variability is impacted by three of the modes just discussed.

Australia is in one of the primary regions affected by the ENSO phenomenon, and Australian scientists from the CSIRO, Bureau of Meteorology (BoM), and universities have been at the forefront of studies into El Niño/La Niña and its impacts, including participating in TOGA. Eastern Australia has a long history of prolonged droughts, broken by devastating floods. One of Australia's best loved poems is *My Country* by Dorothea Mackellar: it refers to Australia as a land 'of droughts and flooding rains'.

During normal conditions, the rising arm of the Walker circulation in the west Pacific brings good rainfall to eastern Australia. However, during a strong El Niño, this rising arm moves to the central Pacific, and Australia experiences drought. Major droughts are well correlated with strong El Niño events. On the other hand, a strong La Niña enhances the region's rainfall, as happened in 2010/2011 when the Brisbane area received its second once-in-a-century flood in 40 years. This was, in fact, Australia's wettest two-year period on record, at least until then. The year 2022 was equally devastating.

13.6.1 The Climate Dogs

The climate modes we have been discussing (plus others) have a significant impact on the productivity of Australian agriculture, and farmers are well aware of the consequences for their decision making, and their viability. As they are not trained scientists (although many study what they can), they look to those who are for guidance. The Departments of Primary Industry (DPI) in several states (New South Wales and Victoria) have worked hand in glove with the BoM to produce such guidance in an easily digestible form: the upshot has been encapsulated in the 'Climate Dogs'. The following is taken verbatim from the NSW DPI website (where you can find short YouTube presentations), with some additional comments where relevant.

RIDGY, otherwise known as the Subtropical Ridge (of the general circulation), is the lead dog of the pack. RIDGY's position and intensity have a significant influence on weather in NSW. Recent changes in RIDGY's behaviour appear to be driving some significant changes to southern NSW rainfall patterns. (This could be consistent with global warming.)

MOJO [the Madden–Julian Oscillation] can have a big influence on Australia's weather and climate, especially during the warmest months of the year. [We have not covered this one, as it is relatively small-scale, and short-term.]

ENSO represents the El Niño Southern Oscillation phenomena. Changes in ENSO's behaviour have a significant influence on rainfall probabilities in inland NSW during the winter and spring period.

INDY represents the Indian Ocean Dipole. Like ENSO, changes in INDY's behaviour also have a significant influence on rainfall probabilities in inland NSW during winter and spring.

SAM represents the Southern Annular Mode and is a complex climate dog. Recent changes in SAM's behaviour increase probabilities of rainfall in spring and summer in some parts of NSW.

EASTIE, the East Coast Low phenomenon, represents the deep low-pressure systems that are an important climate feature along the south-east coast of Australia. [This one is quite localised, and can form with little warning along the coast of New South Wales, often close to Sydney. It drives powerful onshore winds and storm surges, which can cause heavy rainfall, even hail, beach erosion which has caused dramatic property damage, and even driven a bulk carrier aground. Because of its intensity it is sometimes referred to as a 'bomb cyclone'.]

Summary

The oceans play a significant role in both regional and global climate. The surface currents within the major ocean basins are a major component of the transport of heat from tropical to polar regions. The deep ocean circulation – the AMOC, or thermohaline circulation – which includes the Gulf Stream, is vital to the comfort of western Europe and surrounding regions.

Just as important, at least on regional scales, are the various modes of variability that are found in different oceans, with El Niño/La Niña the best known. These are the major source of climate variability around the world, including both droughts and floods. The fact that these seem to be getting worse will be explored in Chapter 19.

Acknowledgements

Table 13.1:

Climate Change 2021: The Physical Science Basis. Contribution of Working Group I to the Sixth Assessment Report of the Intergovernmental Panel on Climate Change [Masson-Delmotte, V., P. Zhai, A. Pirani, S.L. Connors, C. Péan, S. Berger, N. Caud, Y. Chen, L. Goldfarb, M.I. Gomis, M. Huang, K. Leitzell, E. Lonnoy, J.B.R. Matthews, T.K. Maycock, T. Waterfield, O. Yelekçi, R. Yu, and B. Zhou (eds.)]. Cambridge University Press, Cambridge, United Kingdom and New York, NY, USA, 2,391 pp. doi:10.1017/9781009157896.

Figures 13.4, 13.6, 13.7, 13.8, 13.9, 13.10:

Author plot – NOAA data: https://psl.noaa.gov/gcos_wgsp/Timeseries
We wish to thank NOAA for all the valuable information freely available on their website.
We also thank them for the wonderful science undertaken to benefit us all.

FURTHER READING

Physical oceanography is a broad subject, and we have only been able to scratch the surface. For more information the following would be useful:

- *Oceanography: An Invitation to Marine Science* by T. S. Garrison (Cengage, 2015).
- *Essentials of Oceanography* by A. P. Trujillo (Pearson, 2016).
- *Introduction to Physical Oceanography* by J. A. Knauss (Waveland Pr Inc, 2016).
- Abyssal ocean overturning slowdown and warming driven by Antarctic melt water by Qian Li et al., *Nature*, 615, 841–847, 2023.

Of the references noted in previous chapters, one useful book is

- *Elementary Climate Physics* by F. W. Taylor (Oxford University Press, 2005).

An excellent treatment of El Niño can be found in

- *Climate Change and Climate Modelling* by J. D. Neelin (Cambridge University Press, 2011).

REVIEW QUESTIONS

1. Do ocean currents rotate clockwise or counter-clockwise in the Southern Hemisphere?
2. What are the drivers of the AMOC?
3. Explain all four words in the acronym, AMOC.
4. The Southern Oscillation Index involves air pressure measured in what locations?
5. Why does global average temperature usually rise during an El Niño event?
6. Why is the ocean mode in the Indian Ocean referred to as a dipole?

EXERCISE

What guidance can you find for the local agricultural community in your region (either home or school/college) on seasonal climatic conditions? This might be provided by your weather service, or your Agriculture Department. How useful do you think it is?

14 Interactions and Feedbacks in the Climate System

We experience weather in the air around us, and that implies climate: being, after all, the statistics of weather. But if we are to understand climate, and especially how it changes – for any reason, and on any timescale – we need to examine all its interactions with the rest of the planet. Any changes to the flows of energy into or out of the atmosphere have the potential to change the atmosphere, and the climate. Similarly, any flows of chemical species may also change the state of the atmosphere.

In earlier chapters we looked specifically at flows of chemicals such as CO_2, aerosols, and a variety of pollutants. In the previous chapter we looked at the role of sea-surface temperature, and how ocean modes lead to regional climate variability. If we are to go further, both in terms of time and in our ability to understand how climate might change, we need to look at other components of our planet and their atmospheric interactions.

When we see the range of interconnections involved, we realise that we are dealing with a somewhat complex system, with multiple interactions between its components. Whenever we see such a system, we must realise the potential for one interaction between components to end up impacting back on one of these: this is known as feedback, and is ubiquitous in all such systems. If we are to understand climate, we must understand these feedbacks.

In the last sections we will illustrate one of the most important of these processes with a 'relatively simple' model which we invite you to code up and explore.

14.1 Components and Interactions

Our planet and its climate work as a system, consisting of many parts. The most obvious component is the atmosphere. As we discussed in some detail in Part II, the atmosphere is governed by two branches of physics: thermodynamics and fluid mechanics. To these we must add the effects of radiation emission and absorption. Now we have enough science to produce a weather model (which we will discuss in Chapter 15).

In the previous chapter we discussed the contribution of the ocean and its role in climate variability. In this section we will introduce two of the other subsystems of the climate system.

They do not obey 'simple' physical laws, unlike the atmosphere and ocean, so our treatment must be qualitative.

14.1.1 The Cryosphere

The cryosphere consists of water in its frozen state: floating ice; land ice, including glaciers; snow; and permafrost. Although water droplets in the atmosphere may remain liquid even for temperatures 20 or more degrees below freezing, this is not the case on land. So, although temperature is the major determinant of frozen water, we need to remember its large latent heat, meaning that there can be a time lag when temperature changes.

When we consider the cryosphere as a lower boundary of the atmosphere, one of its most important properties is its high reflectivity, or albedo. We touched on this briefly in Chapter 9. Thus, little solar insolation is absorbed at ground level, which contributes to the maintenance of ice and snow. Away from the major ice sheets of Antarctica and Greenland, the cryosphere undergoes a seasonal cycle, and this must have an impact on the atmosphere above. When a layer of ice or snow melts, solar radiation now reaches the land or ocean surface underneath. That surface will certainly absorb more heat, with obvious implications for the local energy budget and the atmosphere above.

Glaciers are frozen rivers, and actually flow, although at barely perceptible rates. This includes the glaciers that drain the Antarctic and Greenland Ice Sheets. Understanding how they flow is a very inexact science. It depends on the details of the underlying bedrock, surface heating, and the penetration of heat from below, especially when they actually extend out into the South or North Atlantic Oceans. We are learning more, but slowly.

There is one component of the cryosphere that deserves an additional comment, and that is permafrost. In the northern polar region, the underlying soil and rock are mostly at temperatures below 0°C, and contain frozen water: a layer known as **permafrost**. In summer, temperatures are usually high enough to melt the top few metres, depending on latitude, so that the tundra turns swampy and muddy. As the globe warms, this thaw zone is extending northward, causing structures that had been 'secure' on solid ground to sink. Global warming is not a blessing for Siberia, despite some claims to the contrary.

14.1.2 The Land Surface

The land surface primarily consists of soils and plants, plus, of course, man-made constructions of roads and buildings. Soils might be sandy – too porous to hold water, so not ideal for plant growth – or clay, which is too impervious. In between lies loam, which is considered excellent soil. Plant cover varies from near zero in desert regions, to multilayered in tropical rainforests. Crops, grasslands, shrublands, and temperate forests lie somewhere between.

The first interaction between the land surface and the atmosphere relates to rainfall, soil moisture, runoff, and evaporation plus transpiration, known as **evapotranspiration**. These processes operate on timescales of days, and require a knowledge of the relevant land-surface characteristics – soil, plant cover, rainfall history – at the local level. In agricultural regions, as well as deciduous forests, these characteristics may be seasonal. We can build relatively simple empirical models, consisting of rainfall, plant uptake, evapotranspiration (which might depend on relative humidity and wind speed), river runoff, and soil moisture, at one or more depths, depending on sophistication. Clearly such models require local data on surface characteristics.

As we saw in Section 7.1, winds blowing over the Earth's surface experience a frictional force, also known as **wind stress**, and hence blow across the isobars. The crossing angle varies, depending on the strength of the frictional force, which depends on the surface's physical characteristics: known as **surface roughness**. Surface roughness may also impact low-level turbulence, which is important in the transport of water vapour up through the atmospheric boundary layer. Again, the relevant characteristics require local data, and may be seasonal.

Vegetation changes due to seasonal growth and decay, or growth and harvesting, are well understood by science. But what of longer-term changes? How might the biosphere change in response to changes in CO_2 content, or rising temperatures, or hydrological cycle changes? Or, more to the point, all three? How might these changes impact different components of the total biospheric system? What might be the impact on an ecosystem if certain key animal species migrate faster than plant species? All of these are important questions that science is now grappling with, and are well beyond the scope of this book.

14.1.3 The Coupled System

Figure 14.1 is a schematic diagram of the main subsystems, and interactions, that make up the full Climate System, or Earth System. (The full Earth System should also include the geosphere, as discussed in Chapter 3 in regard to the long-term carbon cycle, and us, given our current impacts.) As well as the various subsystems we have described so far – atmosphere, ocean, cryosphere, land surface – this diagram indicates how some of the (bio)chemical processes covered in earlier chapters also contribute. (Dashed lines indicate connections that mainly operate on longer timescales.)

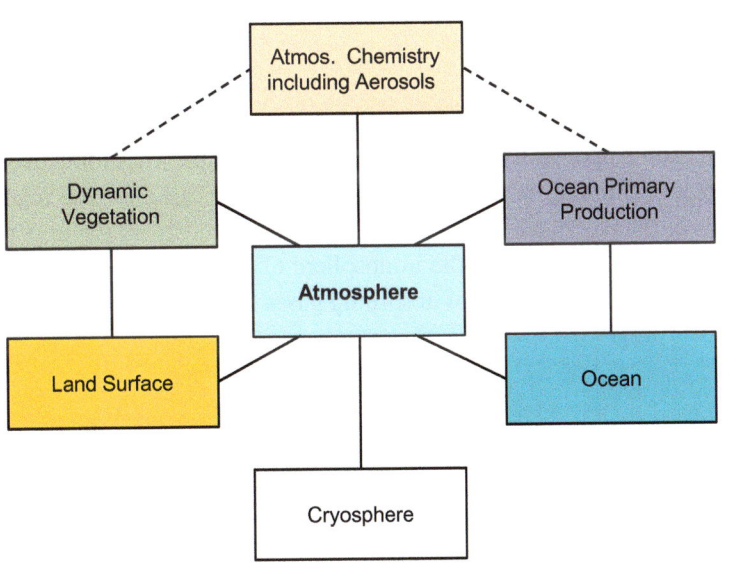

Figure 14.1 The Climate System with its subsystems and interconnections.

14.2 Atmospheric Feedbacks

We turn now to feedback processes. Any time you 'force' a complex system, via some input stimuli, it will respond, and there will most likely be outputs. And some of these may, either directly or indirectly, act as further stimuli to the system. This is what we mean by a feedback process, and is illustrated schematically in Figure 14.2.

The climate system is an example of such a system, and is currently being forced by a range of stimuli, with the emissions of greenhouse gases the prime example. As a consequence, we should be on the lookout for feedback processes. For convenience we have divided these into two groups: those that take place in the atmosphere; and those that take place at the surface.

14.2.1 Water Vapour Feedback

The Clausius–Clapeyron equation tells us that the saturation vapour pressure increases in a roughly exponential manner with temperature – approximately 7% per 1°C – so that a warmer atmosphere will be able to hold more water vapour, trapping additional long-wave radiation, increasing the temperature, and potentially leading to a further increase in water vapour. This is just what is meant by (positive) feedback.

Can we quantify this? If the atmosphere was always fully saturated with water vapour, and could be assumed to remain so, this would be easy. In fact, we know full well that relative humidity varies on many scales, mainly for dynamical reasons: convection in the boundary layer, and the general circulation. This maintains lower tropospheric humidity in the vicinity of 75% for much of the time. The simplest realistic assumption we can make is that average relative humidity (RH) will remain at current values. Evidence to support this idea comes from the fact that RH tends to remain roughly unchanged when temperatures change, for example from season to season, and also from measurements of humidity trends over recent decades. The latest data will be presented in Part V.

14.2.2 Lapse-Rate Feedback

As we saw in Chapter 5, the lapse rate is a quantification of the vertical temperature profile. Why might this – or more importantly, changes to it – be important? The emission of thermal radiation increases with temperature, so that changes in lapse rate can alter the relative emission from different levels in the troposphere. Long-wave radiation emitted in the upper troposphere has a much easier passage to the top of the atmosphere (TOA), and hence to space, for the simple reason that it has a reduced profile of absorbing gases to penetrate (as indicated in Figure 11.4).

Figure 14.2 Schematic diagram of a feedback process.

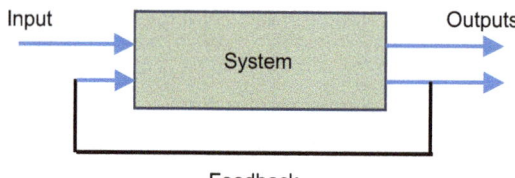

The lapse-rate feedback quantifies the change in long-wave radiation at the TOA due to a non-uniform change in the vertical temperature profile. In the tropics, the prime driver of this profile is moist convection, leading to a profile that is close to moist adiabatic. The result is a warming that is larger in the upper troposphere than lower down. So this results in a larger emission to space, and thus a negative feedback. Observations over the past 20 years confirm these conclusions.

In the extratropics, the vertical temperature profile is primarily driven by a combination of radiation, meridional heat transport, and ocean uptake. Strong temperature inversions in winter lead to a larger lower tropospheric warming (the opposite to the tropics), and a positive lapse-rate feedback in polar regions. On balance, the tropical contribution dominates, leading to a negative global mean lapse-rate feedback.

14.2.3 Cloud Feedbacks

While we have reasonable confidence that RH won't change very much *on average*, even small changes can have significant impacts on cloud cover. The dynamical and thermodynamic processes which control clouds do differ on a wide range of scales, from marine boundary layer turbulence to large-scale frontal systems (for example), so that clouds in different regions may respond quite differently to these changes. As we have seen earlier, these differences may affect their radiative properties.

For many years clouds have presented a challenge to climate modellers. They are smaller than a typical grid square, and so must be parameterised. They have opposite effects on short-wave and long-wave radiation. Low clouds reflect sunlight, contributing to the global albedo. In the infrared, they radiate much like the underlying surface (their temperature is not much lower than the surface temperature), and so have only a small long-wave effect. Hence, an increase in low clouds would constitute a negative feedback. High clouds reflect comparatively little sunlight, but tend to be relatively efficient absorbers and radiators of long-wave radiation, but at a lower temperature. Hence, an increase in high clouds would most likely provide a positive feedback. Thus, the central question is what type of clouds (if any) will change? Alternatively, will cloud fraction increase or decrease?

In an effort to answer these important questions, clouds have been one of the more active areas of research in climate science. Clouds, as we have noted, vary considerably, and for this reason the IPCC now provides different assessments for different types of clouds. Table 14.1 {table 7.9 of AR6} shows how our knowledge in this field has progressed in recent years. It shows the feedbacks for seven different 'types', or regional types, of clouds. The middle column shows the conclusions that had been reached by the time of the previous assessment (AR5), along with the confidence levels. The final column shows the latest conclusions. We see that some new types have been added to the decomposition. We also see that some conclusions have been strengthened, most importantly for the overall effect.

In AR6 we read {page 926}:

The net effect of changes in clouds in response to global warming is to amplify human-induced warming, that is, the net cloud feedback is positive (*high confidence*). Compared to AR5, major advances in the understanding of cloud processes have increased the level of confidence and decreased the uncertainty

Table 14.1 **Assessed sign and confidence level of cloud feedbacks in different regimes in AR5 and AR6. (For some cloud regimes, the feedback was not assessed in AR5, indicated by N/A.) (*Source*: see Acknowledgement at the end of the chapter.)**

Feedback	AR5 (confidence)	AR6 (confidence)
High-cloud altitude	Positive (*high*)	Positive (*high*)
Tropical high-cloud amount	N/A	Negative (*low*)
Subtropical marine low-cloud	N/A (*low*)	Positive (*high*)
Land cloud	N/A	Positive (*low*)
Mid-latitude cloud amount	Positive (*medium*)	Positive (*medium*)
Extratropical cloud optical depth	N/A	Small negative (*medium*)
Arctic cloud	Small positive (*very low*)	Small positive (*low*)
Net cloud feedback	Positive (*medium*)	Positive (*high*)

range in the cloud feedback by about 50%. An assessment of the low-altitude cloud feedback over the subtropical oceans, which was previously the major source of uncertainty in the net cloud feedback, is improved owing to a combined use of climate model simulations, satellite observations, and explicit simulations of clouds, altogether leading to strong evidence that this type of cloud amplifies global warming. The net cloud feedback, obtained by summing the cloud feedbacks assessed for individual regimes, is 0.43 [–0.01 to +0.94] W m^{-2} °C^{-1}. A net negative cloud feedback is *very unlikely* (*high confidence*).

(Note that AR6 attempts to quantify a number of feedbacks in units of W m^{-2} °C^{-1}. We have chosen not to, as we do not feel this information is of any real value to our readers. However, we have included the overall quantification, as it is part of the paragraph we are quoting.)

14.3 Surface-Level Feedbacks

Processes at the surface are much slower than the ones we have just examined. However, if we are interested in how climate might change in the decades and longer ahead, or in how climate has changed in the past, these processes, and any resulting feedbacks, become critical.

14.3.1 Ice–Albedo Feedback

Ice and snow are mainly found at high latitudes and/or altitudes, and for a very obvious reason: if the surrounding temperature is above 0°C the snow/ice will melt. (The actual time to melt will depend on both ambient temperature, and snow/ice depth.) If global temperatures increase, the snow line will be pushed to higher latitudes and altitudes, reducing the planetary snow/ice cover. Snow and ice have a very high albedo, and are second to clouds in their contribution to the overall planetary albedo. Thus, a reduction in cover will entail a reduction in albedo, which again implies a climate warming leading to more melting: again, a positive feedback.

Two small factors need to be kept in mind. Firstly, if the ice or snow surface happens to be overlain by relatively thick cloud, then it won't see much solar radiation to reflect. Secondly, much of the Earth's snow and ice cover is found in the winter hemisphere, at high latitudes, where solar insolation is low anyway. Spring is the key season, as snow/ice cover has reached its peak and solar insolation returns: reductions in spring snow and sea ice have been noted in recent decades.

There is one additional caveat we should add, perhaps in the form of a hemispheric disparity. The high northern latitudes tend to be snow-covered in winter, with thickness being least at the (southern) boundary. It is here that melting will usually begin, and progress to higher latitudes. These are precisely the conditions that suit this feedback. By contrast, the high southern latitudes are dominated by the Antarctic Ice Sheet, which is much thicker, and hence harder to melt. This is a major reason why the high northern latitudes are warming faster than the high southern latitudes.

When we turn our attention to millennial-scale time frames we also need to understand that large-scale melting of ice sheets will lead to a significant rise in sea level. If this rise covers a significant amount of land, then this may also constitute an albedo feedback. Alternatively, of course, the buildup of large ice sheets can only come at the 'expense' of the oceans.

14.3.2 Biospheric Feedbacks

The previous discussion has focused on processes that directly amplify, or dampen, the changes in radiation balances produced by increasing levels of greenhouse gases. On longer timescales there are processes that change the concentrations of these greenhouse gases, most notably biospheric feedbacks. An increase in atmospheric CO_2 should act as a spur to photosynthesis: known as CO_2 fertilisation. However, plants also need sunlight, water, and nutrients: CO_2 is rarely the limiting factor. A number of studies have been undertaken to investigate forest growth in a high CO_2 environment. They have generally shown an initial growth spurt, which has leveled off, suggesting only a small negative feedback.

On longer scales than dynamical changes are potential changes in the carbon cycle resulting from changes in ecosystems. A warming climate places pressures on both plant and animal communities, which may need to migrate to higher latitudes/altitudes. This is far easier for animals, especially birds and insects, than for forests: human actions may have made their migration impossible. As these are major stores of carbon, this is a potentially significant alteration of the biological carbon cycle, and hence a positive feedback. This process could be exacerbated by increased forest fires brought on by hotter and drier conditions.

Modelling has shown that a reduction in rainfall can lead to changes in plant variety with reduction in transpiration, feeding back on rainfall. Changes in plant variety might also affect surface roughness, resulting in changes in low-level atmospheric turbulence, again affecting rainfall. The replacement of taller trees and shrubs with greater leaf area, by low tundra plants during periods of increased snow cover such as an Ice Age glacial advance may have acted as a positive feedback to that process. These are processes on millennial timescales.

14.3.3 Methane Hydrate

There are large deposits of methane in hydrated form, tied to water molecules under pressure (also known as clathrates), trapped in sediments, under permafrost, and under the sea bed, much of it at high latitudes. The CH_4 originates from the decomposition of organic matter present in these sediments, over millions of years. These deposits are currently well contained. (Some nations are considering exploiting these reserves as an energy source.) However, if temperatures rise sufficiently, and they are rising faster at high latitudes, then some of this CH_4 could enter the atmosphere. While clearly a positive feedback, it really only works in one direction. Significant CH_4 release to the atmosphere during this century is considered '*unlikely*'. Longer term is a different matter, and such a change would take centuries to reverse: this is a good example of a **tipping point**.

14.4 Energy-Balance Climate Models

We will now illustrate one of the key concepts in this chapter by looking at a very simplified model which will prove useful in Chapter 16 when we try to understand the Ice Age cycles.

These days, when we think about attempts to understand our climate – to model it – we almost automatically think about the general circulation models that we discuss in the next chapter. But that has not always been the case, of course. The first reasonably reliable models date from the 1980s, and even then, the available computer power was very modest by today's standards. So, what did climate science consist of in earlier decades? While much of it was fundamentally descriptive, some attempts were made to construct models to try to increase our understanding of how energy flows 'create' the climate we observe.

We experience climate at ground level, of course, not the mid-troposphere, and our primary understanding of climate is how it varies with latitude. So, what is it that governs such variation? Ultimately it is energy flows and energy balances. The class of models that have been developed to understand these flows and balances are known collectively as Energy-Balance Climate Models (EBCMs): they are one-dimensional, with latitude the dimension. We will now outline the concepts, and then focus on one of the simpler versions, which we invite you to program for yourself, and play with.

The energy balance at any latitude belt is determined by three contributions: the inflow of solar/short-wave energy; the outflow of long-wave/thermal energy; and the transport of energy to/from other latitude belts. That's the basic concept: we now need to decide how to quantify these pieces. Some EBCMs try to do this as accurately as possible, often using quite sophisticated analysis in order to produce 'accurate' results, while others opt for simplicity and a focus on helping us to understand climate in a reasonably transparent way.

So, let's start to build a model. The energy balance at any latitude will be given by

$$E_B = E_{sw} - E_{lw} - E_T \qquad (14.1)$$

Here, E_{sw} represents the inflow of solar/short-wave radiation, E_{lw} represents the outflow of long-wave/thermal radiation, and E_T represents the transport of energy *out* of that latitude belt, which could be positive or negative, of course. Over a sufficient length of time – say, a year – the balance should be zero, or close to it, which we shall assume. Now comes the challenge: how do we characterise/quantify these three terms?

14.4.1 Insolation

We'll start with the simplest; the inflow of solar radiation. This will be given by

$$E_{sw} = Q(1 - R)s(\lambda) \tag{14.2}$$

where $Q = F_0/4$ (F_0 is the solar constant); R is the fractional reflectivity/albedo (which may or may not be latitude-dependent); $s(\lambda)$ is the latitudinal insolation factor (to account for the latitude variation); and λ is the latitude. We will find it more useful to define $x = \sin \lambda$.

The latitudinal insolation factor may be reasonably approximated by

$$s(x) = 1.241 - 0.723\,x^2 \tag{14.3}$$

Before we progress to the other terms, we should take a closer look at the reflectivity/albedo term, which is currently denoted by the parameter R. There are two basic contributors to the albedo: clouds, and the land surface. In cloudy regions, we assume the reflectivity to be r_c; in cloud-free regions we assume all insolation reaches the surface, which has a reflectivity r_s. We now need to assign a cloud fraction to each latitude (belt): $\eta(x)$, which may or may not be latitude-dependent (remember the hydrologic circulation). So now the reflected fraction is

$$R(x) = \eta(x)r_c + (1 - \eta(x))r_s \tag{14.2'}$$

Because we are interested in the energy absorbed at the surface, we define the absorbed fraction

$$A(x) = 1 - R(x)$$

and finally rewrite Equation (14.2) as

$$E_{sw} = QA(x)s(x) \tag{14.2''}$$

14.4.2 Long-Wave Term

Now we turn to the long-wave term. The first step to understanding this term is to, effectively, split it into two pieces: the upward emission from the surface, and the downward back radiation from the atmosphere to the surface. The first of these is simply given by the Stefan–Boltzmann equation, and depends only on surface temperature. The second part is not so simple, and a number of formulations have been provided: by Brunt, Sellers, and others. Budyko suggested a far simpler formulation for the *net* long-wave radiation (based on analysis of observations), and this is the only one we will provide:

$$E_{lw}(x) = a_1 + b_1 T(x) - \eta(x)[a_2 + b_2 T(x)] \tag{14.4}$$

where $T(x)$ is the surface temperature in °C, η is the fractional cloud cover, and

$$a_1 = 226 \quad a_2 = 48.4 \quad \text{Wm}^{-2}$$
$$b_1 = 2.26 \quad b_2 = 1.62 \quad \text{Wm}^{-2}\text{K}^{-1}$$

For a fixed cloud fraction, this expression simplifies, of course.

14.4.3 Transport

Finally, we turn to the transport term. The transport of energy into and/or out of a latitude belt may be brought about by three mechanisms: transport of latent heat in the atmosphere, transport of sensible heat in the atmosphere, and oceanic transport: formally we may write

$$E_T = L\Delta C_V + \Delta C + \Delta F \tag{14.5}$$

where L is the latent heat of vaporisation of water, ΔC_V represents the mass of water evaporated minus the mass of water which condenses in the air column, ΔC is the net loss of sensible heat carried out of the air column (and hence latitude belt) by air motions, and ΔF is the net loss of sensible heat carried out of the latitude belt by ocean currents.

In principle, we should try to quantify all three, which Sellers did (and for different latitude belts). However, once again, we will turn to the very much simpler parameterisation from Budyko, which brushes it all under the carpet:

$$E_T = c[T(x) - T_{av}] \tag{14.5'}$$

where T_{av} is the global average surface temperature, and $c = 3.8$ W m^{-2} K^{-1}.

14.4.4 Surface Albedo

We now have a set of equations that we can code up, and study how our model behaves. However, there is one key factor in our model that we haven't fully addressed, and that is the surface albedo, which depends on ice cover. Ice cover clearly depends on temperature, which in turn depends on latitude, via the energy balances. This is one of the key foci of these models.

Ice and snow have an albedo of ~0.8, which is much higher than ocean (~0.1) or land (~0.12 for forests, ~0.3 for deserts). While the distribution of land and ocean as a function of latitude and hemisphere may be specified if we so choose, what about ice/snow? This depends on temperature, and several approaches have been suggested. In the simplest we assume an 'ice line' exists at latitude x_s and that

$$r(x) = \begin{array}{ll} r_{ice} & x > x_s \\ r_{if} & x < x_s \end{array} \tag{14.6a}$$

Suggested values are $r_{ice} = 0.6$ and $r_{if} = 0.3$. The question then turns to the location of the ice line. Based on climatological data, it is often set such that $T(x_s) = -10°C$. We have found it better to use $T(x_s) = -2°C$; it all depends on the other numbers you choose.

One smoother transition that we found useful was

$$r(x) = \begin{cases} r_{if} & T > T_{upper} \\ r_{if} - f(T - T_{upper}) & T_{lower} \leq T \leq T_{upper} \\ r_{ice} & T < T_{lower} \end{cases} \tag{14.6b}$$

where

$$f = \frac{r_{ice} - r_{if}}{T_{upper} - T_{lower}}$$

We suggest $T_{upper} = 10°C$, and $T_{lower} = -10°C$.

14.5 Application

We may use the various formulations that we have just discussed to perform a range of modelling exercises to help us understand the variation of temperature with latitude, and how it might be affected by some of the parameters involved. The more detailed the parameterisations, the more 'accurate' the final result is likely to be, but the more opaque the procedure.

These days it is far more informative to use the simpler formulations, and accept that the results are meant to be more 'educational' than real. While we encourage the more adventurous of you to code up a more sophisticated version, much can be learned from a quite simple model. We will use Equation (14.4) for the long-wave radiation, and Equation (14.5′) for the transport term. Assuming a fixed cloud fraction, Equation (14.1) may now be written

$$Qs(x)A(x) - [a + bT(x)] = c[T(x) - T_{av}] \qquad (14.7)$$

This equation may now be solved for the latitudinal temperature distribution:

$$T(x) = \frac{Qs(x)A(x) - a + cT_{av}}{b + c} \qquad (14.8)$$

This only constitutes a true solution if we already know the global mean temperature, and also the ice distribution, which controls $A(x)$. In general, we don't, so we embark on an iterative approach. We start with an initial temperature distribution: that is, a guess. From this we may find $A(x)$, and the average temperature from

$$T_{av} = \int_0^1 T(x)dx \qquad (14.9)$$

We then use these values in Equation (14.8) to determine the new temperature distribution, and from that the new ice line/albedo distribution. This process continues until changes are less than some predetermined tolerance.

We are now in a position to study the effects of varying some of the parameters. For example, let us integrate Equation (14.7) over latitude, which leads to

$$QA_{av} - (a + bT_{av}) = 0 \qquad (14.10)$$

where

$$A_{av} = \int_0^1 s(x)A(x)dx.$$

This is telling us that the global mean temperature is independent of the transport coefficient, c, but obviously does depend on the albedo (i.e. ice) distribution, which most likely does depend on the transport term. The transport coefficient is the key, of course, to the steepness of the temperature gradient: a value close to zero implies little or no transport and a large equator-to-pole gradient, while the gradient decreases as c increases.

14.5.1 The Basic Model

We suggest you start with the simplest case. Set $Q = 342$ W m^{-2}. Assume a uniform cloud fraction of $\eta = 0.5$. We now need to think about the atmospheric transmission term, and from that to the total albedo. The Earth's albedo is close to 0.3, and of that, clouds account for two-

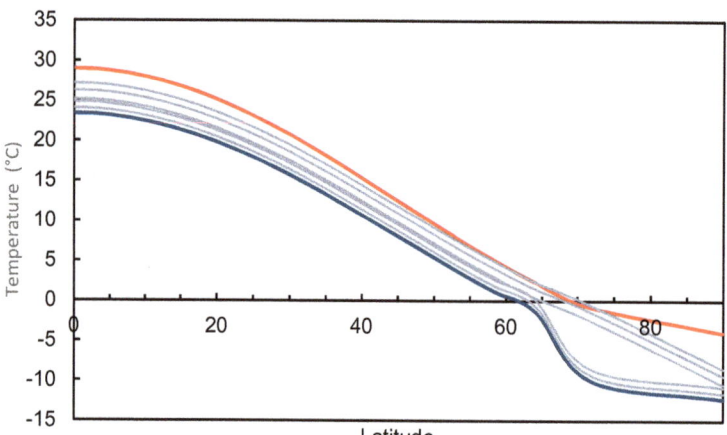

Figure 14.3 Iterative solution of the basic model.

thirds. So assume a cloud albedo of 0.4 (a reasonable value). Use the Budyko ice-albedo idea – Equation (14.6a) – but we have found it better to use $T(x_s) = -2°C$. To run the model in an iterative manner we need some initial guesses in Equation (14.8): we suggest $T_{av} = 15°C$, and, remembering that the global average albedo is $R = 0.3$, we assume that $A(x) = 0.7$. This implies a 'uniform' surface reflectivity as our starting point.

Figure 14.3 shows a typical output. The red line shows the initial guess, and the blue line shows the result after 12 iterations (by which it had effectively converged). Grey lines show some of the intermediate iterations.

Summary

In this chapter we have looked at the remaining two components of the climate system: the cryosphere and the land surface. Both constitute part of the lower boundary of the atmosphere, and thus exert influences on it which may impact our weather and climate. These influences are complex, with some acting on short timescales – for example, albedo and soil moisture – while others act on much longer timescales – for example, the melting of ice sheets, and changes to the carbon cycle.

Whenever we have a multicomponent system, with various pieces exerting 'forces' (however defined) on one another, then we have a clear situation where feedback processes may be involved. This is definitely the case with our climate. To give two key examples: a warming atmosphere holds more water vapour, a powerful greenhouse gas; a warming land surface will lead to a reduction in snow cover, with resulting changes in albedo. These are likely signals of global warming, which we should be looking for: in reality, and in our models.

In the final sections we introduced you to an Energy-Balance Model, based on the work of Budyko and Sellers, half a century ago. While well and truly superseded by the models we will cover in the next chapter, they have much to teach us. We hope you will be able to explore this model, with or without the help of your instructor.

Acknowledgement

Table 14.1:

Climate Change 2021: The Physical Science Basis. Contribution of Working Group I to the Sixth Assessment Report of the Intergovernmental Panel on Climate Change [Masson-Delmotte, V., P. Zhai, A. Pirani, S. L. Connors, C. Péan, S. Berger, N. Caud, Y. Chen, L. Goldfarb, M. I. Gomis, M. Huang, K. Leitzell, E. Lonnoy, J. B. R. Matthews, T. K. Maycock, T. Waterfield, O. Yelekçi, R. Yu, and B. Zhou (eds.)]. Cambridge University Press, Cambridge, UK and New York, NY, USA, 2391 pp. doi:10.1017/9781009157896.

FURTHER READING

- *Climate Change and Climate Modeling* by J. D. Neelin (Cambridge University Press, 2011).
- *Physical Climatology* by W. D. Sellers (University of Chicago Press, 1965).
- A global climate model based on the energy balance of the Earth-atmosphere system by W. D. Sellers, *J. Appl. Meteorol.* 8, 392–400, 1969.
- The effect of solar radiation variations on the climate of the Earth by M. I. Budyko, *Tellus*, 21, 611–619, 1969.
- Seasonal simulation as a test for uncertainties in the parameterizations of a Budyko–Sellers zonal climate model by S. G. Warren and S. H. Schneider, *J. Atmos. Sci.*, 36, 1377–1391, 1979.
- *Global Warming: The Complete Briefing* by J. Houghton (Cambridge University Press, 2015).

REVIEW QUESTIONS

1. In what ways does the cryosphere impact on the atmosphere and climate?
2. In what ways does the land surface impact on the atmosphere and climate?
3. Explain why changes in sea level are slower than other changes in the climate system.
4. Explain some of the interconnections between climate change and the carbon cycle.
5. Explain the ice–albedo feedback mechanism.
6. How do high clouds differ from low clouds in their impact on climate?

EXERCISES

Energy Balance (Budyko–Sellers) models make for a range of interesting exercises. Because they necessarily involve iteration, computation is the key, and any number of languages can be used: Fortran, Python, Matlab, Maple, Excel, or even Basic. (We provided our students with an Excel spreadsheet skeleton, to allow for maximum 'exploration' in the nominal lab time available to them. Many of them simply dropped this onto a USB stick and took it home to play with.)

Now you might work through some or all of the following:

1. Vary the value of the transport parameter c and see what happens: what happens for $c = 0$? Increase the value of c until you produce an ice-free Earth. (You might use the same initial guess distribution each time, or the solution from the previous run.)

2. An increase in greenhouse gases results in more long-wave radiation being trapped by the atmosphere, with some returned to the ground: the effect is a decrease in the *net* upward long-wave radiation from the surface. You can simulate this idea by decreasing the values of the *a* and *b* coefficients in Equation (14.4).

3. The next idea might be to vary the cloud fraction. Here it is important to be consistent in treating both the long-wave emission term – Equation (14.4) – and the albedo.

4. Try the gradual albedo change of Equation (14.6b): what differences does this make? You might have to adjust some of the other parameters to get results similar to those from the Budyko step ice line.

5. While varying one parameter at a time is 'standard practice' in science, you might like to try varying more than one, while predicting the sort of results you will get.

15   Modelling Weather and Climate

The Earth's radiation balances are being altered by a number of changes in the composition of the atmosphere, and as a consequence the climate system is being 'forced', almost certainly in the direction of higher temperatures. In earlier chapters we examined the physics of the processes involved. (We will quantify these changes in Chapter 17, based on the latest IPCC Report, AR6.) We also noted that such warming is highly likely to induce a number of feedback processes, most of them positive, which we discussed in the previous chapter. What are likely to be the effects of such changes?

The only way to answer questions such as this is to endeavour to mathematically model the climate system in sufficient detail. Firstly, of course, we need to model the atmosphere, something we have been doing for half a century to forecast the weather. But weather is not climate, and we need to appreciate the differing expectations we have of each. Because of the significant exchanges of both heat and water between the ocean and the atmosphere, it is clearly necessary to couple an ocean model to our atmosphere model. Ice sheets are likely to be affected by warming, as they are one of the key feedback processes just mentioned. The land surface and the ocean are important players in the carbon cycle, which may be perturbed in response to climate change. So, both of these components of the system must also be modelled.

15.1   General Circulation Models

Today, all weather forecasting, and all 'serious' climate modelling, is done using one or other variant of a General Circulation Model, or GCM. The atmosphere is a dilute gas, which obeys well-established physical laws, starting with Newtonian mechanics, so that determining its time evolution should be, at least in principle, a straightforward task. Unfortunately, Nature rarely equates to the examples in textbooks, so that reality presents a number of key challenges that limit predictability, which we need to understand.

The motion of the atmosphere is governed by seven (scalar) equations:

- Newton's Second Law (force = rate of change of momentum), which is usually referred to as the (horizontal) momentum equation (two scalar equations: zonal and meridional);
- the hydrostatic equation (vertical component of the momentum equation);

- the continuity equation (ensuring conservation of mass);
- the equation of state (the ideal gas equation);
- the First Law of Thermodynamics, also known as the (thermodynamic) energy equation;
- the continuity equation for water vapour (i.e. we need to track water vapour movement).

These equations are then used to predict (that is, they are 'solved' for the time evolution of) seven variables: the three components of the wind vector (u, v, w), the state variables (p, T, ρ), and the specific humidity q.

If our concerns are relatively short term, then we may safely ignore all chemical reactions in the atmosphere, as they generally involve species that are present in such trace amounts that their effects on, say, the radiation flows are miniscule. However, we do need to treat water, in all its phases, with some care, as it undergoes important transformations on a range of scales. In particular, clouds are the dominant variable influence on radiation flows.

These equations are, in general, non-linear (which means that the processes involved may be regarded as 'self-interacting'), and inherently 'messy', so that we can't solve them analytically in sufficient generality to be of use in weather forecasting. Mathematicians do tease them apart to solve them in specific circumstances to better understand how our atmosphere behaves, at least in relatively broad terms. Doing so helps them to understand what the computer later tells them, and potentially alerts them to a coding error.

Instead, we must turn to numerical techniques. This is not unusual: essentially all the analytical problems in physics have already been solved, and numerical approaches are part of the tools of the trade. The standard approach to the numerical solution of differential equations is to convert them to finite difference form, using a suitable algorithm. We then need to employ a spatial grid, and solve via discrete time steps.

15.1.1 Numerical versus Analytical Solutions

Whenever we can, we like to 'solve' the equations that are presented to us in analytical form: this means in terms of mathematical functions such as polynomials ($a + bx + cx^2 \ldots$), or functions such as $\sin(x)$. This gives us a 'once and for all' solution that we can examine and understand. Unfortunately, this is not always possible, especially with many of complex problems that arise in the real world.

In this case we must 'throw the problem to the computer'. Consider a problem (i.e. an equation) where the only variable is x. In this case we break the domain down into a series of points – values of x – and aim to calculate the answer at each point. This will involve converting the original equation from one that is continuous (in x) to one where x takes on discrete values. This process is well understood by mathematicians, although it can still be a matter of judgement. The solution is, of course, only an approximation, but one which – we hope – gets 'better' the more points we have (i.e. the closer together), although this will involve more computation.

In the case of a model of the atmosphere, the output is, of course, a function of all three spatial coordinates, and 'evolves' in time. So we break the atmosphere up into three-dimensional grid boxes, and determine the values of the seven quantities for each grid box.

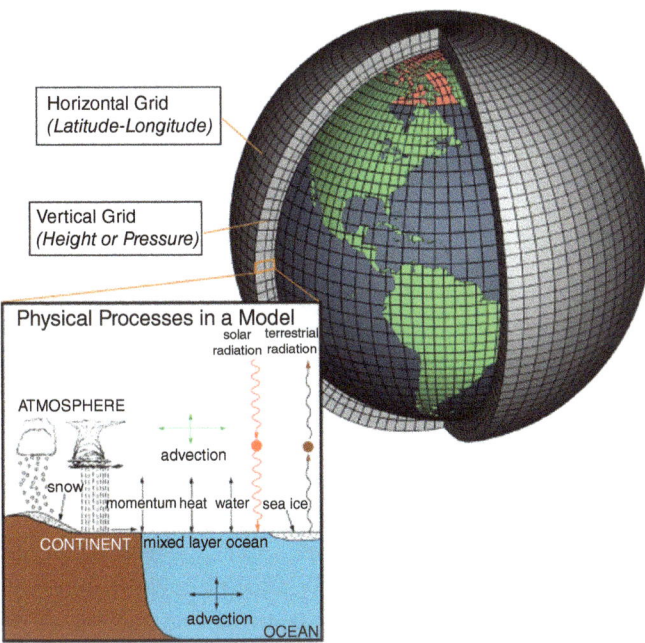

Figure 15.1 Schematic diagram of a weather/climate model. *Source*: NOAA Climate Modelling.

The equations listed above tell us how the air moves in time, carrying with it heat, moisture, etc. The smaller the grid boxes, the more 'accurate' the answer, we hope.

Figure 15.1 from the NOAA website shows a schematic diagram of a weather/climate model. The grid structure is based on latitude and longitude (other options have been used). The boundary and associated processes are also indicated.

15.1.2 Parameterisations

There will always be some processes that we need to incorporate into our model that are too small-scale to be included directly in the system of equations we are solving: such processes (also known as subgrid-scale phenomena) need to be parameterised. Other processes are simply too mathematically complex.

• Turbulent motions near the surface occur on scales much smaller than any grid square, and are responsible for the exchange of heat and moisture between neighbouring grid levels. This is especially important when trying to model the movement of water vapour from the ocean surface up into the boundary layer.
• Surface topography can never be included with enough accuracy, raising a number of issues. A range of low hills, or even one hill, can influence local weather: local weather offices will try to account for this in the forecasts they issue.
• A cumulus cloud has a horizontal scale of ~1 km, distinctly smaller than a grid square, while at the same time stretching upwards through several grid levels. The microphysical processes within such a cloud are orders of magnitude smaller. These clouds must therefore be parameterised in terms of the relative humidity in a grid box, and its instability: these are

then used to predict the 'cloud fraction' within the grid box. (Even a subsaturated grid box may contain some cloud, due to the nature of convection.)

- Computation of radiative transfer, if performed at the sort of accuracy discussed earlier, would consume far too much time, and so this, too, must be replaced by the simplest of two-stream formulations and band models, and even they are usually not performed at each time step.

All of these parameterisations are derived from much effort to understand the physics of the processes, and by running specialised smaller-scale models 'offline', as we discussed in Chapters 9 and 11 for radiation. It can be quite sobering for a scientist to see many years of research work reduced to just a few lines of code.

15.1.3 Boundary Conditions

While weather forecasting is fundamentally an initial value problem (given the state of the atmosphere today, what will it be like tomorrow, etc.?), the boundary conditions at the surface – land, ocean, and cryosphere – are also vital. What surface characteristics are important for our model? They are primarily sea-surface temperature, soil moisture, land-surface roughness, albedo, and leaf area index. From these we may infer the energy and water fluxes to the atmosphere.

Ocean currents move much more slowly than synoptic weather systems, and the ocean's higher heat capacity means that sea-surface temperatures (SSTs) will also change quite slowly, hence we may assume that SSTs will remain constant throughout a weather forecast period (1–2 weeks). The nature of surface vegetation affects surface roughness, which is important in low-level turbulence and convection. Surface conditions are also vital to the surface energy budget: reflection or absorption of solar radiation; sensible versus latent heat. Some means of accounting for soil moisture is needed, of course. Again, these are mostly slowly changing characteristics. Finally, an ice surface has characteristics very different from both land and ocean.

Over the course of a week most of these properties can be assumed to remain fixed, with the possible exception of soil moisture if it happens to rain. Thus their 'input values', which may have been measured along with the atmospheric conditions, can be held fixed throughout the model run. Whenever we turn to longer time frames – seasons, years, decades – we need to enhance our model in some way, by allowing these to vary. Land-surface properties commonly follow a seasonal pattern of growth and decay, or growth and harvest. By contrast, deserts and tropical forests may show very little variability.

15.2 Numerical Weather Prediction

Now the practical questions arise. How much computational power do we have at our disposal? How much time do we have: that is, when do we need the answer? If we are in the business of forecasting weather, for example, we probably need to run our model forward in time for a week, as national weather services provide forecasts at least this far into the future these days. Of course, any forecast is only of use if it is timely, so we want these answers in a matter of hours.

15.2.1 History

Numerical Weather Prediction (NWP) models have evolved over time, starting with the work of Lewis Fry Richardson (1881–1953). While working for the Friends' Ambulance Unit in France during the First World War (he was a Quaker), he devoted his spare time to carrying out the first numerical forecast, using only a slide rule. It took him six months to perform a six-hour forecast, and the results were not very good (to say the least). However, his ideas were sound, and he published them in a book in 1922, imagining a hall full of people carrying out different parts of the calculation together (parallel computing!). The arrival of the first computers turned this dream into the beginnings of a reality.

The UK Met Office has been using computers since the early 1950s for research and, since 1965, for operational forecasting. Their earliest computer had a computing power of ~100 FLOPS (floating point operations per second); today that power is measured in petaflops: an increase of more than 12 orders of magnitude! This growth is a reflection of Moore's law of computing. How much further this can continue remains to be seen, as components are starting to approach the physical scale where quantum effects manifest themselves. Perhaps the solution to such a problem will be provided by quantum computing, but at this stage it is simply too early to tell.

As computing power has increased, so has spatial resolution. The latest NWP models chop the globe up into grid boxes of about 10–20 km on a side, with maybe 70 vertical layers. (Individual national weather services may run 'nested models', with smaller grid boxes down to 1 km over their domain of primary forecast responsibility.) The earliest models used physical height to define the vertical layers: today, pressure is the key.

Numerical stability places upper limits on the time step that may be used: it must be reduced whenever the grid spacing is reduced. How long, for instance, does it take a 'disturbance' to cross a grid square? Current global models use a time step of ~10 minutes, and regional/nested models might use a time step of only 1–4 minutes. If we wish to reduce the grid scale by a factor of 2, that would require 4 times as many spatial grid points, and twice as many time steps. One or two additional vertical layers might also be needed near the surface. All told, this means an order of magnitude increase in the number of computational steps (and data storage).

15.2.2 Predictability

Weather forecasting is essentially an initial value problem: given the current state of the atmosphere, we 'churn the handle' provided by the seven equations to predict what it will be in 24 hours' time, 48 hours' time, etc. How much data are required to specify the initial (current) state? Essentially, we would like to know the values of all the variables – pressure, temperature, wind speed and direction, humidity – at every grid point. These data come from a range of sources. By 2010, the UK Met Office was typically obtaining 23,158 surface observations, 33,640 satellite soundings, 1,586 balloon soundings, and 7,779 satellite wind tracks each day.

All of these data need to be *assimilated* with an absolute minimum of human quality control. These data are measured at a range of times, represent different spatial scales, are sourced from different heights and depths, and take further time to be transmitted to the computers that will process them. In the meantime, the atmosphere is not stationary. Thus, we clearly have a four-dimensional problem, which, because so-called variational techniques are often used, is known as 4D-Var (variational assimilation). Some data will be rejected along the way as they are 'inconsistent' with the predicted state of the atmosphere at the time of its collection: that is, the model is treated as having an intrinsic degree of accuracy, which may be higher than some measured data.

As computer hardware, input data, and algorithms have improved, so has the 'quality' of the outputs; that is, their ability to predict tomorrow's weather. Current generation models are now able to provide valuable forecasts for up to 7 days, and useful indicators beyond that. Further improvements to models, and to the data supplied to them from satellite and ground observations around the world, are likely to extend this time frame even further.

If we were ever to know the state of the atmosphere at a given moment, with the level of precision that we might feel is needed, could we then run our forecast model into the 'indefinite' future? Obviously we could, but would the output be of any real value? The equations involved contain non-linear terms that present the prospect of high sensitivity to initial conditions: the essence of chaos. Twenty days is probably the upper time limit of a forecast model, even assuming 'near perfect' input data.

In the early 1960s, Edward Lorenz (1917–2008) was running experiments on the computer then available, using a simplified atmospheric model, trying to test its forecast range. In those days computers were far less reliable than today, often breaking down. On one occasion he stopped the run, and copied down all its data, before setting it off again. Later, as a test, he decided to restart the program using the numbers he had copied down. To his surprise he found the new output diverged from the previous results. The reason turned out to be the lower precision of the numbers he had recorded compared with those actually stored in the computer's memory. That difference, that sensitivity to the input data, is one of the early signposts to the branch of science we know today as Chaos theory.

15.2.3 Forecast Skill

How might we assess the accuracy of a weather forecast? Is an 'error' of 1°C in the forecast maximum really a failure? On the other hand, to forecast no rain in a city that is predominantly dry at this time of the year is of little value. A good forecast – a good NWP model – needs to outperform both climatology and persistence (tomorrow's weather will be much like today's). We also need to remember that all NWP models are, of necessity, global, and must be assessed on a global basis.

Here's another question to add to the mix. Because weather forecasts now extend out to 7 days, what is important, in terms of a global model, at day 3, in order that the day 5 or 6 forecasts might be acceptably accurate? We said in Chapter 7 that it is the middle levels of the troposphere – say, 500 hPa – that steer weather systems. So, the answer to this new question is that the model must be doing a good job on the bulk of the atmosphere at day 3, not just at ground level where people experience weather. Meteorologists do just such a full-atmosphere comparison between model output and subsequent reality in order to assess forecast quality, and assign a so-called *skill score* based on a detailed statistical assessment.

Some NWP models are run out to 14 days, and available to national weather services, although invariably only the first 7 days are released to the public via forecasts. A recent analysis by a colleague of ours is very interesting. When official 4-day forecasts were first released in Australia in February 1987, they explained 20.1% of the variance (a statistical measure). Today they explain 72.6%: an improvement we have tended to take for granted. Today, experimental 10-day forecasts explain 16.7% of the variance, close to that original 4-day accuracy. What does the future hold?

15.2.4 The Utility of Satellite Data

A few years ago, a team of (mostly) Australian meteorologists carried out an interesting experiment. They ran a weather forecast model both with and without the inclusion of satellite data, and assessed the results using standard analysis tools. More specifically, they looked at the Northern and Southern Hemispheres separately: sample results for the period 15 August to 30 September are shown in Figure 15.2.

Let us focus on the results at 96 hours; 4 days. Without satellite data, the 'accuracy' of the forecast in the Northern Hemisphere was 0.88; or we might say a 12% 'loss of accuracy'. However, in the Southern Hemisphere, the 'accuracy' was only 0.67; or a 33% 'loss of accuracy'. But look what happens when satellite data were included: the 'accuracy' in both hemispheres is close to 0.92, or an 8% 'loss of accuracy'. The reason for the hemispheric difference should be obvious: there are much more ground-based data available in the Northern Hemisphere. When the lead author (a friend of ours) presented these results at a

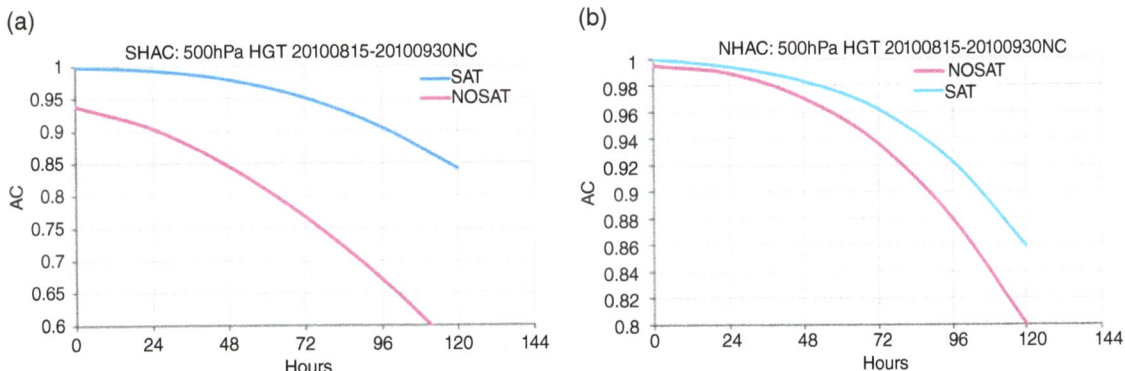

Figure 15.2 500 hPa height anomalies for the control (SAT) and no satellite (NOSAT) analyses for (a) Southern Hemisphere and (b) Northern Hemisphere. *Source*: The considerable impact of earth observations from space on numerical weather prediction by Le Marshall J. et al., *Australian Meteorological and Oceanographic Journal* 63 (2013) 497–500; Bureau of Meteorology.

local conference, he then stood back and remarked that Northern Hemisphere people (mostly) pay for it (i.e. the satellites), but we in the Southern Hemisphere benefit the most. True; and we're grateful.

15.3 Climate Models

While chaos will always place strict limitations on weather forecasting, the situation with longer-term climate prediction is quite different. Climate is fundamentally about statistics, such as average temperature and rainfall in a locality, plus appropriate measures of their variability. It doesn't much matter which day it rains, climate is mainly interested in, say, seasonal rainfall – plus any changes in extremes – and, especially in countries which are strongly impacted by the El Niño/La Niña cycle, the interannual variability. Long-term prediction is thus concerned with the envelope of such parameters, with that spread representing the chaotic effects of natural variability.

Modelling the climate, be it the past or the future, requires simulation runs of months to years to decades to centuries, and even longer, not just a week. Climate models have much in common with weather models, including a vast amount of shared computer code, but there are also a number of key differences. The most obvious is grid size, currently ~100 km or 1 degree of latitude. (Many studies are performed on even coarser grids.) Time steps may be on the order of 30 minutes. This is a direct consequence of the limitations of computing resources, as such models need to be run for a more than 1,000 times longer simulation time. Of course, we don't need the results 'immediately', so some simulations can be run for weeks.

A direct consequence of such a coarse grid is that many aspects of the physics are poorly represented, starting with topography. Many more processes now cannot be included directly and need to be parameterised, with clouds a particular challenge. As a consequence, models such as this are considered less reliable at the local to regional levels, compared with continental to global scales.

15.3.1 Ocean Modelling

We noted above that the most important boundary condition for an atmospheric model is provided by the ocean: in particular, sea-surface temperature has a major influence on heat and water fluxes. So, the next key step in modelling climate is to add an ocean model. The ocean has its own circulation patterns, as we saw in Chapter 13, dictated by the combined effects of continental boundaries, wind stress, density gradients and the Coriolis force. These circulations are major contributors to the transport of heat, and must be reproduced by any model we hope to employ in climate studies. Internal modes of variability such as ENSO are also central to climate, and so we must hope that, in some form, they arise from such an ocean model.

The ocean is also a fluid, and thus subject to the standard laws of fluid mechanics, in some ways similar to, and in other ways quite different from, the atmosphere. In particular, the horizontal momentum equations, the hydrostatic and continuity equations apply. An energy equation is required, but the input/output terms involve simple heat conduction, not radiative transfer. Finally, salinity, rather than humidity, needs to be tracked, as it is the key to the density variations of the ocean. (Evaporation and precipitation will change the ocean surface-layer salinity.)

Once again, boundary conditions are required, starting with the physical boundaries of the continents and ocean bathymetry (especially around continental shelf regions), plus sea ice. They will also include the ocean–atmosphere interface, which governs surface height (which is related to atmospheric surface pressure), plus wind stress, which governs momentum transfer across the interface. The ocean is a much denser, slower-moving fluid than the atmosphere, which dictates differing numerical approaches. We don't need to worry about clouds and similar 'subgrid-scale phenomena', and turbulence is also reduced by the greater density.

As more and more CO_2 enters the atmosphere, we have already discussed the fact that a significant fraction ends up in the ocean, potentially changing its pH and its chemistry. While such effects are largely decoupled from the atmospheric component of the climate at present, this may not always be the case, so that ocean chemistry should be a component of an Earth System Model.

15.3.2 The Land Surface

Land surfaces influence the climate through a number of biogeophysical effects. The nature of a surface – e.g. bare or vegetated – has a direct effect on albedo. Surface roughness affects boundary layer turbulence and cloud formation. Evapotranspiration affects hydrology, and also how energy is partitioned between latent and sensible heat. Such biophysical effects are included in both Coupled Climate and Earth System Models, although they are more fully developed in the latter.

Ice and snow have very different properties from both land and ocean, and so the cryosphere needs its own models. Because of the expected importance of the ice–albedo feedback in any changing climate, these models must include not only static components which might affect the atmosphere above it, but also a dynamic component to account for changes in ice cover in response to influences such as rising temperatures. Sea ice and land ice may have quite different properties, as the former will be directly impacted from underneath if the sea warms. Figure 15.3 presents a simplified schematic diagram of the inputs and outputs of a climate model.

15.3.3 Biogeochemical Interactions

The biogeochemical effects of the land surface mainly comprise the exchange of various chemical compounds with the atmosphere, particularly CO_2. However, plants also exchange organic compounds; for example, isoprene, which is a precursor to secondary organic aerosols. Finally, plants (and their associated root microbes) are also central to the nitrogen cycle. These biogeochemical processes are only included in full Earth System Models.

Predicting how the biosphere may respond to multiple disequilibria of warming temperatures, elevated CO_2 levels, more intense droughts and floods, not to mention the effects of species migration where human constraints still permit it, is a task that science has only recently

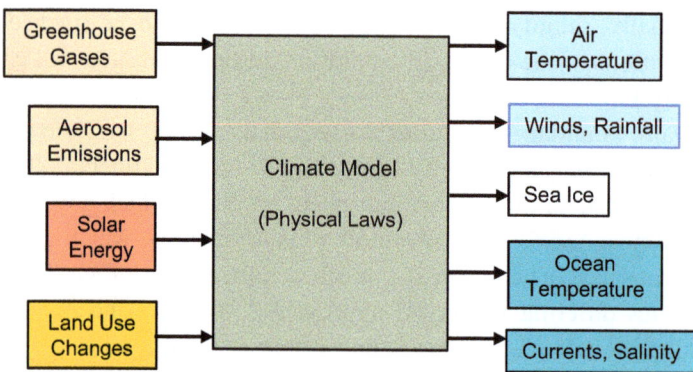

Figure 15.3 Schematic diagram of a climate model.

begun to address. Nevertheless, these interactions, and their potential biogeophysical and biogeochemical feedbacks on the climate, for example via further changes in the carbon cycle, must be included.

These feedbacks take place on much longer timescales than any weather phenomena, and indeed most climate variability, so such models are often run quite separately. It should also be pointed out that much of the content of these models is based on empirical laws that have been derived from observations of ecosystems and communities under a range of conditions they experience. We cannot be absolutely sure what might happen beyond the range of conditions that have already been studied in the field: conditions which might arise due to climate change. These various effects are all required by an Earth System Model.

Climate modelling studies of the glacial–interglacial cycle (Chapter 16) need additional science components. For example, the replacement of taller trees and shrubs with greater leaf area by low tundra plants during a period of glaciation allows more sunlight to reach a snow-covered surface, and hence be reflected, strengthening the ice–albedo feedback. Another important feedback concerns the hydrological cycle. Modelling has shown that reductions in rainfall can lead to changes in plant variety with reduction in transpiration, which then feeds back on rainfall.

15.3.4 Evolution

Figure 15.4 {figure 1 from Box 3 of the IPCC's Third Assessment Report} indicates how climate models have evolved over a 25-year period, starting with just the atmosphere, before adding a basic land-surface model, and then an ocean and sea ice model. You should also note that many of the components that are now 'standard' were initially developed separately, and run 'offline', before being coupled into the package. One example was an atmospheric chemistry model, mainly focused on the fate of SO_2 when it enters the atmosphere, particularly aerosol formation.

What can be expected in the future? There are no major components of the Earth System that are currently excluded from the models, so it comes down to details. Basic primary production biology (i.e. carbon exchanges) is now included, but this could be extended to look at ecology. Ice sheets are difficult to model, especially on complex terrain, so this is an

The Development of Climate Models, Past, Present and Future

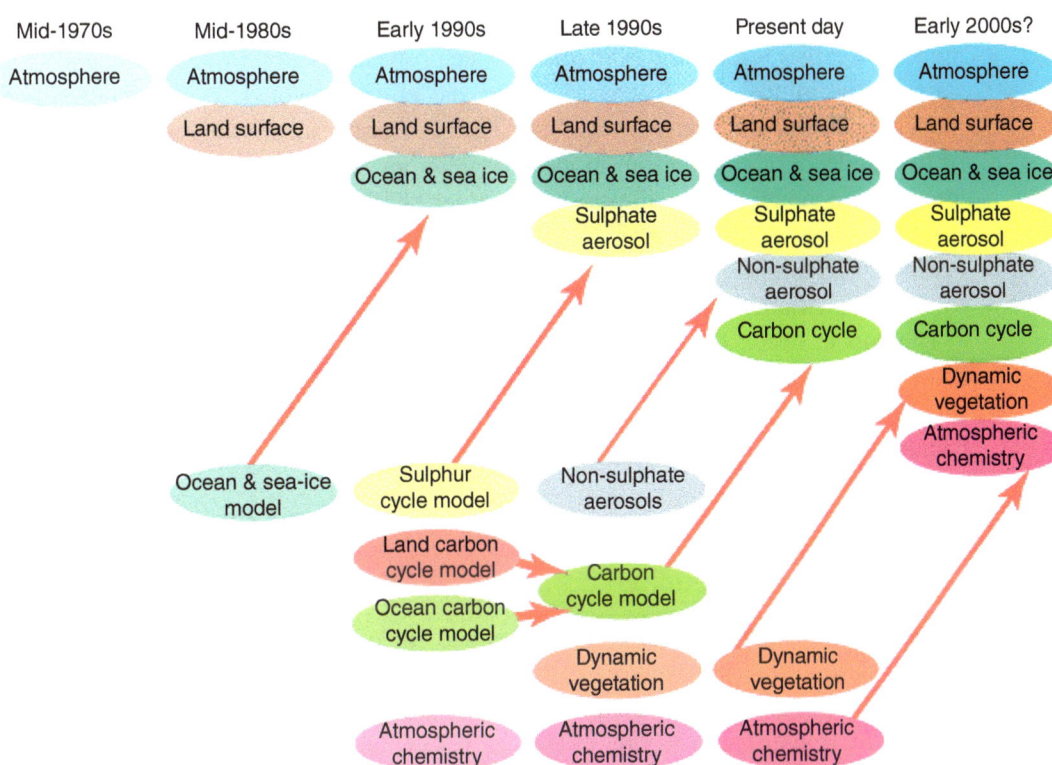

Figure 15.4 The development of climate models from the 1970s to 2000. *Source*: see Acknowledgement at the end of the chapter.

important avenue. And, of course, as computing power improves, so should resolution. This may reduce the need for some of the parameterisations mentioned above, but is very expansive!

15.4 Climate Modelling Studies

The models we have been discussing may be used in many investigations of phenomena, both 'real' and hypothetical. In particular, they may be used to simulate the climate of the twentieth century, or the Holocene, the glacial cycles, or the Earth's distant past. These models may also, of course, be used to anticipate climatic changes in the decades ahead. Models also have capabilities that are either difficult, or even impossible, to achieve in the real world.

Roughly 33 million years ago, South America separated from Antarctica, opening up the Drake Passage, and creating the Circumpolar Current. Using models of varying sophistication, a number of authors have addressed this issue by simply changing the bathymetry: see Further Reading. They conclude that this event had a major effect on Southern Hemisphere climate, starting with a significant cooling around the Antarctic margin, with impacts on bottom water formation.

15.4.1 Running the Models

Today, virtually all climate modelling is done using coupled ocean–atmosphere–land–ice models. Weather forecasting is an initial value problem, as already discussed, and so the initial state of the atmosphere is crucial: hence the efforts devoted to data collection. Climate modelling is the opposite: we want statistically robust results, *independent* of the initial state of the model. For this reason, models are *spun up*, and the early results discarded. For an atmospheric model, a spin-up time of the order of 10 years might be used, as the atmosphere usually undergoes significant change in such a time period. However, the time required for the ocean is much longer. For the upper ocean, the response time may be of the order of 10 years; for the deep Atlantic about 100 years; for the deep Pacific about 1,000 years.

The model needs to be run until it reaches a steady 'climate', and if the atmospheric composition and other relevant parameters are similar to today's conditions, then that 'climate' should be as close as possible to the present-day climate. A crucial factor in the ability of coupled models to simulate the present climate is the fluxes of heat, water, and momentum across the ocean surface.

15.4.2 Natural Climate Variability

Even with fixed atmospheric composition and radiation fluxes – that is, no forcings – the Earth's climate shows variability on a range of time and spatial scales. On short timescales, this is a reflection of the natural variability of the atmosphere. However, on timescales of years and longer, this is mainly a reflection of the internal modes of variability of the ocean. Many such modes have been well studied (Chapter 13). If we want to better understand the effects of such modes on regional rainfall, we might hold the sea-surface temperature fixed in the appropriate pattern, and focus on the results from the atmospheric component. This is one element of how we try to reconstruct the past history of such modes.

An important test of a climate model, and especially the oceanic component, is to see how well it reproduces the major features of such phenomena. Current coupled models are capable of providing the statistics of phenomena such as El Niño, although not of predicting an individual El Niño event some years into the future

15.4.3 Response to Greenhouse Gas Forcing

For over a century, climate scientists have been aware that increasing levels of CO_2 would inevitably lead to an increase in surface temperatures across the planet. Most atmospheric scientists devoted little time to the issue: firstly, because they did not possess the necessary resources to address it; and secondly, because the challenge of understanding and forecasting weather was more pressing. Nevertheless, a few scientists did try to construct (necessarily simple) models of the phenomenon. As models of the climate system have become more sophisticated, and resolution has increased along with computing power, this has been a central theme. However, for many decades the focus was on a somewhat simpler question: what happens to the climate when CO_2 is doubled? This is known as the **climate sensitivity**.

There is an interesting history of scientific attempts to compute the climate sensitivity, some of which are shown in Figure 1.3. The first was by Svante Arrhenius in the late nineteenth century. He estimated a rise of 5–6°C from a doubling of CO_2, which he later revised to 4°C. In 1967, Manabe and Wetherald used a one-dimensional radiative-convective model, with a detailed computation of the radiative transfer, to arrive at a value of 2.3°C (rounded to 2°C to indicate uncertainty). This paper has been dubbed the greatest climate science paper of all time. (Syukuro Manabe shared the 2021 Nobel Prize in Physics 'for the physical modelling of Earth's climate, quantifying variability and reliably predicting global warming': thoroughly deserved!)

In 1979, a committee of the US National Academy of Science (NAS) on anthropogenic global warming, chaired by Jule Charney, estimated the sensitivity to be 3°C ± 1.5°C. Since then, it has been the province of the IPCC Reports to provide such estimates. These successive estimates have largely echoed the US NAS result. However, by the time of the latest Assessment (AR6), the models were suggesting a sensitivity within a *likely* range of 2.5–4.0°C.

15.5 Climate Model Intercomparison Project, CMIP

Climate models, as we have already said, involve many compromises, especially in regard to the parameterisation of processes that cannot be handled explicitly. This raises an immediate and important question: is there an optimal parameterisation for each of these processes? If the answer is 'yes', then all models should incorporate that approach. But if the answer is 'no', or even 'not sure', then it would be most unwise to put all our eggs in the one basket. We should also note that the parameterisations may depend on the resolution of the model, so that models with different resolutions will often have variations in these parameterisations.

For this and many other reasons, different modelling groups around the world run models with subtle differences. This means, of course, that for the same inputs (such as greenhouse gas and aerosol concentrations), the models may produce different outputs. This should not be seen as something negative, but rather an indication of our current uncertainty in the actual phenomena: a form of our collective honesty as scientists.

When the modelling community runs models (with identical inputs and other assumptions), either of the climate of the twentieth century, or a projection for the twenty-first, their outputs will, invariably, show many similar trends, as well as a range of variations about these trends. We can learn valuable lessons from both the similarities (e.g. a rise in temperature in response to a positive forcing) and the differences (which we should treat as a measure of uncertainty), assuming we know what to look for, and ask the right questions.

The Climate Model Intercomparison Project (CMIP) was established in 1995 within the World Climate Research Program, with the aim of better understanding past, present, and future climate, both in response to natural (unforced) variability, and also to prescribed forcings. Comparing model outputs, and especially when we are able to compare these with the historical record, allows us to assess their performance and reliability. One important aspect of CMIP is to make the various model outputs publicly available in a standardised format.

In addition to what might be regarded as 'standard' model comparison exercises such as modelling the climate of the twentieth century, CMIP has also undertaken a number of smaller, focused studies to understand specific phenomena and how the models handle them. The first was to study the response of the Atlantic Meridional Overturning Circulation to surface fluxes of heat and water. A related study focused specifically on the response to a water flux at high latitudes as a result of significant melting of the Greenland ice cap.

15.5.1 CMIP6

Most of the conclusions about our climate, from the recent past to the next century, which will be presented in the remaining chapters, are based on the IPCC's Sixth Assessment Report (AR6), which in turn is informed by the modelling results from CMIP6. So it makes sense to spend a little time getting to grips with the models involved, and the nature of their results. CMIP5 involved 37 models, while CMIP6 involves 50 models (in addition there are 12 high-resolution models: the High-Resolution Model Intercomparison Project; HighResMIP). Figure 15.5 {figure 1.20 in AR6} shows a map with the locations of many of the consortia involved in the CMIP program. (CORDEX is the Coordinated Regional Climate Downscaling Experiment.) Further information on all models is contained in Annex II of AR6. One such consortium shown is eight EU cities, comprising groups from Sweden, the Netherlands, Denmark, Spain, Ireland, Italy, Portugal, and Finland.

A key consideration in any model is its grid size/resolution, for both the atmosphere and the ocean. CMIP models have resolutions ranging from ~250 km down to 50 km, and between 20 and 100 vertical levels. (The high-resolution models, as the name suggests, have grid sizes as low as 10 km.) Average resolution has improved from CMIP5 to CMIP6, but not significantly,

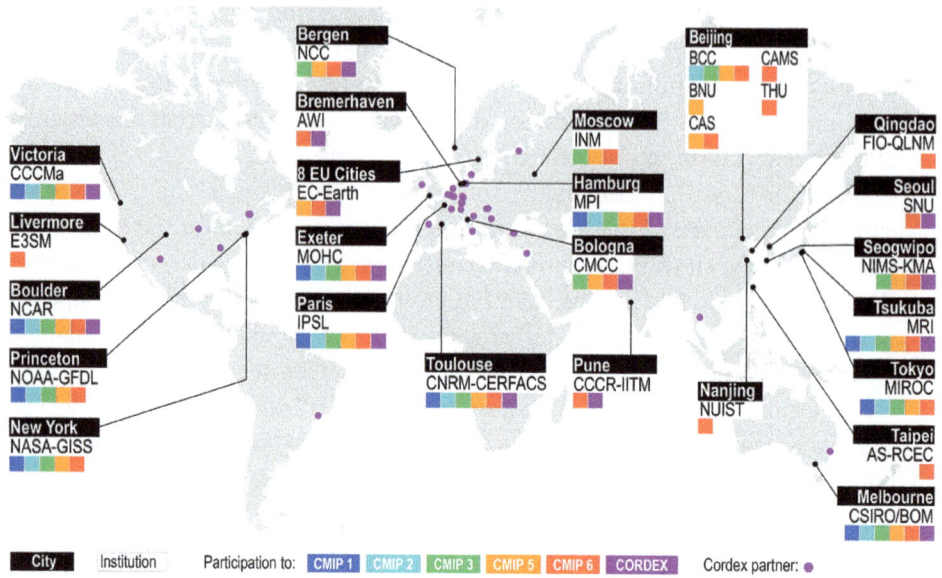

Figure 15.5 World map showing the increased diversity of modelling centres contributing to CMIP and CORDEX. *Source*: see Acknowledgements at the end of the chapter.

and remains at about 150 km for the atmosphere and 75 km for the oceans. The number of vertical levels has increased from ~40 to ~50.

CMIP6 models have included updates in some of the parameterisation schemes mentioned above, with the aim of better representing the physics. Most notable of these developments have been to the schemes involving radiative transfer, cloud microphysics, and aerosols; in particular, how they represent aerosol-induced modification of cloud properties. This, as we will see in Chapter 17, has led to a reassessment of the magnitude of this (negative) forcing of the climate.

AR6 concludes that, for most large-scale indicators of climate change, the simulated mean climate from CMIP6 has marginally improved compared to CMIP5. Some differences still remain, such as regional precipitation patterns. Most Earth System Models, which include biogeochemical feedbacks, perform as well as their counterparts without such complexity. The multi-model mean captures most aspects of observed climate change very well (*high confidence*). The simulation of palaeoclimates has also improved, although some issues still remain.

One detailed study is worth including here. Both CMIP5 and CMIP6 have been put to the test of simulating the climate for the period 1995–2014, and compare their results with the re-analysis of the (atmospheric) data for that period performed by the European Centre for Medium-Range Weather Forecasts (ECMWF). Figure 15.6 {figure 3.3} presents the near-surface temperature data. Panel (a) shows the CMIP6 ensemble average of all the models, while panel (b) shows the multi-model mean bias; the difference between the (average) simulation and 'reality'. Arctic temperature biases are apparent in both ensembles, with errors in clouds, ocean circulation, winds, and surface energy budget frequently cited. Panel (d) shows the corresponding bias for the CMIP5 simulations. Panel (c) shows the multi-model mean of the root-mean-square error, calculated over all months separately and averaged, with respect to climatology. Finally, panels (e) and (f) show the biases from the high-resolution models when run on both high and low resolution. These results provide a hint that improved resolution may reduce biases, although so far improvements are only modest.

15.6 Carbonator.org

Carbonator is a greatly simplified version of a climate model, developed by our good friends at the Climate Change Research Centre of the University of New South Wales, Sydney, Australia. It is publicly available at its website, carbonator.org. To quote from its explanation:

Carbonator is a simple climate model. Unlike a full climate model that can tell us how climate variables evolve at different locations it can only tell us how a subset of variables change on a global scale (e.g. globally averaged temperature). Carbonator uses only 20–30 lines of computer code and takes a few seconds to simulate a few decades of climate system evolution on your computer.

Carbonator is based on the same laws of physics that underpin state-of-the-art climate models (in particular the conservation of energy) and for a limited number of variables will produce very similar results to those models. As such, it is a powerful tool that can be used to explore how the climate system is affected by different factors (like CO_2 emissions, volcanoes or changes in the power output of the sun) how our decisions are likely to affect the climate system in the future.

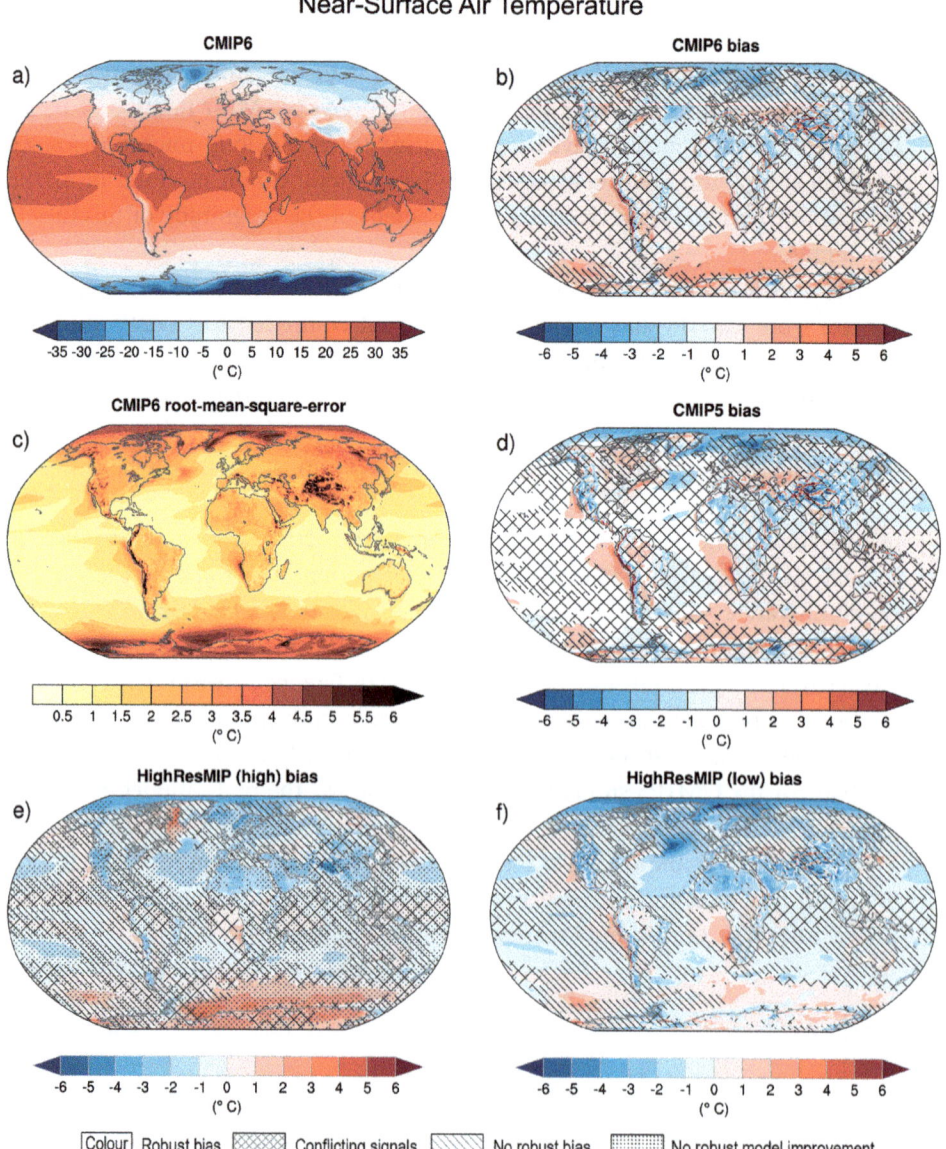

Figure 15.6 Annual mean near-surface (2 m) air temperature (°C) for the period 1995–2014, details as explained in the text. *Source*: see Acknowledgements at the end of the chapter.

Based on a set of six inputs (that you can define):

1. CO_2 emissions (no. of billions of tons of CO_2 released into the atmosphere each year);
2. CH_4 emission;
3. Human aerosol emissions;
4. Volcanic aerosol emissions;

5. Energy reaching the earth from the sun (solar radiation);
6. Reflectivity of the earth surface.

Carbonator will calculate how various climate variables will change over time, including:

- Surface temperature;
- Deep ocean temperature;
- Sea-level and Ocean pH;
- Atmospheric concentrations of CO_2 and CH_4;
- Carbon inventories (land and ocean).

It offers a total of 12 scenarios (and you can also create your own):

1. Rapid Emissions Reduction (RCP3);
2. Moderate Emissions Reduction (RCP4.5);
3. Technology-led Emissions Reduction (RCP6);
4. Business as Usual (RCP8.5);
5. CO_2 pulse;
6. CH_4 pulse;
7. White Roofs;
8. Geoengineering;
9. Geoengineering Failure;
10. Eliminate All Emissions Today;
11. Solar Variations;
12. Large Volcanic Eruption.

Note that the first four are actually IPCC scenarios, which will be discussed in detail in Chapter 20 (the 'RCP' number is their identifier). The two pulse scenarios are quite artificial, as they assume no other emissions of greenhouse gases since pre-industrial times, but are nevertheless quite interesting; as are the three geoengineering scenarios (and especially scenario 9, the failure!). Each scenario consists of the inputs listed above; from 1850 to 2100.

The outputs are, of course, all globally averaged. Nevertheless, they are both valuable and informative. Do check it out. As a pointer, Figure 15.7 shows the temperature evolution under two scenarios: RCP3 and RCP8.5.

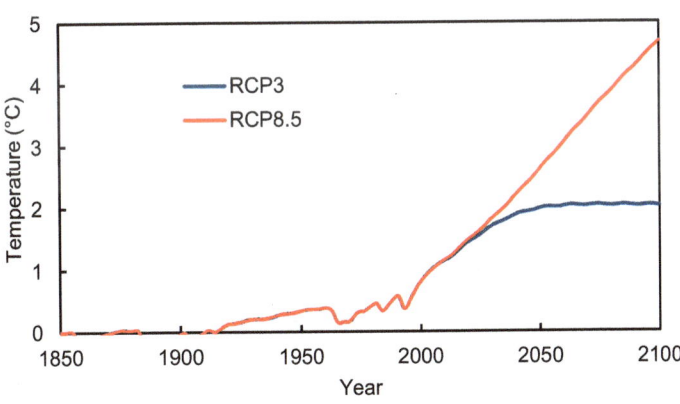

Figure 15.7 Evolution of global average surface temperature under the RCP3 and RCP8.5 scenarios.

Summary

We are more than confident that our climate is changing, and while we think we know why, we really need to be sure, especially if we are to provide any guidance as to humanity's future – surely one of the tasks of science. But the Earth is a huge, and complex, system: how do we address this challenge?

The first step, not surprisingly, is to build a model of the atmosphere; after all, that is where we 'experience' climate (the statistics of weather). We have been doing this for half a century, and are actually getting rather good at it – even if our forecasts do occasionally go dramatically wrong. However, due to chaos, a weather forecast model is only of use for, perhaps, a couple of weeks, so what use are they for climate? Well, because climate is, as we noted, the statistics of weather, that is what we need from such a model. Which day it rains is of no real importance – unless you are planning a picnic – but seasonal rainfall clearly is.

The ocean, being a fluid, obeys essentially the same physical laws as the atmosphere, and so should be capable of being modelled in a similar way, although there are many subtle differences that must be incorporated. Such models are able to handle at least the statistics of phenomena such as ENSO with increasing reliability.

We have seen that atmospheric chemistry, and particularly aerosols, can play a significant role in a changing climate, and we can build models of this. However, the full complexity of tracking all relevant species and their reactions is computationally challenging, and mainly reserved for a specific task before being summarised in a few lines of code for incorporation into larger models.

Finally, if we ever hope to simulate the climate under a high CO_2 scenario, we need to model any interactions and feedbacks with the land surface/biosphere: that is, any perturbations to the carbon cycle. This remains a major challenge.

Acknowledgements

Figure 15.4:

Albritton, D.L., L. G. Meira Filho, U. Cubasch, X. Dai, Y. Ding, D. J. Griggs, B. Hewitson, J. T. Houghton, I. Isaksen, T. Karl, M. McFarland, V. P. Meleshko, J. F. B. Mitchell, M. Noguer, B. S. Nyenzi, M. Oppenheimer, J. E. Penner, S. Pollonais, T. Stocker, and K. E. Trenberth. IPCC, 2001: Technical Summary. In: Climate Change 2001: The Scientific Basis. Contribution of Working Group I to the Third Assessment Report of the Intergovernmental Panel on Climate Change [Houghton, J. T., Y. Ding, D. J. Griggs, M. Noguer, P. J. van der Linden, X. Dai, K. Maskell, and C. A. Johnson (eds.)]. Cambridge University Press, Cambridge, United Kingdom and New York, NY, USA, 21–84 pp.

Figures 15.5 {1.20} and 15.6 {3.3}:

Climate Change 2021: The Physical Science Basis. Contribution of Working Group I to the Sixth Assessment Report of the Intergovernmental Panel on Climate Change [Masson-Delmotte, V., P. Zhai, A. Pirani, S. L. Connors, C. Péan, S. Berger, N. Caud,

Y. Chen, L. Goldfarb, M. I. Gomis, M. Huang, K. Leitzell, E. Lonnoy, J. B. R. Matthews, T. K. Maycock, T. Waterfield, O. Yelekçi, R. Yu, and B. Zhou (eds.)]. Cambridge University Press, Cambridge, United Kingdom and New York, NY, USA, 2391 pp. doi:10.1017/9781009157896.

FURTHER READING

There is, of course, a significant literature on this subject, most of it reasonably technical.

- You will find a very readable chapter on the subject in *Global Warming, The Complete Briefing* by John Houghton (Cambridge University Press, 2015).
- At a more technical level, and certainly more mathematical than our treatment, we suggest *Climate Change and Climate Modeling* by David Neelin (Cambridge University Press, 2012).
- *The Climate Modelling Primer* by A. Henderson-Sellers and K. McGuffie (Wiley, 2014).
- *Climate System Modeling* edited by K. E. Trenberth (Cambridge University Press, 1993).
- See also: Thermal equilibrium of the atmosphere with a given distribution of relative humidity by S. Manabe and R.T. Weatherald. *J. Atmos. Sci.*, 24, 241–259, 1967.
- Effect of the Drake Passage throughflow on global climate by W.P. Sijp and M.H. England. *J. Phys. Oceanogr.*, 34, 1254–1266, 2004.
- Ice–atmosphere feedbacks dominate the response of the climate system to Drake Passage closure by M.H. England et al. *J. Climate*, 30, 5775-5790, 2017.
- The polar ocean and glacial cycles in atmospheric CO_2 concentration by D. Sigman, M. Hain, and G. Haug. *Nature* 466, 47–55, 2010. https://doi.org/10.1038/nature09149
- For a more detailed explanation of the science (including the equations) which underpin the Carbonator model, watch www.youtube.com/watch?v=fkAGwHypMzo

REVIEW QUESTIONS

1. An atmospheric general circulation model tracks the temporal and spatial evolution of which seven variables?
2. What boundary conditions are needed in a numerical weather model?
3. Explain chaos, and its role in weather forecasting.
4. Why do we need to include an ocean model when we try to predict weather/climate on seasonal to interannual timescales?
5. What else do we need?
6. What is meant by 'climate sensitivity'?
7. What are some examples of climate model studies using current models?
8. When we run a climate model, we discard the early results. Why?
9. What does the acronym CMIP stand for?

EXERCISE

Now is the time to start your exploration of Carbonator. We suggest you start with the so-called 'Business as Usual' (RCP8.5) scenario. Click Start. You will see the time series for the six inputs: CO_2 Emissions, CH_4 Emissions, SO_2 Emissions, Volcanic Emissions, Insolation, Albedo, from 1850 to 2100. Now click Run scenario. After a few seconds processing time, you will see the outputs: Temperature, Sea Level Change, pH, CO_2 Concentration, CH_4 Concentration, CO_2/CH_4 radiative forcing, Aerosol radiative forcing, Ocean carbon inventories, Land carbon inventories; again from 1850 to 2100. To add a little interest, you can switch on the internal variability option.

Write a report on the achievements of Nobel Prize winner, Syukuro Manabe.

16 Climates Past

We know that the Earth's climate has changed in the past, including times when humans had yet to put in an appearance. These changes must, therefore, have been the result of 'natural processes'. So how do we know that current changes are not similarly natural? The challenge this poses for science is to understand the processes that were at work in the past, and compare them with the climate drivers, both natural and anthropogenic, at work today.

There are two reasons to try to unravel the climatic history of our planet. The first might be described as its intrinsic interest. The more important reason is to see what lessons we can learn that might help us think about potential future climate change. This is the reason that it is now an important component of the IPCC process. On page 248 {chapter 1} of AR6 we encounter FAQ 1.3: *What Can Past Climate Teach Us About the Future*? We quote:

As scientists seek to refine our understanding of Earth's climate system and how it may evolve in coming decades to centuries, past climate states provide a wealth of insights. Data about these states help to establish the relationship between natural climate drivers and the history of changes in global temperature, global sea levels, the carbon cycle, ocean circulation, and regional climate patterns, including climate extremes. Guided by such data, scientists use Earth system models to identify the chain of events underlying the transitions between past climate states. This is important because during present-day climate change, just as in past climate changes, some aspects of the Earth system (e.g. surface temperature) respond to changes in greenhouse gases on a timescale of decades to centuries, while others (e.g. sea level and carbon cycle) respond over centuries to millennia.

16.1 Methodologies and Lines of Evidence

Palaeoclimatology is, essentially, a branch of geology, but one with significant implications for understanding climate change. The methodologies employed, and the data sources which may be investigated, are quite different from the rest of climate science. As such, it is a specialised field of study, and mostly well beyond both the level, and the requirements, of this book. Our treatment will, by any objective measure, be somewhat superficial. For more information on this field of study we direct you to the Further Reading.

Despite that caveat, we do need to give you a brief outline of some of the key forms of data, etc. that palaeoclimatologists are using to help us understand the climate of the past, before the advent of accurate instruments, or written records, or even folklore. We quote three paragraphs from IPCC AR6, page 178.

Certain geological and biological materials preserve evidence of past climate changes. These 'natural archives' include corals, trees, glacier ice, speleothems (stalactites and stalagmites), loess deposits (dust sediments), fossil pollen, peat, lake sediment and marine sediment. By the early 20th century, laboratory research had begun to use tree rings to reconstruct precipitation and the possible influence of sunspots on climatic change. Radiocarbon dating, developed in the 1940s, allows accurate determination of the age of carbon-containing materials from the past 50,000 years; this dating technique ushered in an era of rapid progress in paleoclimate studies.

On longer time scales, tiny air bubbles trapped in polar ice sheets provide direct evidence of past atmospheric composition, including CO_2 levels, and the ^{18}O isotope in frozen precipitation serves as a proxy marker for temperature. [Covered in Section 2.5.3 of this book.] Sulphate deposits in glacier ice and as ash layers within sediments record major volcanic eruptions, providing another mechanism for dating.

Global reconstructions of sea surface temperature were developed from material contained in deep-sea sediment cores, providing the first quantitative constraints for model simulations of ice-age climates. Paleoclimate data and modelling showed that the Atlantic Ocean circulation has not been stable over glacial–interglacial time periods, and that many changes in ocean circulation are associated with abrupt transitions in climate in the North Atlantic region.

Figure 16.1 {figure 1.7 from AR6 chapter 1} shows the sources of information related to our climate, both instrumental (important for Chapter 18) and palaeoclimatic.

16.2 Interesting Tales from Planetary History

Until comparatively recently, the long-term history of our planet has been the province of geologists. The first reason for this is obvious: you have to dig through layers of rock and examine the evidence provided – clearly geology. A second reason seemed to follow from those endeavours: while the Earth is about 4.5 billion years old, the evidence suggested that it was a lifeless place until ~540 million years ago; about seven-eighths of its existence. So, the role of biology appeared to be quite limited. Discoveries during the past half century have turned that thinking on its head, and we now recognise that microbial life forms first appeared around 3.7 billion years ago. Thus, biology has been a contributor for almost the Earth's entire history.

We won't go into all the details of the geological timescale, but a few terms will be useful. The first ~700 million years are referred to as the **Hadean Eon**, during which the Earth was under continual bombardment from small planetesimals. After the Hadean came the **Archaean**, from 3.8 to 2.5 billion years ago, followed by the **Proterozoic**, which lasted until 542 million years ago. This marked the start of multicellular life, and the remaining eon – the **Phanerozoic** – is divided into the **Palaeozoic**, **Mesozoic**, and **Cenozoic Eras**, each of which is further subdivided into the periods and epochs some of you may be (vaguely) familiar with: Figure 16.2.

16.2.1 The Early Atmosphere

The Earth's first atmosphere is likely to have mainly consisted of H_2 and He, which are the most abundant elements in the universe. However, this would not have lasted, as these gases are

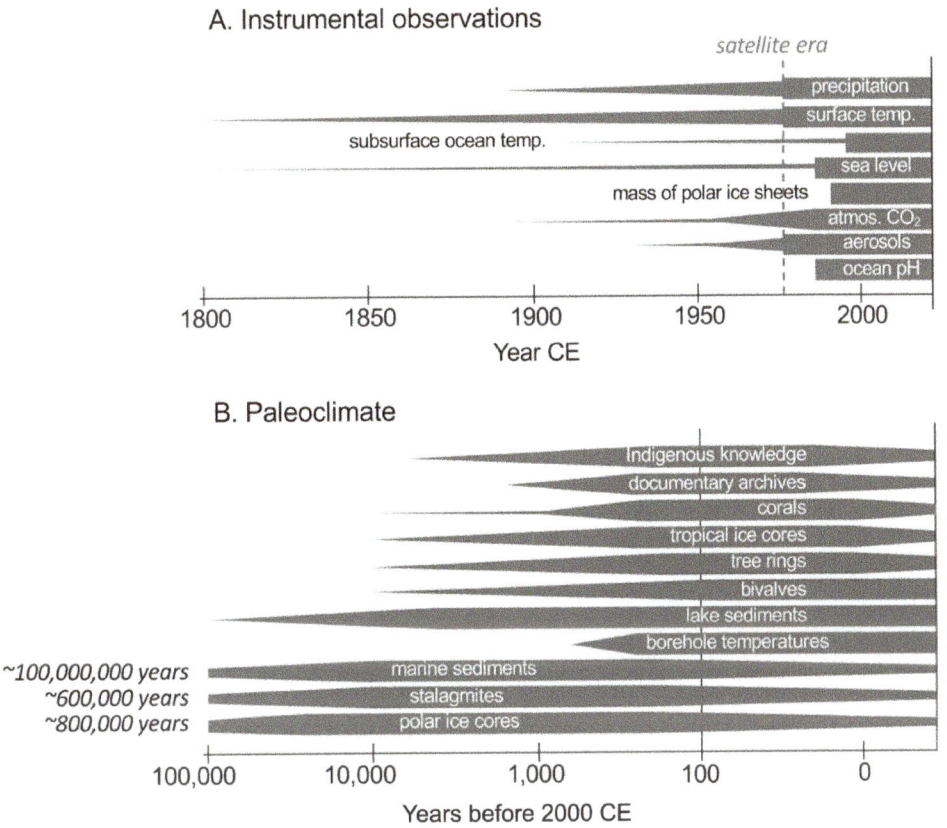

Figure 16.1 Schematic diagram of temporal coverage of (a) selected instrumental climate observations, and (b) selected palaeoclimate archives. The satellite era began in 1979 CE. The width of the taper gives an indication of the amount/extent of available record. *Source*: Acknowledgements at the end of the chapter.

too light to be gravitationally retained by the Earth. In addition, the much stronger solar wind than exists today would have helped strip this early atmosphere. During the Hadean, heating and outgassing caused by continual bombardment would have released water vapour and other volatiles, helping to form the primordial atmosphere and oceans. While such bombardment largely ceased around 3.8 billion years ago, bolide impacts have continued throughout the Earth's history, although now at widely spaced intervals. Perhaps the best known is the impact that is now widely accepted to have ended the reign of the dinosaurs, 65 million years ago, and known as the *K-T* **impact** (*K-T* for Cretaceous–Tertiary).

Evidence for the composition of the Earth's second atmosphere is mostly indirect, and incomplete. Present-day volcanic emissions are made up of 80–90% H_2O, 6–12% CO_2, 1–2% SO_2, with traces of H_2, CO, H_2S, CH_4, NH_3, and N_2. While this might constitute a good guess for earlier times, it is likely that proportions of the reduced gases (H_2, H_2S, CO, and CH_4) would have been distinctly higher when the Earth's mantle was less oxidised than it is today. Nitrogen is relatively unreactive, and is likely to have steadily accumulated. Some oxygen would have been liberated by the photodissociation of water vapour, with the hydrogen again

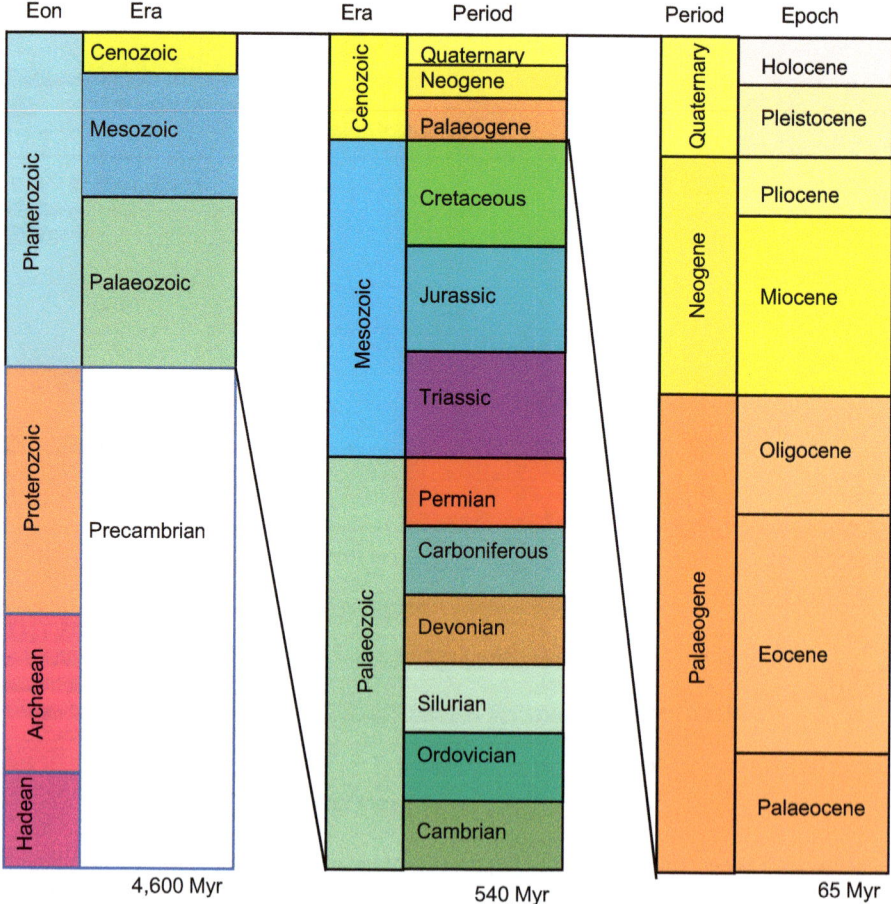

Figure 16.2 The geological timescale (at the level of detail needed for this chapter).

escaping to space. Most of the oxygen would then have reacted with crustal elements (Si, Al, Fe, Ti, etc.) to form oxides.

The earliest bacterial life forms did not have oxygen to use as an energy source, and so a number of exotic **metabolic pathways** were employed. Photosynthesis has two main steps. In the second, electrons are added to CO_2 to help convert that molecule into sugars. The first step is getting the electrons, which are stripped from a source molecule and used to create an electrochemical potential which powers the second step. Today the source of electrons is water, with oxygen a waste product, which can then be used by a different life form: animals. Before photosynthesis appeared, electrons may have been obtained by splitting H_2S, with sulphur as a waste product.

By around 2.4 billion years ago, some bacteria (cyano, or blue-green bacteria) began to develop a new pathway, using the oxygen in water to form basic organic molecules such as formaldehyde. However, this did not lead to a steady rise in the atmosphere's oxygen content:

instead, it went to oxidise large amounts of iron, one of the more abundant metals in the Earth's crust. Many of the world's iron ore deposits are thus very ancient.

Eventually the rate of oxygen production exceeded the rate at which geological processes were able to bring fresh iron to the surface, and its concentration in the atmosphere began to rise. By around 2.0 billion years ago its concentration reached 1% of its modern value. However, it was not until around 700 million years ago that it reached 10%, which was enough to allow an **ozone shield** to form, offering protection from deadly UV radiation to any land-based life forms. For the first time in its history, life was able to move out of the oceans and onto the land. The availability of new ecological niches led to the so-called **Cambrian explosion** of multicellular life forms around 550 million years ago.

16.2.2 Temperature History

One of the most intriguing questions about the early atmosphere is the strength of its greenhouse effect. Models of stellar evolution clearly imply that, around the time that life made its first appearance, the Sun's output was smaller than its present value by around 25%. This is referred to as the **faint young sun paradox**. As there is no convincing evidence to suggest that the Earth was ever frozen, this can only imply a significantly larger greenhouse effect than today. This clearly demands higher levels of greenhouse gases such as carbon dioxide than in the recent geological past.

Carbonate rocks, which are known from all periods of Earth history, were precipitated from oceans containing CO_2 that was in equilibrium with atmospheric CO_2. The characteristics of sedimentary rocks vary in response to a number of environmental factors at the time of their formation, including water depth. Some rocks from the Archaean suggest a water depth of a few kilometres, implying that there were substantial oceans at this time. From other evidence, including biological, it is likely that our planet has always been at least partially water covered, and with average surface temperatures around 7°C, although variations of a few degrees from this are certainly possible. (Average surface temperature today is ~15°C.)

Other information that can be read in the rock record is evidence of the grinding action of glaciers. This record suggests that the Earth has experienced a number of periods of glaciation – or **Ice Ages** – although it is doubtful that the planet was ever fully ice-covered. The earliest Ice Age appears at around 2.7 to 2.3 billion years ago, after which the Earth was comparatively warm (ice-free) for ~1 billion years. A second glaciation occurred around 900 million years ago, and two others followed at 820–730 million years ago and 640–580 million years ago.

We should point out that some authors have suggested that the Earth may have been fully ice-covered at times in the past: known as **Snowball Earth**. They believe that this occurred sometime before 650 Mya, during the Cryogenian Period (720–635 Mya). They argue that this best explains sedimentary deposits of glacial origin at tropical latitudes (at that point in time). Energy-Balance Climate Models (Chapter 14) imply that, once fully iced, the high albedo would make it near impossible to thaw. Perhaps 'Slushball Earth' is more likely.

What factors might account for these temperature variations? One of a number of factors which have been suggested is the movement of continents. At certain times there have been substantial land masses at or near the poles, which can provide for the accumulation of ice and snow, increasing the planetary albedo: the ice–albedo feedback process. At other times, the positioning of the continents facilitated the easier circulation of the oceans from equatorial to polar regions, substantially reducing the equator-to-pole temperature gradient. (Check the role of the transport parameter, c, in the Energy-Balance Model from Chapter 14.)

As solar luminosity steadily increased, we must assume that the major greenhouse gas, CO_2, was being steadily removed from the atmosphere. As we saw in Chapter 3, plate tectonics can affect the geological carbon cycle, impacting on the rates of volcanism and weathering. We looked at the deep carbon cycle, and it is similar processes that we must assume now, namely weathering of calcium silicate rocks:

$$CaSiO_3 + CO_2 \rightarrow CaCO_3 + SiO_2$$

Note that the white cliffs of Dover are $CaCO_3$; sand is SiO_2. We know that at times in the past there has been extensive volcanism (over millions of years), which would have increased atmospheric CO_2, leading to global warming. One associated consequence is increased rainfall, leading to increased weathering of silicate rocks.

16.2.3 The Gaia Hypothesis

The interactions between the biosphere, and the physical parts of the Earth System, may seem minor, and the impact of the former on the latter inconsequential, but that would (probably) be a mistake. James Lovelock was one of those called in by NASA when planning its first Mars landing, in the early 1970s. The question he and others were asked was, 'what were the tell-tale signs of life that the Viking lander should look for?' Lovelock studied the available data on the composition of the Martian atmosphere, and compared it to Earth's. He noted that Earth's atmosphere is a long way out of chemical equilibrium, kept that way by life processes (Table 2.1): Mars' atmosphere was not. He thus concluded that Mars was, at least today, a dead planet.

This caused him to think deeply on the role that life plays in the chemical balances of our environment as a whole, not just its atmosphere, and led him to put forward the Gaia hypothesis, which essentially states that life acts to control its environment through a range of interactions and feedback processes, in order to maximise the habitability of planet Earth. 'Life, or the biosphere, regulates or maintains the climate and the atmospheric composition, at an optimum for itself.' Stated in its strongest form, 'the Earth acts as a single super-organism', in accordance with the rules of **geophysiology**.

While few scientists would admit to taking it literally, there are aspects of this hypothesis that can be illuminating. Perhaps the most useful lesson (to us) is the idea that Nature must find ways to deal with all by-products of biochemical processes, or else they will build up and choke the planet. The rise of oxygen was just such an example, as it was toxic to almost all life forms at the time. Fortunately, that rise was sufficiently slow to allow for the evolution of new life forms that would employ oxygen as a fuel. (An element of the Gaia hypothesis is that the biosphere can evolve fast enough.) Similarly, we need to be very conscious today of our dependence on the natural cycles of vital elements (e.g. Chapter 3). Manifestly we are currently altering some of these in ways we have so far chosen to ignore.

It is also worth pointing out that Gaia – the ancient Greek Earth Mother – is interested in the preservation of life in general, not any specific life form. If one particular life form appears to be a threat, then maybe Gaia will find a way to eliminate it. The human species is not special in Gaia's eyes.

Largely in response to criticisms, even outright hostility, to the Gaia hypothesis, Watson and Lovelock introduced 'the **parable of Daisyworld**'. The model consists of a (water-free) planet populated with two types of daisies, plus bare ground, all with distinctly different albedoes. Dark daisies absorb most of the incoming solar radiation, and thus warm themselves and their surrounds. Light daisies reflect most of the solar radiation, and thus are able to 'cool' themselves. Bare ground has an intermediate albedo. Both daisy types are able to live and grow within a temperature range of 5–40°C, with an optimum growth rate at 22.5°C. Outside this range they die. (Note that the model has no geography.)

We then assume that insolation steadily increases, and the planet warms. Initially the planet is sterile. Then a temperature is reached that is warm enough for the daisies to grow, and they expand to cover more of the planetary surface. (This, of course, assumes that they can grow fast enough.) The low albedo of the dark daisies helps to increase their temperature, closer to the optimal – a positive feedback – so they grow faster. Some of this warmth is transported from dark daisy patches to the light daisies, lifting their temperature towards the optimal. Nevertheless, the planet is initially dominated by dark daisies.

Solar insolation continues to increase, pushing the daisies towards, and then past, their optimal temperature. Now the light daisies have an advantage, as their higher albedo means they are always cooler than their competitors. So, the fractional coverage by light daisies grows, while that of the dark daisies declines. Finally, of course, as the insolation continues to increase, a temperature is reached where no daisies can survive, and the planet is, once again, sterile. The lesson from the parable is that, between them, the two daisies are able to maintain the planet's (surface) temperature in a relatively stable range, hospitable for life, for a significant period of time. This is known as **homeostasis**.

A web search will provide lots more information on Daisyworld. Check it out.

16.3 Tertiary Climate

The **Cenozoic Era** covers the time from the dinosaur extinction, 65 Mya, to today, and comprises the Palaeogene, Neogene, and Quaternary Periods (the latter started ~2.6 Mya): see Figure 16.2. (Until quite recently, the Palaeogene and Neogene Periods were collectively known as the Tertiary Period.) Its climate gradually cooled, and CO_2 concentration reduced: the reasons are not fully understood.

Plate tectonics was certainly active. The impact of the Indian plate hitting the Asian continent ~40 Mya would have had a significant impact. The Tibetan uplift changed atmospheric circulation and created the Indian monsoon. This, when combined with the exposure of fresh rock, allowed enhanced weathering, facilitating the reaction above. The closing of Panama, between 13 and 2.6 Mya, strengthened the Gulf Stream/AMOC. By contrast, the Drake Passage opened up, as we discussed in the previous chapter.

In recent years, palaeoclimatologists have taken a considerable interest in the Tertiary, as more and more of the data mentioned at the start of this chapter have been located and carefully analysed. One of the main reasons for this activity is to understand any significant changes and their drivers, and to learn any lessons that may have relevance for our currently changing climate, as noted in the introduction to this chapter. For example, what was the planet like during any extended periods when temperatures were just a few degrees warmer than they are today, but where we appear to be heading?

16.3.1 Early Eocene Climatic Optimum

The first two Epochs of the Palaeogene were the **Palaeocene** (66 Mya to 56 Mya) and the **Eocene** (56 Mya to 34 Mya). Around the boundary occurred a notably warm period known as the **Palaeocene–Eocene Thermal Maximum** (PETM), when the global average temperature rose by an estimated 5–8°C. This was closely followed by the **Early Eocene Climatic Optimum** (EECO), dated at 53–49 Mya, with temperatures 10–18°C above the 1850–1900 average. AR6 chose to focus on the EECO.

The onset of the PETM/EECO was most likely the result of volcanism, which caused significant changes to the Earth's carbon cycle, with a corresponding rise in temperature. There was a sizeable reduction in the $^{13}C/^{12}C$ ratio in both marine and terrestrial carbonates and in organic carbon. Changes in other isotopic ratios (e.g. $^{18}O/^{16}O$) are also noted. These data have been interpreted as showing that ~12,000 Gt of carbon were released over roughly 50,000 years: an average of 0.24 Gt per year. (Human activity currently emits about 10 Gt of carbon per year, but we've been doing this for only a century, or so. At this rate, we'd equal the total just quoted in a 'mere' 1,200 years.)

In chapter 9 of AR6 we find their table 9.6, which compares temperatures, CO_2 concentrations, and mean sea level from a number of palaeo periods, starting with the EECO. We have reproduced that table as our Table 16.1. In particular, it lists the global mean sea level as +70 to +76 m at this time. However, it needs to be noted that at this time there was no Antarctic ice cap.

Table 16.1 Reference ranges of age, global mean surface temperature (GMST), atmospheric CO_2 concentration, and global mean sea level (GMSL) for the palaeo periods discussed in this chapter (*source*: based on IPCC AR6 table 9.6: see Acknowledgements at the end of the chapter).

Palaeo period	Years	GMST relative to 1850–1900	CO_2 (ppm)	GMSL (m)
Early Eocene Climatic Optimum	53–49 Mya	+10°C to +18°C	1150–2500	+70 to +75
Mid-Pliocene Warm Period	3.3–3.0 Mya	+2.5°C to +4°C	360–420	+5 to +25
Marine Isotope Stage	~424–395 kya	0.5°C ± 1.6°C	265–286	+6 to +13
Last Interglacial	~129–116 kya	+0.5°C to +1.5°C	266–282	+5 to +10
Last Glacial Max.	21–19 kya	–5°C to –7°C	188–194	–125 to –134
Last Deglacial Trans.	18–11 kya	n/a	193–271	–120 to –50
Early Holocene	11.65–6.5 kya	n/a	250–268	–50 to –3.5
Mid-Holocene	6.5–5.5 kya	+0.2°C to 1.0°C	260–268	–3.5 to +0.5
Last Millennium	850–1850 CE	–0.14°C to +0.24°C	278–285	–0.05 to +0.03

16.3.2 Mid-Pliocene Warm Period

The **Pliocene** Epoch, from 5.333 Mya to 2.58 Mya, is the final epoch of the Neogene Period. Recently, considerable research has focused on the **Mid-Pliocene Warm Period** (MPWP), 3.3–3.0 Mya. The CO_2 concentration was 360–420 ppm, higher than the pre-industrial value of 280 ppm, but comparable to today. Global mean surface temperature is estimated to have been between 2.5°C and 4.0°C, relative to 1850–1900 (*medium confidence*): Table 16.1. Model simulations show a multi-model surface temperature of 3.2 [2.1 to 4.8] °C warmer than control simulations. The consistency between these two approaches gives us confidence in other conclusions.

The important conclusion about this event is sea level, which was between 5 and 25 m higher than today (*medium confidence*), 25 m being regarded as a 'plausible' upper bound. So, what/ where was the source? Ice-sheet models have been used in an effort to answer this very important question. They indicate that Northern Hemisphere glaciation was limited to high elevations in Greenland, with the more recent glaciation appearing in the cooler late Pliocene. An Antarctic contribution is clearly required, with the suggested range of 5.4–17.8 m, mainly from the East Antarctic Ice Sheet.

Given the clear relevance of this event to our current and (potentially) future climatic conditions, ongoing research is vital. Of course, it should be pointed out that ice sheets have long timescales, so that any rise of this magnitude could take thousands of years. There is also an interesting technical point to add. As the Greenland Ice Sheet was forming, it was the temperature at its base that was critical. Today, for that ice sheet to melt, it is the temperature at elevations for 1–2 km up the plateau, which will, of course be colder, that are important.

16.4 Pleistocene Climate

The Quaternary Period started 2,588,000 years ago, and contains the Earth's current Ice Age. It is divided into the **Pleistocene** and **Holocene Epochs**, with the latter, which started 11,700 years ago. Note that an Ice Age typically contains glacials and interglacials: we are currently in an interglacial; that does not mean this Ice Age has finished. Understanding the glacial–interglacial cycle of the Pleistocene, and also the remarkably benign climate of the Holocene, are both important scientific challenges.

In the early nineteenth century, geologists began to study the glaciers in the European Alps, and concluded that some of the erratic pieces of alpine rock that were 'randomly' scattered on the slopes had been moved there by the actions of glaciers. In 1837, Louis Agassiz proposed that the Earth had been (largely) covered by ice in the relatively recent past. He suggested that Switzerland had been like Greenland, containing not just glaciers, but one vast sheet of ice. Evidence from other parts of the world was then added; for example, landscapes that had been scraped smooth in some way.

It took many more years, and the development of new scientific techniques such as radioisotope dating and stable isotope ratios, before a full picture of the current Ice Age emerged. We now know that ice sheets covering large parts of Eurasia and North America have advanced and retreated many times in the past ~2.5 million years, with perhaps the best evidence coming from studies of ice cores from both Greenland and Antarctica, which now provide data covering the past 800,000 years.

Figure 16.3 {figure 5.3 from IPCC AR5} contains a number of plots based in part on this ice core (and related) data. Part (g) shows δ ^{18}O (the variation in the ^{18}O/^{16}O ratio) from benthic foraminifera from marine sediments (a proxy for global ice volume), while parts (e) and (f) show the inferred tropical and Antarctic sea-surface temperature (respectively) based on this and other data such as the deuterium ratio. Also shown is the mixing ratio of CO_2 (d) and reconstructed sea level (h). (The other traces will be discussed below.) The temperature data show a clear cycle with a period of roughly 100,000 years, with additional structure hinting at other possible cycles. What might be the cause of such behaviour?

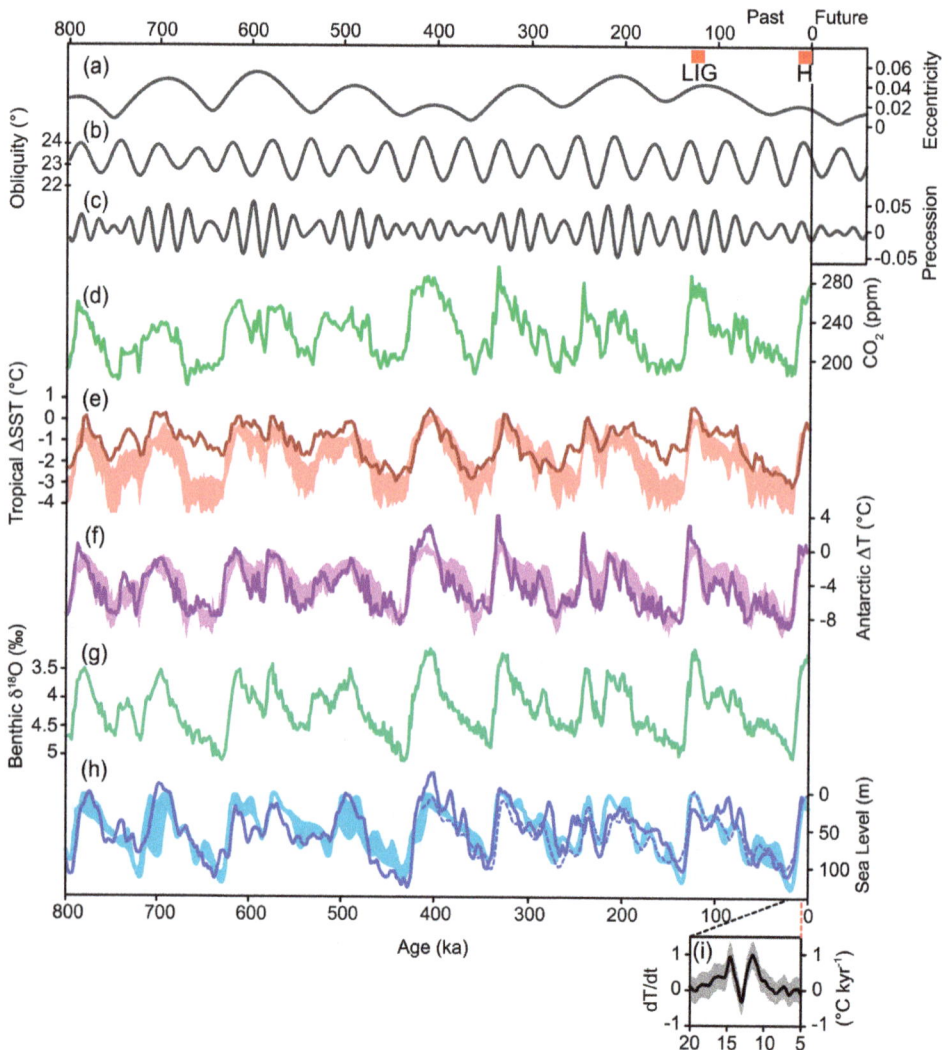

Figure 16.3 Orbital parameters and proxy records over the last 800 kyr. *Source*: see Acknowledgements at the end of the chapter.

16.4.1 Orbital Perturbations

In the 1920s and 1930s, Serbian astronomer Milutin Milankovic suggested an explanation could be found in the small changes in the Earth's orbital parameters, which are the result of the gravitational effects of the other planets, especially the largest, Jupiter. At that time there was little or no reliable data which could be used to verify this theory (the cycles just mentioned were yet to be established), so it was largely ignored. Today it is clear that these orbital changes are the key to the glacial–interglacial cycles throughout the Quaternary.

As Kepler first noted, the Earth's orbit is elliptical rather than circular. The **eccentricity** of the Earth's orbit varies between 0.0001 (very close to circular) and 0.068 (noticeably elliptical), and back, with a period close to 100,000 years: trace (a) in Figure 16.3. (In fact, the variation in eccentricity consists of multiple cycles with different periods and amplitudes, but the dominant period is ~100,000 years.)

Figure 16.4 illustrates this variation with two cases: an ellipticity of 0.0 (i.e. a circle) in black, and 0.2 in blue. (This is larger than the largest actual value, but was the smallest that clearly shows the differences.) The Sun, which sits at the focus of the ellipse, is in yellow, and the centre of the ellipse is marked. The distance between the two is given by $c = ea$, where a is the length of the semi-major axis, and e is the eccentricity (see Figure 9.3).

A technical, and personal, aside. We undertook a search for suitable diagrams to illustrate this variation, and could not find one we considered adequately accurate. So we need to stress a few salient points. The Sun is always at a focus of the ellipse, never the centre. The length of the major axis remains constant: this means that the orbit, effectively, 'slides sideways'. The constancy of the major axis also means that the orbital period (the year) remains constant – this is effectively Kepler's third law – and the total mechanical energy (i.e. kinetic plus potential) also remains constant (this is more or less implied by Kepler's second law).

The tilt of the Earth's axis (its **obliquity**) – currently 23.5° – varies between 22° and 24.5° with a period of 41,000 years; trace (b) in Figure 16.3. The larger this angle, the more extreme is the summer–winter insolation variation for mid and high latitudes. Finally, this axis **precesses**, with a period of 23,000 years: trace (c) in Figure 16.3. This affects which hemisphere is in summer at

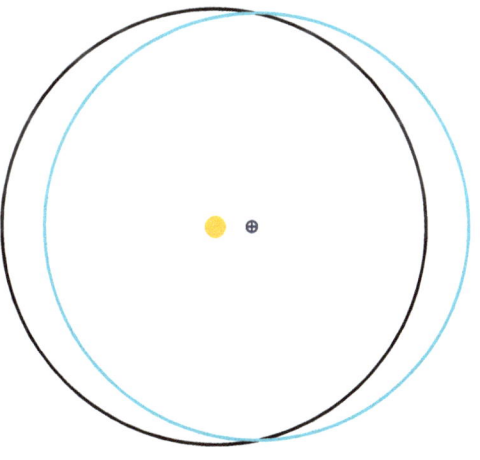

Figure 16.4 Earth's orbit for two values of the eccentricity.

perihelion (the time of the year when we are closest to the Sun). The Earth's orbital ellipse also rotates (apsidal precession), as does Mercury's (the explanation of which is one of the early triumphs of Einstein's General Relativity), and this slightly affects the 'effective' value of the precession of the Earth's axis.

How might such orbital changes – or orbital forcings – lead to the significant climatic effects we see in the glacial–interglacial cycle data? The eccentricity cycle corresponds remarkably well with the dominant glacial–interglacial period. At times when the eccentricity is close to maximum, the Earth–Sun distance varies by ~5% between perihelion and aphelion, and solar insolation varies by 10%. However, the total annual insolation is essentially unaffected (the difference is on the order of 0.2%), so while we might expect effects at the regional (even hemispheric) level we should not – at first sight – expect a global signal. Yet a glacial maximum clearly has a global signal, which is evident in the Antarctic ice core data.

As we know that the ice sheets build up on high-latitude land masses, and that the majority of such land is found in the Northern Hemisphere, this is where we must focus our attention. Surprising as it may seem, it is not cold northern winters that are the trigger. Even when the Northern Hemisphere is in winter at perihelion, snow and ice will still cover the high latitudes. However, six months later when summer comes, and the Earth is at aphelion, there will not be sufficient insolation to melt it all. Come the next winter there will still be some ice and snow to help cool the local atmosphere, directly and indirectly, encouraging more to accumulate. Now we have the starting conditions for a slow but steady ice–albedo feedback.

16.4.2 Carbon Cycle Feedbacks

Both the $\delta^{18}O$ data and the sea-level data show a very asymmetric pattern: how can we explain this? Modelling studies based just on orbital forcing, plus the ice–albedo feedback mechanism, are unable to produce the magnitude of global climate change attested to by the ice core data: clearly, additional processes are required. The ice core data also indicate that CO_2 levels dropped, more or less in sync with temperature. This obviously led to a reduced greenhouse effect, amplifying/driving the temperature drop.

So, what has driven this carbon cycle feedback? While a number of factors are almost certainly at work, the primary factor is the greater solubility of CO_2 in the colder oceans. This would also increase the biological carbon sequestration in the ocean depths. Plant growth on newly exposed continental shelves would also have made a contribution. However, this is likely to be offset by the expanding ice sheets that were replacing the boreal forests.

16.4.3 Sea-Level Variations

As we have already indicated, one of the big questions that climate science is facing is what will happen to global sea levels in the decades to centuries, and longer, into the future. At present, the global community is aiming to hold global warming to 2°C above pre-industrial values: and preferably 1.5°C. The latter seems highly unlikely given current emissions trajectories, and even the former may be optimistic. But let's say we do level off around the +2°C mark; what then? As noted in the quote from FAQ 1.3 at the start of the chapter, both sea level, and the carbon cycle, are much slower to respond to greenhouse forcing than global surface temperature.

Why might this be? The carbon cycle is slow to respond, as the main mechanism for its drawdown from the atmosphere is the establishment of new ecosystems, starting with plant (and microbe) communities, which will then draw in various animal species. Sea-level rise and fall is mainly driven from the ice sheets. Water has a latent heat of 334 kJ/kg, but a specific heat of 'only' 4.182 kJ/kg/°C. To put that in perspective, the amount of heat that would melt 1 kg of ice would raise the temperature of that melted water by 80°C. That is thermal inertia!

There have been periods during past interglacials when temperatures have been just a few degrees warmer than today, with CO_2 concentrations no more, and often less than today. These periods typically lasted for a few millennia or more, long enough for these components of the system to equilibrate. We will briefly discuss a couple of these periods, and the current IPCC conclusions.

Approximately 424–395 kya is a period with the somewhat odd name of **Marine Isotope Stage** (MIS) 11. The CO_2 concentration was 265–286 ppm, much the same as the oft-quoted pre-industrial value of 280 ppm. Global mean surface temperature, relative to 1850–1900, is given as +0.5°C ± 1.6°C. Yet sea level was 6–13 m higher: Table 16.1. This would require the loss of much of the Greenland and West Antarctic Ice Sheets, and possibly also a contribution from East Antarctica. This clearly demonstrates the high sensitivity of both ice sheets to extended periods of thermal forcing, even at warming levels comparable to today.

One of the best-studied periods is the Last Interglacial (known as the **Eemian**), about 129–116 kya. The CO_2 concentration was very similar to MIS 11, and the temperature was only 1.0°C ± 0.5°C above 1850–1900: Table 16.1. The IPCC states 'it is *virtually certain* that global mean sea level was higher than today, likely by 5–10 m'. Again, this can only be achieved by substantial contributions from the Greenland and Antarctic Ice Sheets, whose long-term stability must now be considered questionable.

The final period worth studying is the Last Glacial Maximum (LGM), 19–21 kya. The CO_2 concentration was 188–194 ppm, and temperatures 5–7°C cooler than 1850–1900: Table 16.1. What do we know about sea level, and its association with the ice sheets? From geological proxies, and models, our best estimates are that sea level was 125–134 m below present. Various studies have attempted to 'partition' this water loss among various ice sheets. A reasonable estimate is that 76 ± 7 can be said to be from the North American Laurentide Ice Sheet, 18 ± 5 m from the Eurasian Ice Sheet, 10 ± 2 m from Antarctica, 4 ± 1 m from Greenland, 5.5 ± 0.5 from glaciers, and 2.4 ± 0.3 m from an increase in ocean density (greater salinity). The rest may be allocated to groundwater, land ice reservoirs, and lakes (still uncertain).

Finally, there are two small points which need to be taken into account. Clearly, thermal expansion will have an impact on sea level when we are looking at temperatures that differ from today. The second is isostatic rebound. Parts of the high northern latitudes were under 2 km (and more) of ice at glacial maxima. When that is removed, these land masses will slowly rise under a form of hydrostatic pressure from other parts of the planetary surface, transferred underneath by the mantle, which is not quite a solid.

16.5 Holocene Climate

16.5.1 Younger Dryas

One of the most remarkable, and dramatic, examples of climate variability occurred at the very end of the Pleistocene, as the world was coming out of the Last Glacial Maximum. It is actually named after an arctic flower, the Younger Dryas, which shows up in pollen and ice core records, indicating a sudden drop in temperatures, back towards glacial levels. This relatively sudden 'cold snap' appears to have been largely confined to continental areas around the North Atlantic – that is, Canada and western Europe – with much smaller impacts elsewhere. It took around a century to kick in, lasted just over a thousand years (12,900 to 11,700 years bp), and then, in the space of a few decades, temperatures rose sharply towards the typical Holocene temperature pattern.

All of this occurred at a time of wide-scale glacial retreat, when the ice sheets that had covered large areas of North America (most of Canada, and large parts of northern USA) and Europe were giving way to lower albedo land surfaces which should have been aiding the warming process. To find an answer to this paradox we need to look for more subtle physical processes that we have so far not included in our discussion.

The 'standard' answer (although not universally accepted), as might be guessed, lies in the Atlantic Ocean, and more specifically the Atlantic Meridional Overturning Circulation (AMOC) – the conveyor belt discussed in Chapter 13. As the North American ice sheets began to melt, about 14,000 years ago, melt water was collecting in south-central Canada and northern USA, and flowing down the Mississippi River, with a flow comparable to the Amazon today. At this time glaciers were blocking any flow to the east. This created a significant influx of fresh water into the central Atlantic.

However, there came a time when those glaciers had retreated sufficiently that much of this melt water was now able to flow across what would soon become the Great Lakes and down the St Lawrence River, providing a large influx of fresh water to the North Atlantic. Now we remember that the driver of the AMOC is the sinking of cold, dense waters in the North Atlantic, caused in part by the increase in salinity following evaporation. With this huge influx of fresh water into the North Atlantic Ocean, the salinity was too low to allow the water to become dense enough to sink, and the conveyor belt stopped. And that stopped the flow of the Gulf Stream, with its warming effects on Canada and Europe.

16.5.2 Early Civilisation

Human civilisation has largely arisen during the relatively stable – even benign – climate of the Holocene, which started 11,700 years ago. For this time period the data available to us include tree-ring data, written records, and archaeological evidence left by some early farmers and urban dwellers.

In the early Holocene, the Sahara turned green. Grass grew and lakes filled, grassland animals moved in and humans followed. Mud cores off West Africa indicate the Sahara's wet period ended abruptly about 5,000 years ago. In a few hundred years the landscape dried dramatically, with some people moving to the Nile valley, where the Egyptian civilisation arose. (The Early Dynastic Period is dated at ~3150–2686 BCE, followed by the Old Kingdom, 2686–2181 BCE, during which the Great Pyramids were built.) A green Sahara recurs because of the axial precession, which affects how much of the monsoon rain arrives.

We know that some of the earliest city states flourished and then vanished. While conflict is one of the causes, regional-scale climate change is also considered to be an important factor in some cases. Whether this is a result of climate variability or of local-scale feedbacks resulting from land-use change is an interesting question.

A recently analysed speleothem (limestone stalagmite) from a cave in north-eastern Iraq has helped shed light on the rise and dramatic fall of the Assyrian Empire. Researchers were able to reconstruct a high-resolution 4,000-year record of precipitation and temperature (using ratios of uranium and oxygen isotopes), and align this with archaeological and cuneiform records, with remarkable correlation.

The rise and zenith of the Assyrian Empire, from 920 to 630 BCE, occurred during a period of higher-than-average rainfall. The rapid fall, between ~620 and 600 BCE, lies within a drought period from 675 to 550 BCE. (Ashurbanipal, the last great Assyrian king – 'king of Assyria, king of the world' – ruled from 669 to 631 BCE.) Nineveh, the largest city of its day, was not resettled for over a century. (This turned out to be a blessing for archaeologists 2,500 years later, who found an absolute treasure trove of cuneiform tablets buried in the rubble.) The Assyrians relied on seasonal rainfall; the Babylonians to the south, who conquered them, used irrigation.

16.5.3 The Past 2,000 Years

We do have a good picture of climate over the past 2,000 years, with reasonably tight constraints on global mean temperature. The period from 950 to 1250 CE is known as the **Medieval Warm Period** (or Medieval Climatic Anomaly), and was characterised by warmer temperatures in the North Atlantic region, and possibly also China. This was the time of the Viking settlement in Greenland. We cannot be too exact on numbers, but it is likely that temperatures then were at least 0.1°C below the period 1960–1990.

This was followed by a period known as **The Little Ice Age** (1350–1850 CE), when European temperatures dropped by 1°C or possibly a little more. (The Viking settlement in Greenland is believed to have ended some time between 1450 and 1500.) In the middle (1645–1715) is the '**Maunder Minimum**', when sunspot numbers were anomalously low. Decreased solar activity also allowed for a higher creation rate for ^{14}C, which is seen in tree rings. Hence the Sun is considered a primary factor in this climatic anomaly, but it may not be solely responsible.

Volcanic activity is another cause of climatic variability, and is an important input into modelling efforts for the past 2,000 years. However, two issues remain. Firstly, while it is often easy enough to find the evidence for a significant eruption, dating it with the level of precision required is often difficult. (Tree-ring data are capable of providing useful information on several climate variables on a year-by-year basis; sulphate deposits are now extracted from ice cores.) Secondly, we need to estimate the resulting stratospheric loadings, which must always be considered approximate.

Many of the world's most active volcanoes are found around the so-called Pacific Ring of Fire, and especially the Indonesian archipeligo, so that European knowledge of them really only began with the 'Age of Exploration', starting around 1500. The 1883 eruption of Krakatoa was well documented; however, the 1815 Tambora eruption – around 10 times larger, and causing 'the year without a summer' in Europe – took some unravelling. More recently, scientists were able to pin down key details of an eruption of the Samalas volcano, on the island of Lombok, in 1257. This blast had widespread impacts across Europe, including excessive rain that ruined crops and led to major famine.

16.5.4 Sea-Level Rise during the Holocene

About half of the GMSL rise since the Last Glacial Maximum (LGM) – between 50 and 60 m – occurred during the early Holocene at a sustained rate of about 15 m per thousand years, from around 11.4 to 8.2 kya, possibly punctuated by melt-water pulses. An abrupt rise of about 1.1 m was associated with drainage of Lakes Agassiz (in central North America) and Ojibway (north-east Canada), associated with the collapsing Laurentide Ice Sheet, which covered (most of) Canada and the northern USA, which contributed 27 m to sea-level rise. The Scandinavian Ice Sheet contributed about 2 m. The Greenland Ice Sheet contributed about 4 m. Finally, estimates of the Antarctic contribution during the early Holocene vary from about 1.2 m to 8.5 m. Further data can be found in Table 16.1.

16.6 The Anthropocene

For some years now, a growing number of (environmental) scientists have been suggesting that human activity is having such a profound impact on our planet that we may have entered a new epoch, the Anthropocene. However, such determinations normally belong to the domain of geology, and its professionals. Generally speaking, they look for stratigraphic evidence of a 'golden spike': a Global Boundary Stratotype Section and Point (GSSP). A group of scientists called the Anthropocene Working Group believe they have found such a spike in Lake Crawford, in Ontario, Canada (with supporting evidence from other locations).

This highly stratified lake allows sediment to steadily collect at the bottom, which contains the evidence they have been seeking. In particular, since 1950 these sediments show the arrival of fly ash from fossil fuel burning, nitrogen from fertilisers, microplastics, and plutonium from nuclear bomb detonation. They propose this date as the start of the new epoch. This, of course, is in line with the decision of the carbon dating community to use 1950 as 'present' when quoting carbon dates as so many years bp, and for related reasons.

Summary

This chapter has explored a number of facets of the environmental history of our planet, from its beginnings. Our knowledge of the past composition of our atmosphere is, of course, mostly indirect: our knowledge of the past climate even more so. We may divide this subject into two specific topics. The first is the deep past – a challenge that a number of scientists have spent time addressing – and the second centres around sea level in the 'more recent' past (the Tertiary and Quaternary); including the current Ice Age.

Within the first of these, there are two very interesting issues. The first is the so-called faint young sun paradox: the fact that solar insolation was distinctly lower in the distant past. This clearly calls for a stronger greenhouse effect, but also calls for processes to reduce this, as solar output increased over time. The second is the rise of oxygen, a chemically and biologically fascinating story. The Gaia hypothesis provides some food for thought.

Many aspects of our climate, such as temperature, respond very quickly – decadal time-scales – to changes in the atmospheric concentration of CO_2. Sea level does not. As sea-level rise may well be a serious issue for humanity in the decades to centuries to come, we look to the past to guide us. Information is now available for two time periods during the Tertiary Period where we have reliable data: the Early Eocene Climate Optimum, and the Mid-Pliocene Warm Period. (Note that the word 'period' is being used with multiple meanings here.)

The Earth is currently in an Ice Age, consisting of a glacial–interglacial cycle, starting around two and a half million years ago: right now we are in an interglacial. We understand that this cycle is driven by the gravitational attractions of the other planets. Modern humans appeared during the previous interglacial, and endured the Last Glacial Maximum (as did most of the biosphere, of course). Human civilisation has arisen during the relatively benign Holocene, with quite stable CO_2 concentration, temperatures, and sea level.

Acknowledgements

Figure 16.1 {figure 1.7}, Table 16.1 {table 9.6}:

Climate Change 2021: The Physical Science Basis. Contribution of Working Group I to the Sixth Assessment Report of the Intergovernmental Panel on Climate Change [Masson-Delmotte, V., P. Zhai, A. Pirani, S. L. Connors, C. Péan, S. Berger, N. Caud, Y. Chen, L. Goldfarb, M. I. Gomis, M. Huang, K. Leitzell, E. Lonnoy, J. B. R. Matthews, T. K. Maycock, T. Waterfield, O. Yelekçi, R. Yu, and B. Zhou (eds.)]. Cambridge University Press, Cambridge, United Kingdom and New York, NY, USA. doi:10.1017/9781009157896.

Figure 16.3 {figure 5.3 from AR5}:

Masson-Delmotte, V., M. Schulz, A. Abe-Ouchi, J. Beer, A. Ganopolski, J. F. González Rouco, E. Jansen, K. Lambeck, J. Luterbacher, T. Naish, T. Osborn, B. Otto-Bliesner, T. Quinn, R. Ramesh, M. Rojas, X. Shao and A. Timmermann, 2013: Information from Paleoclimate Archives. In: Climate Change 2013: The Physical Science Basis. Contribution of Working Group I to the Fifth Assessment Report of the Intergovernmental Panel on Climate Change [Stocker, T. F., D. Qin, G.-K. Plattner, M. Tignor, S. K. Allen, J. Boschung, A. Nauels, Y. Xia, V. Bex and P. M. Midgley (eds.)]. Cambridge University Press, Cambridge, United Kingdom and New York, NY, USA, pp. 383–464. doi:10.1017/CBO9781107415324.013.

FURTHER READING

- *Earth. Evolution of a Habitable Planet* by J. I. Lunine (Cambridge University Press, 1999).
- *Atmospheric Change. An Earth System Perspective* by T. E. Graedel and P. J. Crutzen (W. H. Freeman, 1993).
- *Paleoclimatology. Reconstructing Climates of the Quaternary* by R. Bradley (Elsevier, 2013).
- *Paleoclimatology. From Snowball Earth to the Anthropocene* by C. P. Summerhayes (Wiley-Blackwell, 2020).

REVIEW QUESTIONS

1. What was the earliest atmosphere composed of?
2. Why did it not last?
3. Approximately when did the earliest forms of life appear on Earth?
4. What evidence do we have to constrain temperatures during the Archaean?
5. What happened to the oxygen produced by photolysis of water vapour?
6. Why could terrestrial life not emerge before the Earth had an oxygen atmosphere?
7. What evidence prompted geologists to posit that the Earth had undergone a recent Ice Age?
8. Why is the Northern Hemisphere the key focus of the Milankovic hypothesis?
9. Explain the feedback processes that are necessary to make the Milankovic hypothesis actually 'fit the data'.
10. Why might the Earth have been gradually cooling during The Little Ice Age?

EXERCISE

Your Energy Balance (Budyko–Sellars) model from Chapter 14 can provide for a very informative exercise for this chapter. Most of the parameters in that model are meant to be based on the physics of the relevant process, even though this has been grossly simplified. However, the Sun is 'external' to the problem, and we have seen that its output has steadily increased during its life. What happens if we reduce Q? Inspection of Equation (14.10) quickly shows that T_{av} must decrease: this invariably means that the entire temperature distribution is reduced, which in turn implies that the ice line moves to lower latitudes. This will cause the average temperature to lower a little further: the positive ice–albedo feedback has set in. Depending on the actual choice of the various parameters, it may be found that only a modest decrease in Q is needed to produce a fully iced Earth.

What happens to this planet if we steadily return Q to its current value, or even higher? It turns out that the planet remains iced, because of the high albedo! This suggests that our current, partially iced, climate is an unstable state. The simpler formulations are obviously the best place to start. Set $Q = 341$ W m^{-2}, and use Equation (14.3). The step ice line of Equation (14.6a), and the long-wave formulation of Equation (14.4), are also logical. Set the cloud fraction at 0.5, at least to begin. We would also suggest you use equal steps in x (say, 0.05) rather than latitude. Start with an initial guess for the temperature distribution, compute T_{av} and iterate using Equation (14.8). (You might use some form of convergence criterion, or simply perform 20 iterations.) When you are steadily increasing (or decreasing) the solar constant, always use the temperature distribution of one run as the initial guess for the next.

Part V

Our Changing Climate

In the first four parts of this book we focused on the first half of its title: the science. We looked at the chemistry of our environment, and how it has been changing. Then we looked at the atmosphere as a physical system, and the basic laws that govern it, and its circulation. In Part III we looked at radiant energy, the ultimate driver of climate, and some of the factors that have at least the potential to alter either the inflow, or the outflow, of that energy. Finally, in Part IV, we pulled the various pieces together, to see how well we understood the climate system. We know that we can build climate models, with various degrees of accuracy, while also understanding their limitations and uncertainties. This understanding is central to good science.

In the final four chapters we turn our attention to the second half of the book's title: our changing climate. Are we in a position to ask, and answer, the all-important questions that those words denote? We are confident that the answer is yes, provided we ask the right questions, and understand any caveats. Note that this is a rapidly evolving 'space'. Much of the material in these chapters has been extracted from AR6, the most recent IPCC Report. However, in a few years it will be superseded: we hope our readers will endeavour to keep abreast of any developments in the years ahead.

Chapters 17 and 18 focus on the climate of the past century or so, and its changes. Firstly, we look at the changes in atmospheric composition – greenhouse gases, aerosols, etc. – plus any changes at the surface, especially its albedo. All of these changes are likely to alter the balance of energy flowing either into, or out of, the Earth–atmosphere system. We need to quantify these forcings. We then look at the quantifiable changes observed in our environment, not just temperature. Can we connect the changes in the energy fluxes to the changes in the environment? Using the simulations provided by our models, we will see that the answer is yes.

Finally, in Chapters 19 and 20, we focus on the twenty-first century. In the first we look not so much at climate but at weather, and the rise of 'extreme events'. What is really happening? Are we able to make sense of it? And what of the decades ahead? As we clearly cannot know what decisions our leaders will make, we have to make some educated guesses: scenarios of how our world might 'evolve'. Then we use our models to simulate the climatic consequences: many of the resulting projections (the technical term) are far from comforting.

17　Driving Climate Change

We saw in the previous chapter that climate has changed in the past, and some of these changes can be described as dramatic. However, when we focus on the *rate of change*, most of these changes were, in reality, quite gradual. Over the past century or so, and especially since the end of the Second World War, our climate has actually been changing at an unprecedented rate. And we believe we know why.

Human activity has been changing the composition of our atmosphere, and has also been responsible for significant changes to the land surface, such as cutting down forests to grow crops or graze livestock. And we have seen that the Earth's surface temperature has risen by around 1°C, with dramatic impacts which are beginning to be felt. (Some of these will be examined in Chapter 19.) Are these changes connected?

There are two steps in answering this question. The first of these is to quantitatively connect the changes just mentioned with any changes in the radiant energy fluxes into and out of the Earth–atmosphere system: the fluxes that ultimately govern our climate. This is the thrust of this chapter. The second of these steps, determining the response of the Earth System to any flux changes, is somewhat more challenging, and requires a 'full-scale' 3D model of the system: that will be the subject of the next chapter.

In this chapter we will be using some of the technical terminology, and acronyms, used in AR6. We have collected these in a box towards the end of the chapter. [In all cases where we use an AR6 figure or table, etc. we will indicate the source in curly brackets: {..}.]

17.1　Effective Radiative Forcing

It is relatively straightforward to define the radiative forcing produced by the increase in a greenhouse gas such as CO_2: run a radiative transfer code 'before' and 'after', and compare the additional radiation absorbed. But what do we mean by before and after? For this chapter, the answer is relatively simple: 'before' means with the situation in the time frame of 1750–1850 (gas concentrations, aerosols, land use, etc.); 'after' usually means today, unless we are attempting to understand climate change as a function of time. This, naturally, implies that the forcing is likely to change from one IPCC Assessment Report to the next, but that is simply a consequence of the desire that each report be as up to date as possible.

However, the IPCC makes use of several variants of radiative forcing, which we now outline. **Instantaneous radiative forcing** (IRF) is the simplest: it is the net change in the top of atmosphere radiative flux following a 'perturbation'. For example, for the greenhouse gases (CO_2, CH_4, N_2O), this involves running high spectral resolution models to obtain the most accurate results. Some special care is needed due to the spectral overlap between the different gases. In AR5, the IPCC used the **Stratospheric-Temperature-Adjusted Radiative Forcing** (SARF), to allow for the response to any forcing by the stratosphere, which is radiatively decoupled from the troposphere.

AR5 and AR6 take this one step further, by introducing the **effective radiative forcing** (ERF), which also includes responses in the troposphere and effects on clouds and circulation; collectively referred to as 'adjustments'. Specifically, these include 'adjustments in both tropospheric and stratospheric temperatures, water vapour, clouds, and some surface properties, such as surface albedo from vegetation changes, that are uncoupled from any changes in surface air temperature'. (This requires running a full Earth System Model.)

Adjustments are processes that are independent of **global surface average temperature** (GSAT) changes (and generally occur on timescales of hours to months), while feedbacks refer to changes caused by GSAT change (on longer timescales). These definitions split the feedbacks we considered in Chapter 15 into two categories. The full details of the ERF can be found in Chapter 7 {box 7.1} of AR6, and are somewhat technical: definitely beyond the scope of this book. For consistency, we will be quoting ERF throughout this chapter.

17.2 Natural Drivers

While our primary focus is on the anthropogenic drivers of climate change – after all, these are the ones we may (or may not) be able to mitigate – we must not lose sight of the fact that there are other drivers of climate change: the natural drivers, some of which were a key focus of the previous chapter. And they are still – most likely – in play today. So we should examine these first.

17.2.1 Solar and Orbital Forcing

We have seen that solar output is not constant, but varies both by virtue of stellar evolution, and the sunspot cycle. In round numbers, stellar evolution is probably increasing insolation by somewhere between 5% and 10% per billion years: 0.005% to 0.01% per million years; or 0.000005% to 0.00001% per thousand years. I think we can safely eliminate that from further consideration. The sunspot cycle appears to be a regular, cyclic fluctuation, without any long-term (multidecadal) trends.

However, any longer-term – or 'low-frequency' – fluctuations would, of course, be a very different matter. AR5 assessed solar variability over a range of timescales, and concluded that total solar irradiance (TSI) fluctuations over the past 9 kyr were <1 W m^{-2}, but with no assessment of confidence. AR5 also concluded that TSI variations show little change ($<0.1\%$) since the Maunder Minimum (1645–1715). Further work has strengthened these conclusions, with the current claim that, since the late nineteenth century, the global mean solar forcing is in the range of –0.06 to +0.08 W m^{-2} (*high confidence*).

The slow, periodic changes in the Earth's orbit that we discussed in the previous chapter mainly cause variations in seasonal and latitudinal insolation. Calculations of such orbital variations are now available, going back tens of millions of years. Over the past million years, average insolation over boreal summer at 65° N varied by as much as 83 W m^{-2}, but only 3.2 W m^{-2} during the past thousand years, with no substantial effect on globally averaged radiative forcing (0.02 W m^{-2} during the past millennium).

17.2.2 Volcanic Aerosol Forcing

A large volcanic eruption will inject large quantities of sulphur gases into the stratosphere, which will then be converted to a sulphate aerosol layer, as we saw in Chapter 4. Such a layer will enhance the planetary albedo, with significant impacts on insolation, lasting a few years. As such, they are the dominant natural driver of climate variability. One of the Carbonator scenarios focuses on a large volcanic eruption: check it out.

The eight largest volcanic eruptions in the past two centuries are believed to be: Tambora (April 1815), Krakatoa (August 1883), Tarawera (June 1886), Pelee/Soufrier/Santa Maria (May–October 1902), Katmai (June 1912), Agung (March 1963), El Chichon (April 1982), and Pinatubo (June 1991). Can we trace such impacts back, beyond the records of the past few centuries?

In recent years, scientists have found signatures in sulphate layers in the Greenland ice cap, and have produced reconstructions going back 2,500 years, and even further, but with less certainty. The time between large eruptions varies between 3 and 130 years, with an average of 40–45 years. Averaged over the period 950–1250 CE, stratospheric aerosol optical depth (SAOD) was 0.012, and similar to the period 1850–1900 (0.011). It was higher in 1450–1850, at 0.017. AR6 concludes that volcanic aerosol forcing since 1900 is 'not unusual' in the context of at least the past 2,500 years (*medium confidence*).

17.3 Greenhouse Gases

Greenhouse gases (GHGs) are, of course, dominated by the 'big three', collectively referred to as **well-mixed greenhouse gases** (WMGHGs). All have atmospheric lifetimes of around a decade or more, as we saw in Chapter 2. In addition, there is a growing list of other gases that are radiatively active; that is, greenhouse gases.

17.3.1 WMGHGs

Figure 17.1 shows the growth in these three gases since 1850, based on the latest IPCC data {annexe III 1a}. AR6 also provides a table of the CO_2 growth rate for a number of time periods {table 2.1}. For the period 1000–1750, its concentration rose slightly from 278 ppm to 285 ppm, which, allowing for measurement uncertainty (e.g. different ice cores), gives a growth rate that is

Figure 17.1 Concentrations of CO_2, CH_4 and N_2O since 1850. *Source*: Acknowledgements at the end of the chapter.

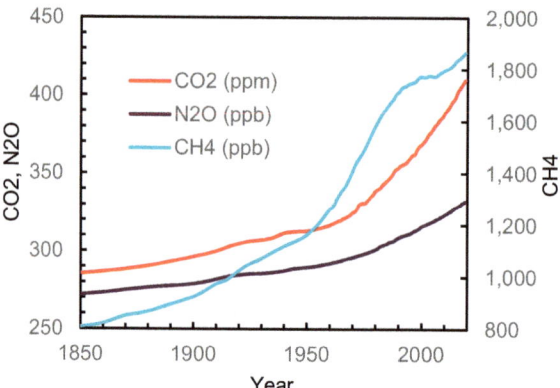

consistent with zero (i.e. anywhere from a very small decline to a very small increase). For the pre-industrial baseline period of 1850–1900, it rose from ~285 ppm to ~296 ppm, for a growth rate of ~20 ppm per century. Over the past 100 years (1919–2019) it rose from ~304.4 ppm to ~410 ppm, a growth rate of ~100 ppm per century. And finally, between 1995 and 2014, it rose from 360 ppm to 397 ppm, a growth rate close to 200 ppm per century. The trend is clear. We should also point out that there is substantial year-to-year variability in CO_2 growth rate, with peak growth occurring during the El Niño events of 1997–1998 and 2015–2016.

Methane, CH_4, is the second most important greenhouse gas. Its concentration has more than doubled since 1850, despite its short atmospheric lifetime of around a decade (before being 'degraded' to CO_2, as we saw in Chapter 2). On a per molecule, or per kilogram, basis, it is a far more potent greenhouse gas than CO_2, with the result that its ERF (which, remember, is based on the recent increases) is roughly a quarter that of CO_2. Recently, many nations have signed up to a pledge to cut methane emissions by 35% by 2030.

17.3.2 Halogenated Greenhouse Gases

Many industrial chemical species are also greenhouse gases, invariably those containing one or more of the halogens. The key reason they are of concern is that they have an absorption feature in the atmospheric window region, ~10 μm, primarily due to the vibration frequency of the carbon–halogen bond. Many of these are also ozone-depleting substances (ODS), as those same bonds are hard to break in the troposphere, with the main group being the chlorofluorocarbons (CFCs). Others have been produced as replacements for these gases, as they have a C–H bond which is vulnerable to OH attack in the troposphere. Figure 17.2 {figure 2.6 from AR6} shows the growth, and sometimes decline, of the main species of concern. (Note that all concentrations are expressed in parts per trillion, ppt.)

17.3.3 Effective Radiative Forcings

Table 17.1, a condensed version of {table 7.5} of AR6, summarises the effective radiative forcings of the gases we have just discussed. The AR6 table lists many of the halogenated species

Table 17.1 Present-day mole fractions in parts per trillion, except where specified, and effective radiative forcing (ERF, in W m⁻²) for the well-mixed greenhouse gases. (*Source*: see Acknowledgements at the end of the chapter.)

Greenhouse gas	Year	Year	Year	ERF (W m⁻²)	ERF (W m⁻²)
	1850	2011	2019	1850–2011	1850–2019
CO_2 (ppm)	285.5	390.5	409.9	1.738	2.012 ± 0.241
CH_4 (ppb)	807.6	1,803.3	1,866.3	0.473	0.496 ± 0.099
N_2O (ppb)	272.1	324.4	332.1	0.177	0.201 ± 0.030
Sum of HFCs (HFC-134a equivalent)	0.0	128.6	237.1	0.022	0.040
Sum of CFCs+HCFCs +ODSs (CFC-12 eq.)	0.0	1,050.1	1,031.9	0.362	0.354
Sum of PFCs (CF_4 equivalent)	34.0	98.9	109.4	0.006	0.007
Sum of halogenated species				0.394	0.408 ± 0.078
Total				2.782	3.118 ± 0.258

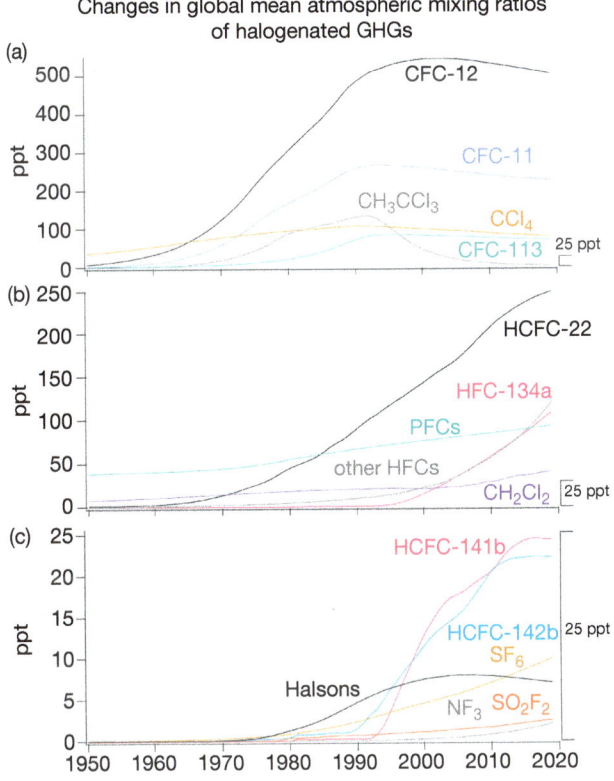

Figure 17.2 Global mean atmospheric mixing ratios of select ozone-depleting substances and other greenhouse gases. Data shown are based on the CMIP6 historical data set and data from NOAA and AGAGE global networks. (Note that the *y*-axis is different for each panel.) *Source*: see Acknowledgements at the end of the chapter.

individually; we have chosen to group them, based on the summaries that AR6 also chose. We have selected the concentrations in 1850, 2011, and 2019, plus the ERF for 2011 versus 1850, and 2019 versus 1850. While the halogenated species have largely stabilised (mainly due to the decline of the CFCs, governed by the Montreal Protocol), the forcings due to the WMGHGs have continued to rise, giving an overall ERF increase of 12% in the past 8 years.

17.3.4 Ozone

Ground-level ozone is a key component of air quality assessments, and has been monitored in select urban settings for many decades. However, due to its relatively short photochemical lifetime, such data are of limited value in determining any long-term global trends. There have been large-scale increases in tropospheric ozone at rural sites across the Northern Hemisphere from 1970 to 2010, and a doubling of European surface ozone during the twentieth century. Surface ozone likely increased in East Asia, but has levelled off or decreased in the eastern USA and western Europe. We also have some valuable ozonesonde and aircraft data through the troposphere. These show positive trends in most, but not all assessed regions.

What about pre-industrial levels? Some useful information can be gleaned from oxygen isotope data from air trapped in polar ice, along with chemical modelling.

So, we have limited data, with significant regional (and temporal) variability, which is understandable, given the nature of ozone photochemistry. AR6 states that given this limited evidence, global tropospheric ozone has increased by less than 40% between 1850 and 2005 (*low confidence*), with significant regional variation.

Stratospheric ozone decreased for a period last century, but has remained stable from the mid-1990s: at about 3.5% less than the 1964–1980 reference period. Forcing actually depends on the altitude of any declines. As we saw in Chapter 10, the strongest stratospheric ozone loss occurs in the austral spring, over Antarctica, although there are signs of a recovery this century. Overall, compared to the reference period, stratospheric ozone outside the polar regions declined by about 2.5% from 1980 to 1995 and stabilised after 2000. There is no consensus as to the resulting forcing, except that it is much smaller than that due to changes in tropospheric ozone. The total ozone ERF from 1750 to 2019 is estimated as 0.47 [0.24 to 0.70] W m^{-2}, and this is dominated by tropospheric ozone.

17.3.5 Stratospheric Water Vapour

The IPCC looks at several other gases which have potential forcings. The first is stratospheric water vapour (SWV). Two data sources have been studied. Balloon-borne measurements over Boulder, Colorado (40° N) from 1980 to 2010 show an average net increase of 1.0 ± 0.2 ppm (27 ± 6%). Meanwhile, merged satellite data since the 1980s suggest little change. AR6 concludes that the ERF from SWV is 0.05 ± 0.05 W m^{-2}, but with *low confidence*.

17.4   Aerosol Forcings

The interactions between atmospheric particulates and radiant energy are many and varied, as we have discussed in Chapter 4. Aerosols – with the key exception of black carbon – scatter solar radiation, some of it back to space, thus contributing to the planetary albedo. This is known as the aerosol direct effect, or **aerosol–radiation interaction**. Black carbon, on the other hand, is a strong absorber of solar radiation, potentially reducing the albedo, depending, in part, on the albedo of the underlying surface. However, the details are complex, as aerosol particles come in a great range of sizes, as well as chemical composition (and hence optical properties). (If you skipped Chapter 4, go back and read Section 4.7.)

We also know that every cloud droplet has an aerosol particle as a seed (Chapter 6), and there are good reasons to believe that an increased number of suitable aerosols in a volume of air that has reached saturation is likely to see a higher number of droplets – which must, of course, be smaller – than would have been the case otherwise. Such a cloud will be more reflective: again implying a forcing of the planet's energy fluxes. This is known as the aerosol indirect effect, the Twomey effect, or the **aerosol–cloud interaction**.

Both of these effects act to reduce the inflow of solar radiant energy, potentially a negative forcing. However, they will only constitute a radiative forcing if their contribution is changing in time, particularly over the past century or so. So, what information – i.e. data – do we need to decide this question, and to quantify any actual forcing? As we discussed in Chapters 4 and 12, aerosols are studied, and at least partially quantified, using a variety of both ground-based and space-based instrumentation. Given the wide geographic variability of aerosol concentrations and properties, both are vital. Some proxy data are also available from sulphate and black carbon in ice cores, which is very useful for earlier times.

17.4.1 Aerosol Trends

As we noted in previous chapters, industrial emissions of SO_2 have increased as a result of economic/industrial changes over the past century or so, some of which ends up as sulphate aerosol. Similarly, increasing demand for food, especially in tropical regions, has led to an increase in slash-and-burn agricultural practices, which inevitably create soot. Hence it is these two aerosol types that have been the main focus of investigation.

Sulphate in ice cores increased by a factor of 8 from the end of the nineteenth century to the 1970s in continental Europe, and by a factor of 4 from the 1940s to the 1970s in Russia. In all regions studied, concentrations have since halved from their peaks. Black carbon also increased in the twentieth century. Time series of sulphate and black carbon from a number of ice core data sets are presented in Figures 17.3(a) and (b), respectively {AR6 figure 2.9}.

For the past 20 years we have extensive global data from satellite observation. Figure 17.3(c) shows the trends in aerosol optical depth based on data from MODIS and MISR, with additional constraints from the AERONET network. We know that the anthropogenic contribution to aerosols is predominantly from the submicron (fine) range. This data set is presented in Figure 17.3(d), showing a significant decline of more than 1.5% per year, from 2000 to 2019, has occurred over Europe and North America, with positive trends over India and East Africa. Globally, there is a trend of –0.03% per year.

Overall, AR6 concludes that the data show positive trends from 1900 until the last quarter of the twentieth century, and decreases thereafter for the northern mid-latitudes (*high confidence*). However, there is *low confidence* in systematic changes in other parts of the world. Remote-sensing data indicate that aerosol optical depth exhibits predominantly negative trends since 2000 over Northern Hemisphere mid-latitudes and Southern Hemisphere continents, but increases over south Asia and East Africa (*high confidence*). A global decrease is thus assessed with *medium confidence*.

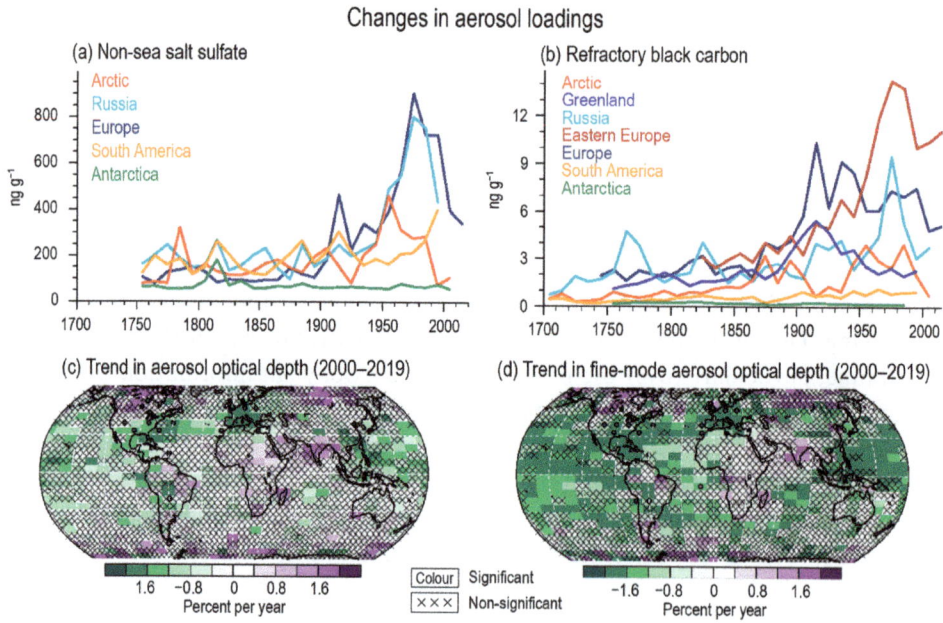

Figure 17.3 Aerosol evolution. Changes are shown as 10-year averaged time series (a, b) and trends in remote-sensing aerosol optical depth (AOD) and AODf (c, d). (a) Concentrations of non-sea-salt (nss) sulphate (ng g^{-1}). (b) Black carbon (BC) in glacier ice from the Arctic, Russia, Europe, South America, Antarctica, and BC from Greenland and eastern Europe. (c) Linear trend in annual mean AOD retrieved from satellite data for the 2000–2019 period (% year^{-1}). The average trend from MODIS and MISR is shown. (d) Linear trend in 2000–2019 as in (c), but for fine mode AOD, AODf, and using only MISR over land. 'x' marks denote non-significant trends. *Source*: see Acknowledgements at the end of the chapter.

17.4.2 Radiative Forcing

This time we are going to have to go through a number of steps to get a final answer. Step 1 is the instantaneous radiative forcing at TOA due to changes from direct aerosol–radiation interactions, which AR6 denotes as IRFari. When the required adjustments are made, we obtain the ERFari: the ERF due to such changes. Similarly, the ERF due to interactions between aerosols and clouds can be divided into the instantaneous component, which AR6 denotes by IRFaci, and the subsequent changes in cloud water content or extent. We will now examine these steps in turn.

In recent years, our understanding of the processes that govern aerosol radiative properties, and hence IRFari, has improved, as a result of better observations and modelling. As already noted, determination of any forcing requires a reasonable estimate of changes that have occurred during the industrial era. In the case of aerosol processes, this is derived from global aerosol model simulations for both aerosol optical depth and the absorption component.

The observational data come from broadband radiative flux measurements from CERES and aerosol optical depth from MODIS, combined with modelling to determine the anthropogenic

Table 17.2 Present-day effective radiative forcing (ERF) due to changes in aerosol–radiation interactions, aerosol–cloud Interactions, and total aerosol ERF from CMIP6 (2014 relative to 1850) and CMIP5 (per year relative to 1860). An additional 5% is applied to the CMIP5 and CMIP6 model results to account for land-surface cooling. (*Source*: see Acknowledgements at the end of the chapter.)

	ERFari (W m^{-2})	ERFaci (W m^{-2})	ERFari+aci (W m^{-2})
CMIP6 (2014–1850)	-0.25 ± 0.40	-0.86 ± 0.57	-1.11 ± 0.38
CMIP5 (2000–1860)	-0.27 ± 0.35	-0.96 ± 0.55	-1.23 ± 0.48

fraction. From this it is estimated that the clear-sky IRFari is -0.6 W m^{-2}, which, because global cloud cover is typically 0.5, implies an all-sky value of -0.3 W m^{-2}. Other studies have come to a slightly higher value, leading to consensus value of -0.35 W m^{-2}, but with a sizeable uncertainty. Improvements and reassessments have also been made in the determination of the contribution to IRFari by black carbon, although there is still considerable uncertainty.

To go from the instantaneous forcing to IPCC's current ERF numbers requires the use of global models. AR6 lists the results from a broad range of Earth System Models and model runs {table 7.6}. We summarise those results in our Table 17.2.

Determining the various forcings due to aerosol–cloud interactions is even more challenging than the direct forcing, and we have chosen to skip those details. However, when all factors are examined and combined it turns out to be the larger of the two forcings, by a significant amount. These results are also included in Table 17.2, along with the combined ERF due to both the aerosol–radiation interaction (ERFari) and the aerosol–cloud interaction (ERFaci). The results shown are firstly those from CMIP6 (2014 relative to 1850), and secondly CMIP5 (2000 relative to 1860). In each case we quote the average of the models used, and the 5–95% confidence range. These results are also shown in Figure 17.4 {figure 7.5}, which also shows individual model results.

17.5 Other Forcings

The IPCC recognises a number of other 'agents' that may be driving climate change, of which the most important is land-use changes. Some very minor agents will also be noted.

17.5.1 Land Use

Land-use changes can force the climate in a number of ways, the most obvious of which is via the surface albedo. Deforestation typically replaces darker forested areas with lighter cropland or pasture, increasing the albedo: a negative forcing. Any efforts at reforestation will have the opposite effect. The details will depend on the nature of the forest, crops, and underlying soil.

Land-use changes also affect the amount of evapotranspiration by the vegetation. Irrigation directly affects evaporation, causing a global increase estimated at 32,500 m^3 s^{-1}. Such changes

Figure 17.4 Net aerosol effective radiative forcing (ERF) from different lines of evidence. The headline AR6 assessment of –1.3 [–2.0 to –0.6] W m^{-2} is highlighted in purple for 1750–2014 and compared to the AR5 assessment of –0.9 [–1.9 to –0.1] W m^{-2} for 1750–2011. The CMIP5 and CMIP6 model results are depicted by blue and red crosses respectively. Uncertainty ranges are represented by black bars for the total aerosol ERF and depict *very likely* ranges. *Source*: see Acknowledgements at the end of the chapter.

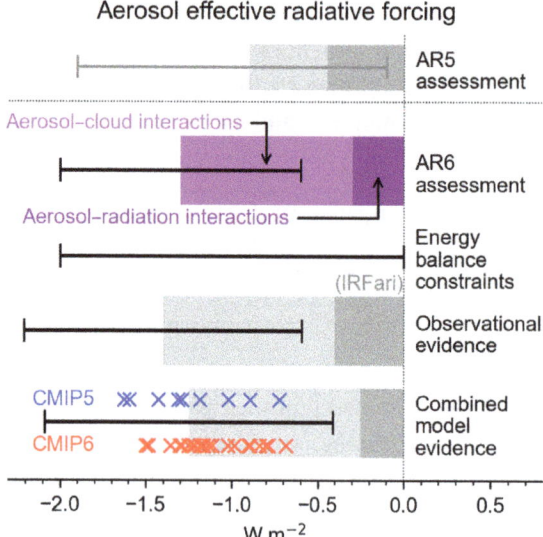

Aerosol effective radiative forcing

affect the latent heat budget, but do not directly affect the TOA fluxes, because the lifetime of water vapour is so short. However, evaporation can affect low-cloud amounts, and hence the ERF through adjustments.

Land management can also impact on the fluxes – emissions and removals – of the major greenhouse gases to the atmosphere. While it is these emissions that have the greatest impact on climate, they are already included in the GHG inventories. Land-use change may also affect the emissions of dust aerosols and biogenic volatile organic compounds (BVOCs), which may form secondary aerosols, and may also affect the atmospheric concentrations of ozone and methane.

The definition of ERF currently used by the IPCC excludes adjustments in land-surface temperature, but not changes in vegetation and snow cover. Assessing the ERFs of these various processes is clearly a challenge, often requiring the use of an Earth System Model with a well-tested land-surface component. Few models have attempted this exercise in full detail, although many useful contributions have been made. The consensus from CMIP6 is an IRF of –0.15 ± 0.12 W m^{-2}, and an ERF of –0.11 ± 0.09 W m^{-2}. Albedo changes have also been assessed from satellite data.

The contribution from irrigation (low-cloud amount) has been assessed as –0.05 [–0.1 to 0.05] W m^{-2}. The indirect effects through biogenic emissions are very uncertain, and it is not yet possible to make an assessment. Combining the irrigation effect with the previous result, AR6 concludes that the ERF from land-use changes is –0.2 ± 0.1 W m^{-2} (with *medium confidence*), for the period from 1750 to 2019.

17.5.2 Minor Agents

We will finish with a couple of very minor, although interesting, potential climate forcers. The emission of water vapour (from the burning of hydrocarbon fuels) by aircraft at cruising altitudes often produces contrails, or cirrus clouds. The ERF (from 1750 to 2019) has been assessed as 0.06 [0.02 to 0.10] W m^{-2}, with *low confidence*.

When light-absorbing particles such as soot get deposited on snow and ice, they will absorb solar radiation and warm, helping to melt the snow/ice, hence reducing the local albedo. This will constitute a positive forcing. (This may actually constitute a potential threat to the stability of the Tibetan Plateau glaciers, which feed many important rivers.) While such effects are largely regional, some progress has been made as to the global impact. The overall assessment presented in AR6 is an ERF of +0.08 [0.00 to 0.18] W m^{-2}.

17.6 Synthesis

Figure 17.5 {figure 7.6} shows the synthesis of the various ERFs investigated by the IPCC, based on changes between 1750 and 2019. We see a total forcing due to anthropogenic activities of 2.72 [1.96 to 3.48] W m^{-2}. We may compare this value with that from the previous report (AR5) of 2.3 [1.1 to 3.3] W m^{-2} for the period 1750–2011. The main reason for the increase is, of course, the increased emissions of greenhouse gases in that eight-year interval. This is partially offset by an increased estimate for the negative forcing (i.e. cooling) caused by aerosol–cloud interactions, which has been revised upwards this round.

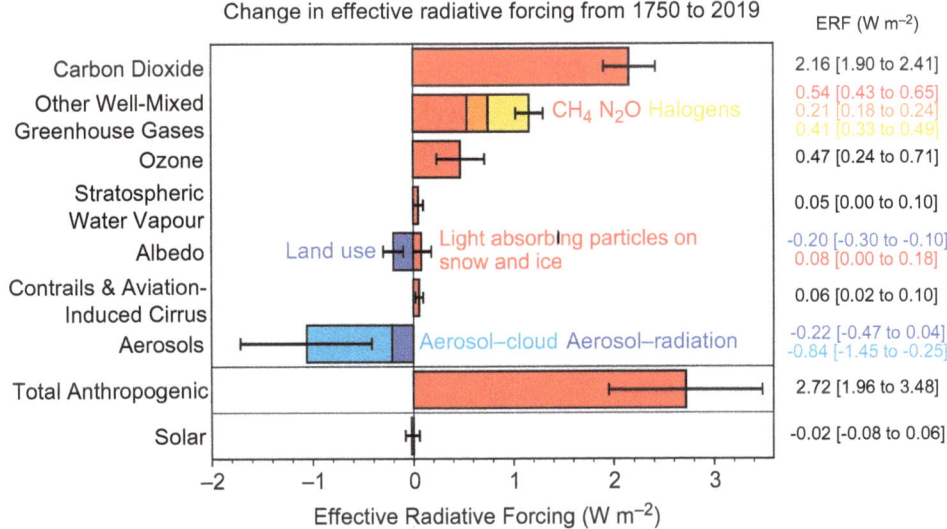

Figure 17.5 Effective radiative forcing (ERF) from 1750 to 2019 by contributing forcing agents (carbon dioxide, other well-mixed greenhouse gases (WMGHGs), ozone, stratospheric water vapour, surface albedo, contrails and aviation-induced cirrus, aerosols, anthropogenic total, and solar). Solid bars represent best estimates, and *very likely* (5–95%) ranges are given by error bars. Non-CO_2 WMGHGs are further broken down into contributions from methane (CH_4), nitrous oxide (N_2O), and halogenated compounds. Surface albedo is broken down into land-use changes and light-absorbing particles on snow and ice. Aerosols are broken down into contributions from aerosol–cloud interactions and aerosol–radiation interactions. *Source*: see Acknowledgements at the end of the chapter.

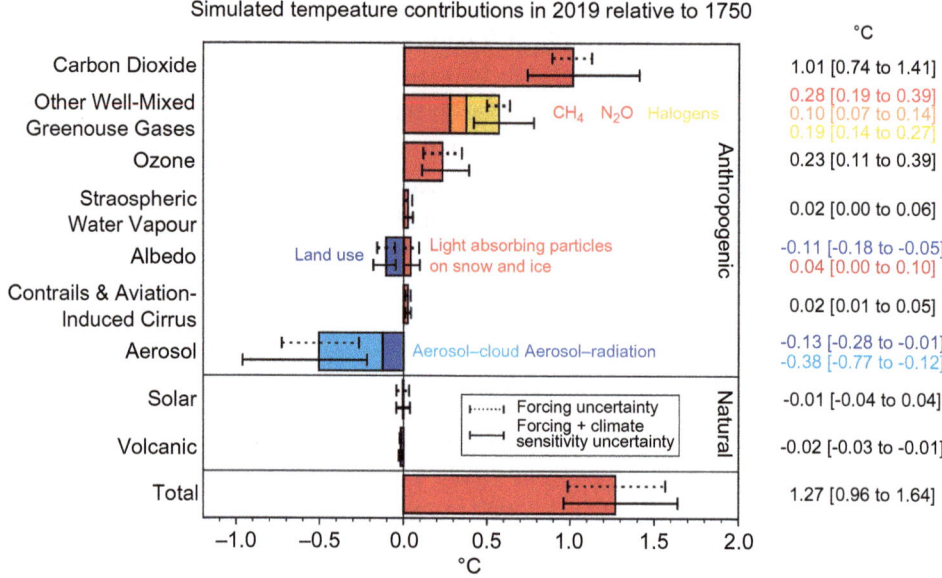

Figure 17.6 The contributions of forcing agents to 2019 temperature change relative to 1750 produced using the two-layer emulator. *Source*: see Acknowledgements at the end of the chapter.

Can we convert these 'sciency' numbers into a temperature change, something much easier to relate to? This, of course, requires a suitable climate model, with its own set of uncertainties. Nevertheless, AR6 provides such an assessment, and it is presented in Figure 17.6 {figure 7.7}. Two sets of uncertainty ranges are provided: one due to the uncertainty range due to the ERFs (Figure 17.5), and a larger range which adds in the uncertainty in the climate sensitivity: the direct connection between ERF and the change in temperature.

17.6.1 Planetary Energy Budget

Figure 17.7 {figure 7.2} shows the most recent estimate of the planetary energy budget, for both 'all-sky' and 'clear-sky' conditions. This figure also includes a very important number, the net imbalance of 0.7 [0.5, 0.9] Wm^{-2}.

17.6.2 Evolution of ERF

Until now we have focused on the effective radiative forcing today, compared with the atmosphere and climate of the pre-industrial era. However, in the next chapter we will be examining the evolution of our climate over that time frame. To do this accurately, we need to know how the ERF has evolved in that interval. Figure 17.8 {figure 2.10} provides the temporal evolution of the various climate drivers from 1750 to 2020.

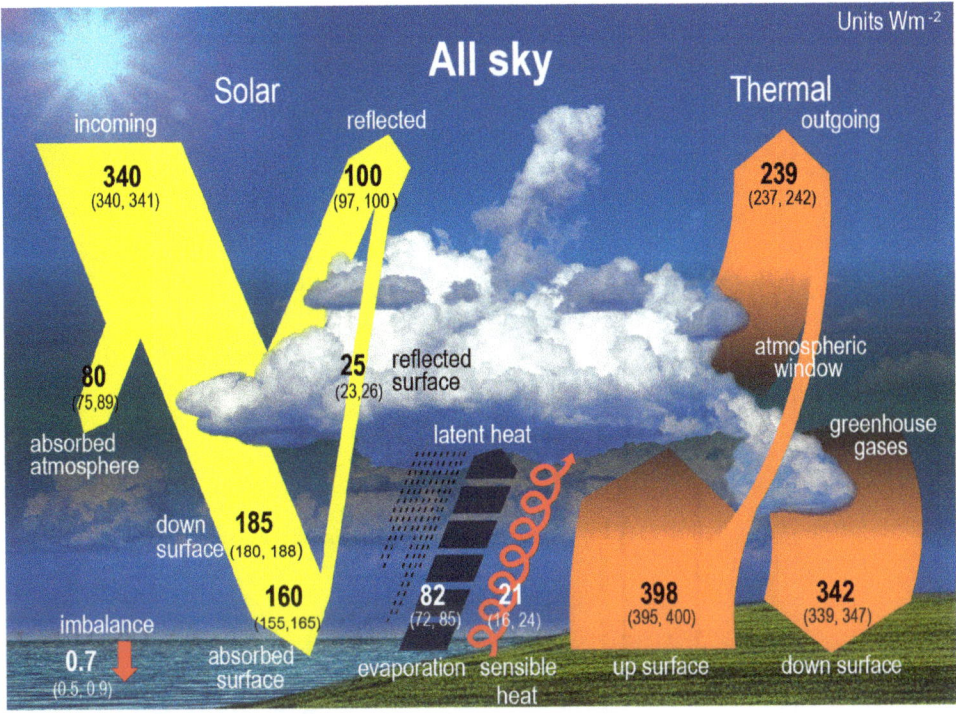

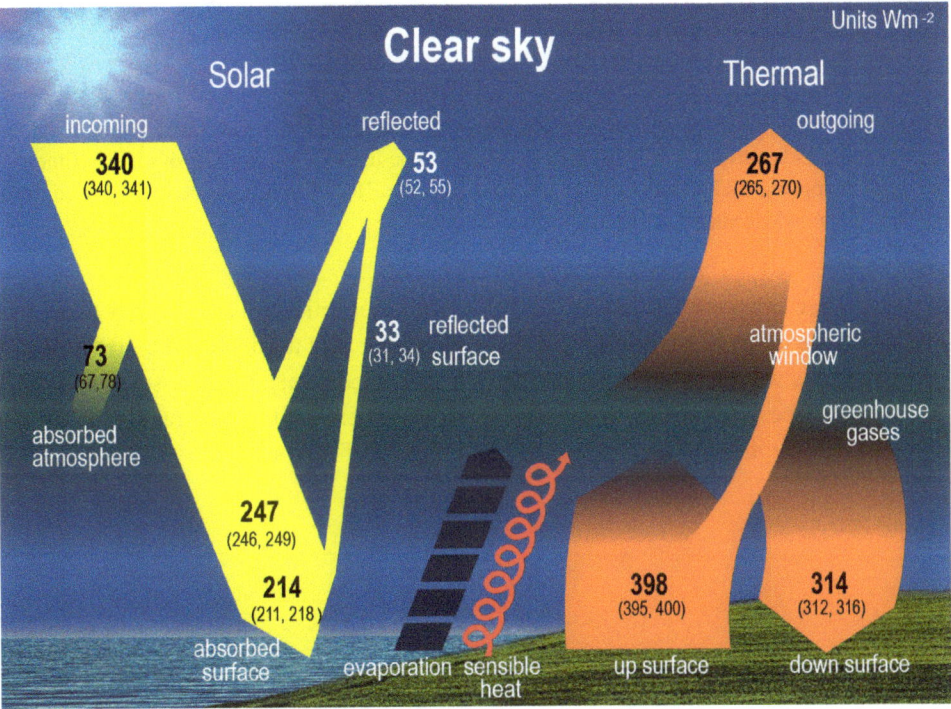

Figure 17.7 Schematic representation of the global mean energy budget of the Earth (upper panel) and its equivalent without consideration of cloud effects (lower panel). Numbers indicate best estimates for the magnitudes of the globally averaged energy components in W m^{-2} together with their uncertainty ranges in parentheses (5–95% confidence range), representing climate conditions at the beginning of the twenty-first century. *Source*: see Acknowledgements at the end of the chapter.

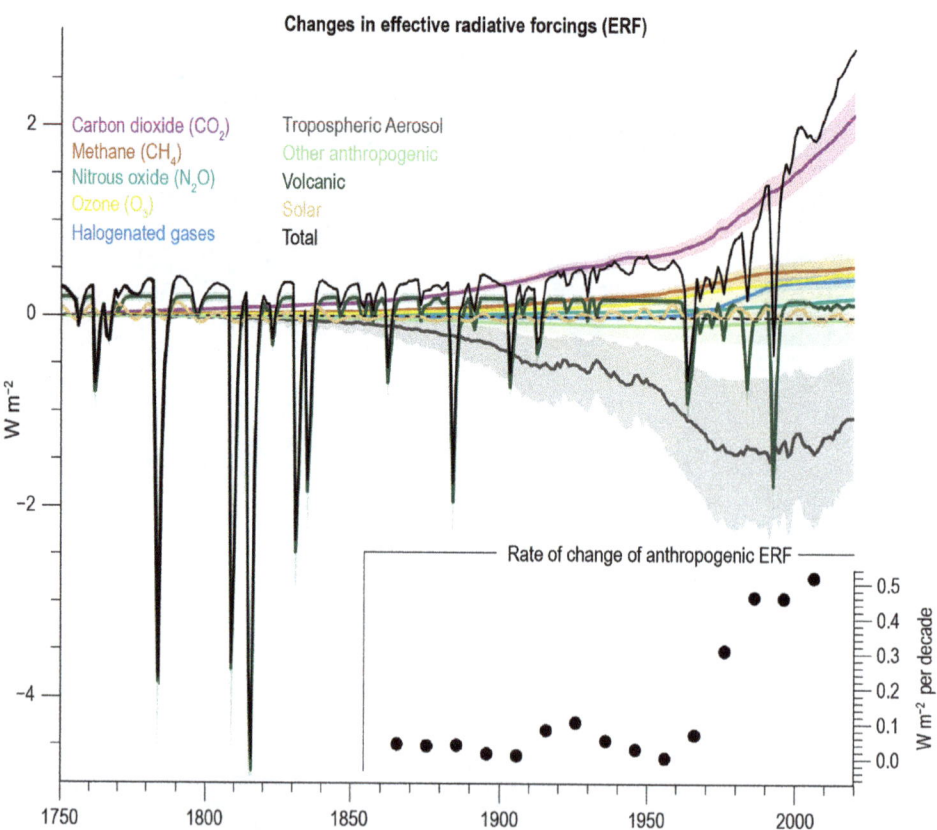

Figure 17.8 Temporal evolution of effective radiative forcing (ERF) related to the drivers assessed above. Shading indicates the 5–95% uncertainty range. The inset shows the rate of change (linear trend) in total anthropogenic ERF (total without TSI and volcanic ERF) for 30-year periods centred at each dot. *Source*: see Acknowledgements at the end of the chapter.

We see immediately the impact of volcanic eruptions – strong, but short-lived – many of which we noted in Section 17.2.2. You should be able to identify the dips in this figure with the volcanoes we listed. We also see the sizeable uncertainty range associated with the (negative) forcing due to aerosols. The panel in the lower right – the rate of change of ERF – is particularly interesting, illustrating the urgency of the problem.

Summary of IPCC Terminology and Acronyms used in this Chapter

Instantaneous radiative forcing (IRF) the net change in the top of atmosphere (TOA) radiative flux following a 'perturbation'.

Stratosphere Temperature-Adjusted Radiative Forcing allows for the response to any forcing by the stratosphere, which is radiatively decoupled from the troposphere.

Effective radiative forcing (ERF) also includes responses in the troposphere and effects on clouds and atmospheric circulation; collectively referred to as 'adjustments'.

AOD aerosol optical depth.

AODf fine-mode aerosol optical depth.

aci aerosol–cloud interaction [the modification of cloud reflectivity due to increased aerosol amounts: see Section 6.6.1].

ari aerosol–radiation interaction [reflection of solar radiation by aerosols: see Section 4.4.3]

ERFaci the ERF due to aerosol–cloud interaction.

ERFari the ERF due to aerosol–radiation interaction.

ERFari+aci the total aerosol ERF.

GSAT global surface average temperature.

TSI total solar irradiance.

WMGHGs well-mixed greenhouse gases [CO_2, CH_4, N_2O].

Summary

In this chapter we have taken the key first step in attempting to understand how our climate is changing. This has been a two-stage process. Firstly, we have looked at the available data on any changes in the properties and composition of our atmosphere (and land surface) that might, potentially, lead to changes in the radiation fluxes either into, or out of, our planet. Secondly, we have computed the resulting changes in those fluxes (using the AR6 concept of effective radiative forcing). What can we conclude?

Firstly, we looked at the natural drivers; a partial carryover from the previous chapter. Solar insolation variations appear to be far too small to be of concern. Volcanic eruptions can have significant consequences, although most are local. However, when significant quantities of sulphur gases enter the stratosphere, the resulting aerosol layer may spread over large parts of the globe, exerting a measurable cooling effect which lasts a few years. Such eruptions are sporadic, and there does not appear to be any long-term trend.

Secondly, we looked at the increasing levels of greenhouse gases, major and minor, widely acknowledged as the major culprits in this whole saga. This turns out to be one of the more straightforward chapters of our story, as both the data, and the physics, are there for all to see. AR6 concludes that, over the past 170 years, the increase in GHGs has forced the radiation budget by about +3.1 W m^{-2}: i.e. a planetary warming.

After this, we turned our attention to aerosols. At the start of Chapter 4 we justified devoting an entire chapter to aerosols by claiming they were important climate players. Section 17.4 is the proof. Aerosols cool our planet, both by directly reflecting radiation back to space, and – more significantly – by increasing cloud reflectivity. Their net effect is to cancel out more than one-third of the forcing (warming) due to greenhouse gas increases: major players indeed. Because this contribution is masking such a large fraction of the GHG warming (masking our misdeeds?), have we been deluding ourselves all these years?

The land surface, and especially any changes in land use, can have range of small but subtle environmental impacts, including to biogeochemistry, the hydrological cycle, and albedo. The recent changes due to human decisions (direct and indirect) have had a small negative impact on the radiation balance; i.e. a cooling effect.

Acknowledgements

Figures 17.1 {Annexe III Ia}, 17.2 {2.6}, 17.3 {2.9}, 17.4 {7.5}, 17.5 {7.6}, 17.6 {7.7}, 17.7 (2.10), 17.8 {7.2}; and Tables 17.1 {7.5}, 17.2 {7.6}:

Climate Change 2021: The Physical Science Basis. Contribution of Working Group I to the Sixth Assessment Report of the Intergovernmental Panel on Climate Change [Masson-Delmotte, V., P. Zhai, A. Pirani, S. L. Connors, C. Péan, S. Berger, N. Caud, Y. Chen, L. Goldfarb, M. I. Gomis, M. Huang, K. Leitzell, E. Lonnoy, J. B. R. Matthews, T. K. Maycock, T. Waterfield, O. Yelekçi, R. Yu, and B. Zhou (eds.)]. Cambridge University Press, Cambridge, United Kingdom and New York, NY, USA, 2391 pp. doi:10.1017/9781009157896.

REVIEW QUESTIONS

1. What are the two natural drivers of climate change?
2. Are they important in today's changing climate?
3. What is the most important contribution that aerosols make to radiative forcing?
4. What are the main ways that changes in land use can 'force' the climate?
5. By how much did the addition of greenhouse gases between 2011 and 2019 'force' the climate?
6. Currently the Earth's radiation budget is unbalanced. Why?
7. How might that balance be restored?

EXERCISE

The primary focus of this chapter has been the positive forcing (i.e. warming) produced by all the greenhouse gases that have been added to the atmosphere in the past century or so. (The negative forcing by aerosols is an important secondary focus, of course.) Although we often only talk about CO_2, we realise that there are many other gases that contribute. An important question, especially for policy makers looking for ways to reduce their impacts, is how might we compare other gases to CO_2? If it is going to cost \$x to eliminate 1 tonne of CO_2, and \$y to eliminate 1 tonne of CH_4, which will give the better bang for the buck? In order to provide useful guidance, a number of **metrics** have been introduced. This exercise is designed to guide you through a couple of these so you can 'speak the language'.

We saw in Chapter 11 (Section 11.5) that the radiative forcing produced by any increase in a greenhouse gas depends on the level of saturation of its absorption bands. We also noted that the CO_2 bands were already heavily saturated, whereas the bands of other gases were less, and in some cases significantly less, saturated. It is for this reason that, 'pound for pound', the extra CH_4 that has been added to our atmosphere is having a greater warming effect than the extra CO_2. This is the starting point for the concept of CO_{2e}, or **CO_2 equivalent**.

But there is a second piece of science that needs to be borne in mind. We saw in Chapters 2 and 3 that atmospheric chemistry, and especially the OH radical, has a way of 'removing' some of the gases we add to the atmosphere. We enclosed the key word in the last sentence in quotes, because although CH_4 has an atmospheric lifetime of 'only' about a decade, it is then converted into CO_2: still a greenhouse gas. So, while a gas may have an instantaneous warming effect many times larger than CO_2, what might that be in 20 years', 50 years', 100 years', or even 500 years' time? This is the basis of the **Global Warming Potential**, and the related Global Temperature Potential.

For this exercise we invite you to find the formal definitions of the metrics just discussed, and also have a look at some of the (possibly surprising) results of these quantifications. This is a topic that has been addressed in the IPCC Assessment Reports, so that is one place to start. AR6 has a relatively small section: {section 7.6}. AR5 has a somewhat longer treatment: {section 8.7, plus appendix 8A}. Another useful source is Wikipedia (last viewed August 2023), which has an up-to-date article with many references to the original science. You might also like to research the origins/uses of some of these gases: almost all are man-made.

18 How, and *Why*, is Our Climate Changing?

The title of this chapter is in the form of two questions, although it is the second that is the more crucial. You may be tempted to scratch your head as a result. After all, the theme of the entire book is *Our Changing Climate*, and we have referred to the rise in atmospheric CO_2 content over the past century or more, and the rise in temperature of the past 75 years, multiple times. So, aren't these self-answering questions?

No. Firstly, scientists do not stop at self-answering questions: they delve deeper. But the key reason is that global average surface temperature is only the crudest, and most reported, evidence of a changing climate. In this chapter we will dive into AR6 in order to find many more indicators of a changing climate: this is referred to as detection. We will also interrogate our CMIP6 simulations to see if we really do understand the science behind such changes. That is to say, how much of the observed change(s) can we attribute to human actions?

18.1 Large-Scale Indicators of Climate Change

We will set the scene for this chapter by drawing on the contents of cross-chapter box 2.2 of AR6: *Large-Scale Indicators of Climate Change*. It is a 'cross-chapter' box because the subject matter is not just relevant to chapter 2 {Changing State of the Climate System}, but also chapters 3 {Human Influence on the Climate System} and 4 {Future Global Climate}. For now, we will be focusing on chapters 2 and 3 (our 'How' and '*Why*', respectively). Chapter 4 largely focuses on the future: we will draw from it in Chapter 20.

18.1.1 Domains and Indicators

Firstly, we need to understand 'large scale', as defined by AR6.

Understanding of large-scale variability and change requires knowledge of both the responses to forcings and the role of internal variability. Many forcings have substantial hemispheric and continental scale variations. Modes of climate variability are generally driven by ocean basin scale processes. The climate system involves interactions from the micro- to the global-scale and, as such, any threshold for defining 'large-scale' is arbitrary, but, within these chapters on the basis of these considerations large-scale is defined to include ocean basin and continental scales as well as hemispheric and global scales.

Key climate indicators should constitute a finite set of distinct variables and/or metrics that may collectively point to important overall changes in the climate system that provide a synthesis of climate system evolution and are of broad societal relevance. Key indicators have been selected across the atmospheric, oceanic, cryospheric and biospheric domains, with land a cross-cutting component.

These indicators are {cross-chapter box 2.2, table 1}:

1. Atmosphere and surface
   - Surface and upper air temperatures;
   - Hydrological components (i.e. the water cycle);
   - Atmospheric circulation (Hadley cell, storm tracks).
2. Cryosphere
   - Sea-ice extent/area, seasonality and thickness;
   - Terrestrial snow cover;
   - Glacier mass and extent;
   - Terrestrial permafrost temperature and active layer thickness.
3. Ocean
   - Temperature/ocean heat content;
   - Salinity;
   - Sea level;
   - Circulation;
   - pH and deoxygenation.
4. Biosphere
   - Seasonal cycle of CO_2;
   - Marine biosphere (distribution, primary production, etc.);
   - Terrestrial biosphere (distribution, growing season, etc.).

We will devote a section to each of these four domains, although we will not address all indicators in equivalent detail.

18.1.2 Lines of Evidence

This chapter addresses two questions, and for the moment we'll focus on the first: *how* is our climate changing? To answer this we need data, and perhaps other types of information which might support or enhance that data. So, what data do climate scientists examine?

To focus first of all on the increase in global mean surface temperature, we need temperature data from around the world, over an extended period of time. Similar rainfall data might also tell a story. But we need to remember that the direct surface measurements available are far from uniform, with little or none over the oceans, which account for most of the Earth's surface. So we must make the best use of what is available, and use our judgement and experience.

Now we strike a second challenge. Over the extended periods we are interested in, many changes occur: in the types of instruments used; calibration issues; their location (and perhaps the effects of changes in their local environment; e.g. the heat island effect), to name a few. Remember that almost all the data they collected were intended for weather forecast purposes, not the detection of long-term trends. Again, scientific judgement is needed.

Satellites have been providing steadily increasing quantity, and quality, of data since the launch of Nimbus-7 and TIROS-N in October 1978: 1979 is often considered the start of the 'satellite era'. Part (a) of Figure 16.1 shows the temporal history of instrumental observations, with the beginning of the satellite era marked. As just one example, this technology step is the key to reliable data on Arctic sea-ice extent. Again, we must be conscious of calibration issues, and the minor or major changes in instrumentation between one mission (5–10 years, but getting longer) and the next. Ground-based data – 'ground truthing' – are vital here.

Measuring the concentrations of CO_2 and other greenhouse gases is now routine, although we only do this at a small number of sites worldwide. Monitoring aerosol properties is, as we trust you know by now, a greater challenge. Ocean salinity, pH, and oxygen content all require specialised technology, especially if we are to probe the ocean depths. Over recent decades this has been a productive area of research and technological development.

Finally, one indicator on our list is atmospheric circulation. This is meteorology – day-to-day weather – starting with the data fed into numerical weather models. However, as we have already noted, those data were collected for immediate use, with little time for quality control. Models are steadily being improved, asking differing questions of such data, so that there can be some inconsistencies over the extended periods of time that is our current focus.

For this reason, a number of meteorological agencies have decided to reprocess these data in a fully consistent way, to provide a more reliable picture of the dynamics of our atmosphere – a full, four-dimensional picture of the wind, pressure, and temperature fields as a function of time – a process known as **re-analysis**. Examples include the Modern-Era Retrospective analysis for Research and Applications, version 2 (MERRA-2), the ECMWF re-analysis product ERA, and JRA-55 from the Japanese Meteorological Agency. IPCC uses these products when and where appropriate.

18.1.3 Simulation and Attribution

The second half of the challenge we have set ourselves is the *why*; and more specifically, how much of the why can be put down to humanity's actions. Clearly the necessary answers will come from our climate models, and how we interrogate them. In Section 15.5 we discussed the current models, with particular emphasis on CMIP6: it is these models which provide the basis for most AR6 conclusions.

As should have been clear from that chapter and section, there is no such thing as a perfect model: otherwise, why run so many? So the scientists involved in the IPCC process must look at the results from all models – each can be regarded as a form of data, like the observations – and come to appropriate conclusions. The techniques used to analyse this set of data are inherently statistical, and specialised, and the details are well beyond the scope of this book. However, we do need to emphasise that it is through such analysis that the IPCC is able to quantify the confidence and likelihood language mentioned in Section 1.5.2.

These techniques evolve from report to report, as does virtually every aspect of the science, and our understanding of it. AR6 states:

these new fingerprinting and other probabilistic methods for detection and attribution as well as efforts to better incorporate associated uncertainties have addressed a number of shortcomings in previously applied detection and attribution techniques. They further strengthen the confidence in attribution of observed large-scale changes to a combination of external forcings.

Among the words quoted in the previous paragraph, one deserves special attention: **attribution**. We run models using a variety of forcings (drivers): natural only (volcanic eruptions and solar variability); greenhouse gases; aerosols; and all forcings. Then we compare the results to observations, and see which simulations most closely match up. Attribution of a particular climate change signal to human activity requires that we can show three things:

1. that the signal is unlikely to be entirely due to internal variability;
2. that it is consistent with the estimated responses to the given combination of natural and human forcings;
3. that it is not consistent with alternative, physically plausible explanations.

18.1.4 Structure

In the remainder of this chapter we will firstly present the observational evidence for how our climate is changing, along with the relevant conclusions from Chapter 2 of AR6. This will be followed, in each case, by the relevant conclusions from Chapter 3 as to the human contribution: starting with 'It is unequivocal that human influence has warmed the atmosphere, ocean and land surface since pre-industrial times.' Many of the conclusions we provide will be quoted verbatim. While this may seem like laziness, we will argue that the scientists involved spend significant effort in choosing their words, and any attempt to improve on their efforts would almost certainly be counterproductive. All the figures presented in this chapter are from AR6, chapters 2, 3 and 9, with the AR6 figure number in curly brackets: {..}.

18.2 Atmosphere and Earth's Surface

This section is based on {section 2.3.1} from AR6, which is, not surprisingly, by far the longest on the four domains.

18.2.1 Surface Temperature

Before we start, we need to define our terms. Most of the time when we talk about temperature, we mean some form of global average: IPCC recognises two approaches to this averaging. Global mean surface temperature (GMST) is a combination of land-surface air temperatures (LSAT) and sea-surface temperatures (SSTs). Global surface air temperature (GSAT) is a combination of LSAT and marine air temperatures (MATs). While closely related, they are physically different. It is important, of course, to always understand what the data mean.

Section 2.3.1.1.1 focuses on the 'deep past', 65 Mya to 8 kya, which we covered in Chapter 16. Section 2.3.1.1.2 then focuses on the past 7,000 years, which we have also touched on. Nevertheless, it is worth summarising, as it helps to put the recent temperature rise into context. The so-called Holocene thermal maximum, between 10,000 and 5,000 years ago, saw regional temperatures up to 1°C higher than 1850–1900. However, peak warming was regional, occurring at different times. After that time, GMST generally decreased, with the coldest multi-century interval occurring between 1450 and 1850 (*high confidence*). This cooling trend was reversed in the second half of the nineteenth century. Since the mid-twentieth century, GMST has increased at an observed rate that is unprecedented for any 50-year period in at least the last 2,000 years (*high confidence*).

Figure 18.1a {figure 2.11} summarises changes in surface temperature during the Holocene, through to the present day. For the mid-Holocene, and also the last ~1,000 years, the uncertainty range is also indicated, reflecting the increasing uncertainty the further back in time we try to make such determinations (as expected). (All temperatures are relative to the 1850–1900 reference period.) Figure 18.1b shows the geographical distribution of temperature rise, firstly between 1900 and 1980, and secondly between 1981 and 2020. The rise over the past 40 years is clearly dramatic! Finally Figure 18.1c shows how temperature has increased faster over land than over the oceans, a reflection of the fact that the oceans are able to sequester the heating (radiation imbalance; see Figure 17.8) to greater depths.

The IPCC has spent much time and effort (or, more accurately, taken note of the time and effort of the scientists involved) in carefully re-analysing how the global temperature data are assessed. The resulting estimates show only minor differences, although the latest estimates do now imply a slightly stronger trend in temperature change, which has implications for how much more greenhouse gases humanity can emit if we are to stay below the hoped-for 1.5°C. Figure 18.2 {cross-chapter box 2.3, figure 1} shows the key details. Part (a) shows the change in the assessed global surface temperature increase from the 1850–1900 baseline. The top bar shows the AR6 assessment: the 'whiskers' show the 90% (*very likely*) ranges. The results of the reassessment are shown next, followed by the contribution from the additional years of data between AR5 and AR6. Part (b) shows the assessed global surface temperature anomalies (i.e. relative to the reference period 1961–1990) from both AR5 and AR6, along with the trend line. It is the trend line that has implications for the remaining carbon budget.

18.2.1.1 Simulation Results

All CMIP6 models simulate the rise in temperature in the first part of the twentieth century, and the small decrease after the Second World War, followed by the rapid increase in the past three decades. Individual models may differ from the observed temperature by a few tenths of a degree, but the ensemble average is good. One reason for the model-to-model variations may be how well they handle internal variability: getting ENSO statistics right is a good, even a necessary, requirement for a reliable model, but getting the phase right is an extra challenge. This is just one of the many reasons we make use of a range of models.

Figure 18.3 {FAQ 3.1, figure 1} shows the observed warming (1850–2019), and the modelling results assuming only natural causes/forcings (green), greenhouse gases (red), aerosols and other human drivers only (blue), and all forcings (grey), with the observations in black. (Thick/dotted lines show the model means, while the shading covers the 5th to 95th percentile ranges.) Clearly the observed warming is only reproduced by including all forcing agents.

Changes in surface temperature

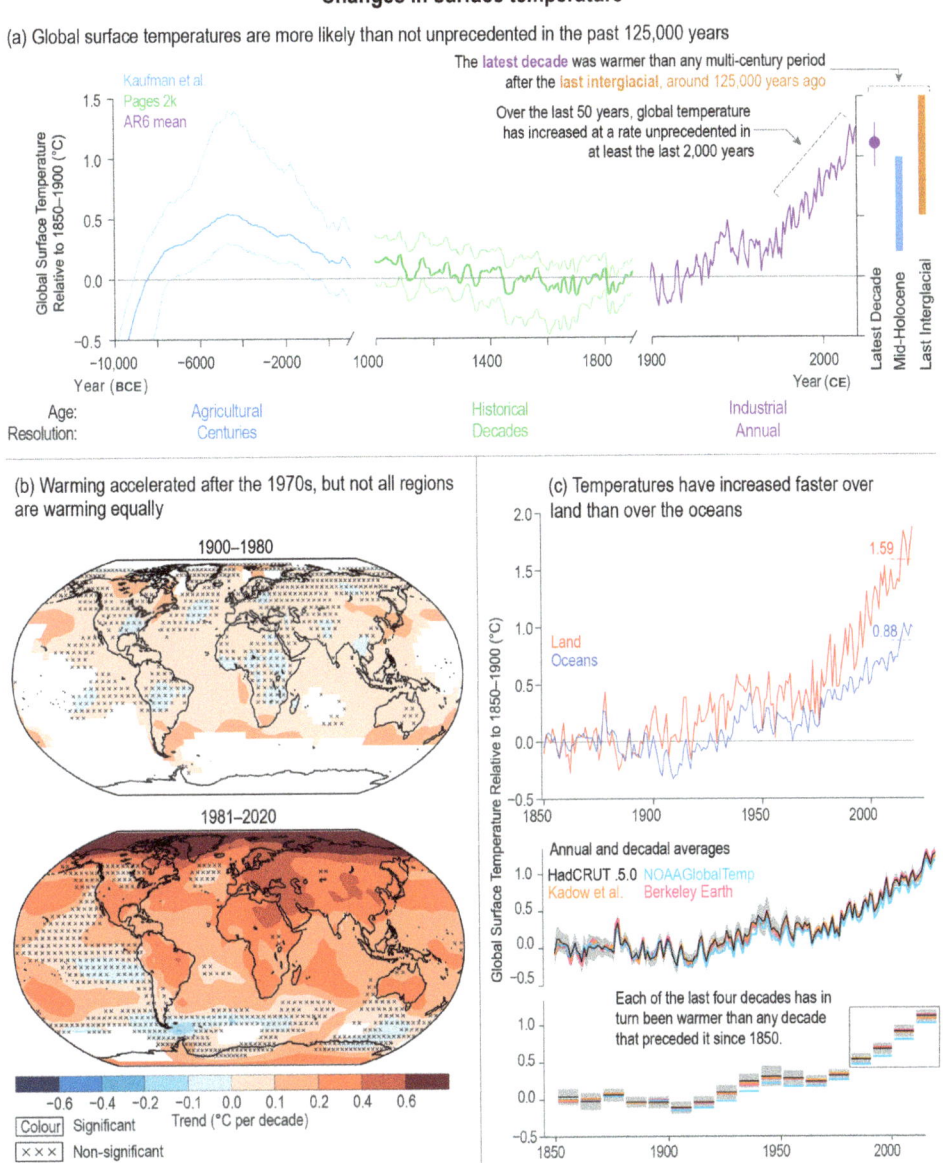

Figure 18.1 Earth's surface temperature history with key findings annotated with each panel. *Source*: see Acknowledgements at the end of the chapter.

AR6 has also chosen to separate this combination of observations and simulations into regions, as the models now do a good enough job at this scale. Figure 18.4 {figure 3.9} shows this breakdown. In each case we see the simulation results for all forcings (brown), natural forcings only (green), greenhouse gases only (grey), and aerosols only (blue): the observations are black. The all-forcings simulations encompass the temperature changes for all regions,

(a) Change in assessed historical global surface temperature estimates since AR5

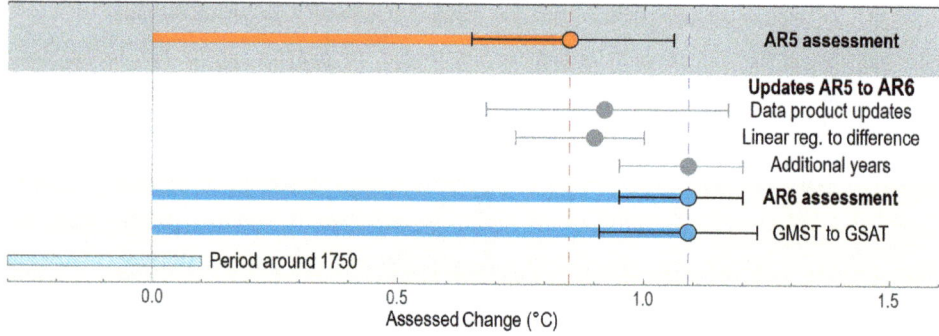

(b) Assessed global surface temperature anomalies

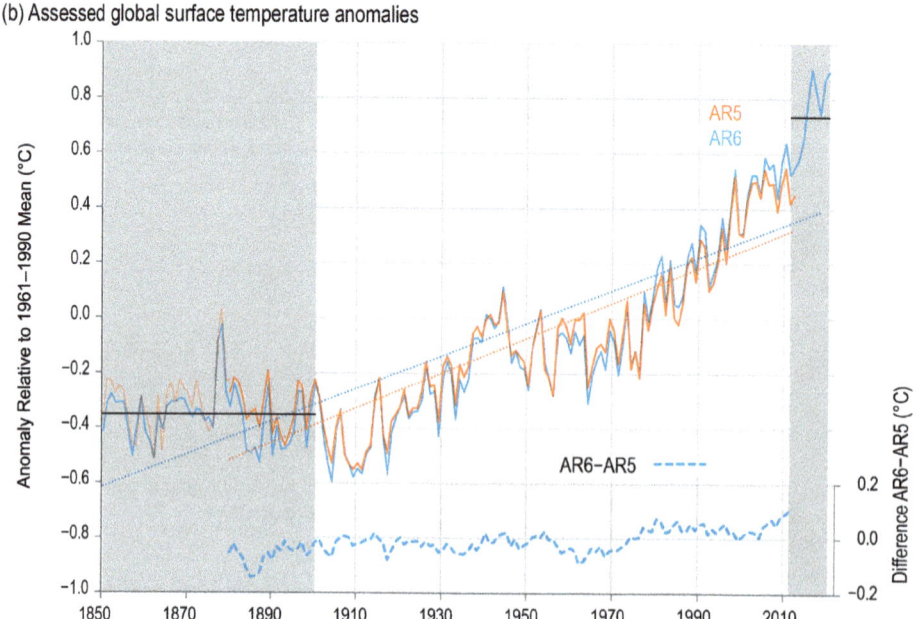

Figure 18.2 Changes in assessed historical surface temperature changes since AR5. *Source*: see Acknowledgements at the end of the chapter.

while the natural-only simulations fail to do so in recent decades, except for Antarctica. The warming due to greenhouse gases is partly offset by aerosol cooling, and that offset is stronger over land than over ocean.

18.2.1.2 Human Impact

We can do no better than to quote from the relevant paragraphs of the Executive Summary to chapter 3.

The *likely* range of human-induced warming in global-mean surface air temperature (GSAT) in 2010–2019 relative to 1850–1900 is 0.8°C–1.3°C, encompassing the observed warming of 0.9°C–1.2°C, while the change attributable to natural forcings is only –0.1°C to +0.1°C. The best estimate of

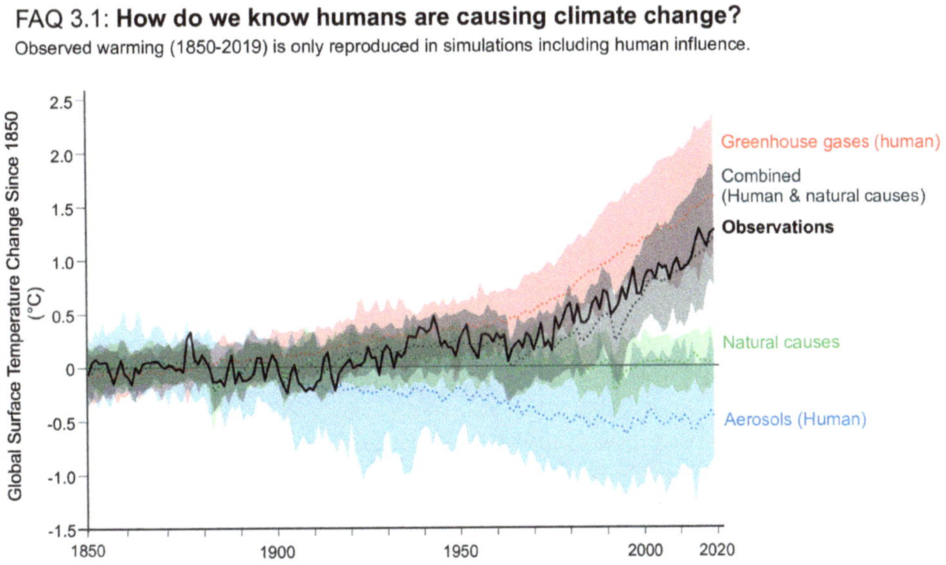

FAQ 3.1: **How do we know humans are causing climate change?**
Observed warming (1850-2019) is only reproduced in simulations including human influence.

Figure 18.3 Observed warming (1850–2019) is only reproduced in simulations including human influence. *Source*: see Acknowledgements at the end of the chapter.

human-induced warming is 1.07 °C … Over the same period, forcing from greenhouse gases *likely* increased GSAT by 1.0°C–2.0°C, while other anthropogenic forcings including aerosols *likely* decreased GSAT by 0.0°C–0.8°C. It is *very likely* that human-induced greenhouse gas increases were the main driver of tropospheric warming since comprehensive satellite observations started in 1979, and *extremely likely* that human-induced stratospheric ozone depletion was the main driver of cooling in the lower stratosphere between 1979 and the mid-1990s.

The CMIP6 model ensemble reproduces the observed global surface temperature trend and variability with biases small enough to support detection and attribution of human-induced warming (*very high confidence*) … CMIP6 models broadly reproduce surface temperature variations over the past millennium, including the cooling that follows periods of intense volcanism (*medium confidence*).

The slower rate of GMST increase observed over 1998–2012 compared to 1951–2012 was a temporary event followed by a strong GMST increase (*very high confidence*). Internal variability, particularly Pacific Decadal Variability, and variations in solar and volcanic forcings partly offset the anthropogenic surface warming trend over the 1998–2012 period (*high confidence*). Global ocean heat content [Section 18.4.1 below] continued to increase throughout this period, indicating continuous warming of the entire climate system (*very high confidence*). Since 2012, GMST has warmed strongly, with the past five years (2016–2020) being the warmest five-year period in the instrumental record since at least 1850 (*high confidence*).

18.2.1.3 Post AR6

On 8 August 2023, the WMO released a summary of climatic observations for July, and we will include those that are relevant.

The global average temperature for July 2023 is confirmed to be the highest on record for any month. The month was 0.72°C warmer than the 1991–2020 average for July, and 0.33°C warmer than the previous

Anomaly of Near-Surface Air Temperature

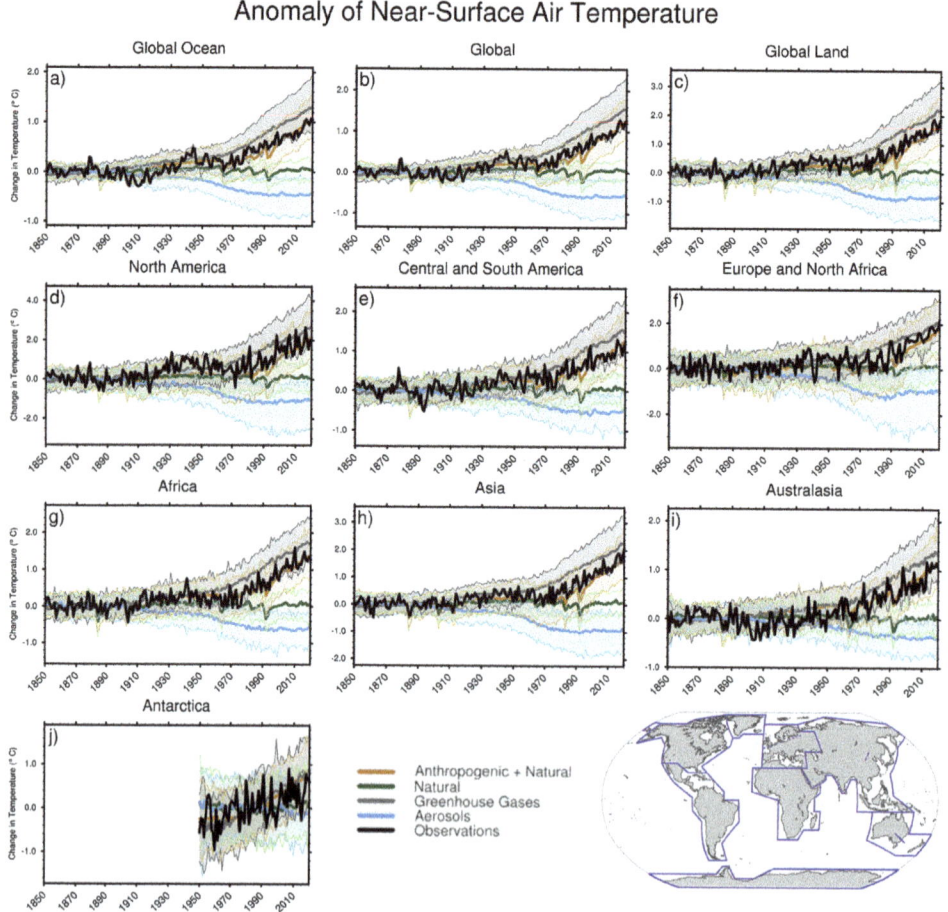

Figure 18.4 Global, land, ocean, and continental annual mean near-surface air temperature anomalies in CMIP6 models and observations. *Source*: see Acknowledgements at the end of the chapter.

warmest month, July 2019. The month is estimated to have been around 1.5°C warmer than the average for 1850–1900. Heatwaves were experienced in multiple regions in the Northern Hemisphere, including southern Europe. Well above-average temperatures occurred over several South American countries and around much of Antarctica.

The report also looked at sea-surface temperature.

Global average sea surface temperatures continued to rise, after a long period of unusually high temperatures since April 2023, reaching record levels in July. For the month as a whole, global average sea surface temperatures were 0.51°C above the 1991–2020 average. The North Atlantic was 1.05°C above average in July, as temperatures in the northeastern part of the basin remained above average, and unusually high temperatures developed in the northwestern Atlantic. Marine heatwaves developed south of Greenland and in the Labrador Sea, in the Caribbean basin, and across the Mediterranean Sea. El Niño conditions continued to develop over the equatorial eastern Pacific.

18.2.2 Hydrological Cycle

18.2.2.1 Humidity

Firstly, we will look at the indicators surface humidity and total water vapour column (TWVC). Near-surface humidity has been measured using in-situ instrumentation and satellite-derived estimates, plus various data re-analysis products. The in-situ data are, of course, poorly sampled, especially in the Southern Hemisphere. Strong El Niño events are known to enhance surface moisture levels, so that ENSO trends will impact moisture trends. Increasing temperatures, both of the surface and (lower) atmosphere, will likely lead to increased specific humidity, but potentially to a decrease in relative humidity (RH).

Figure 18.5 {figure 2.13} shows the assessed changes since 1973. Panel (a) shows the global distribution of changes in specific humidity, q, while panel (b) shows the global average as assessed by different analysis products (anomalies relative to a base period of 1981–2010). The impact of the 1997–1998 El Niño is evident. Panel (c) shows the trends in relative humidity, while panel (d) shows the global average anomalies. The colours in this figure are not ideal for easy eyeballing of the information, but the patterns of the different contributions are obviously similar, and this should help you. (Trends are calculated using ordinary least squares regression not covered in our book).

Again, we quote:

In summary, observations since the 1970s show a *very likely* increase in near surface specific humidity over both land and oceans. A *very likely* decrease in relative humidity has occurred over much of the

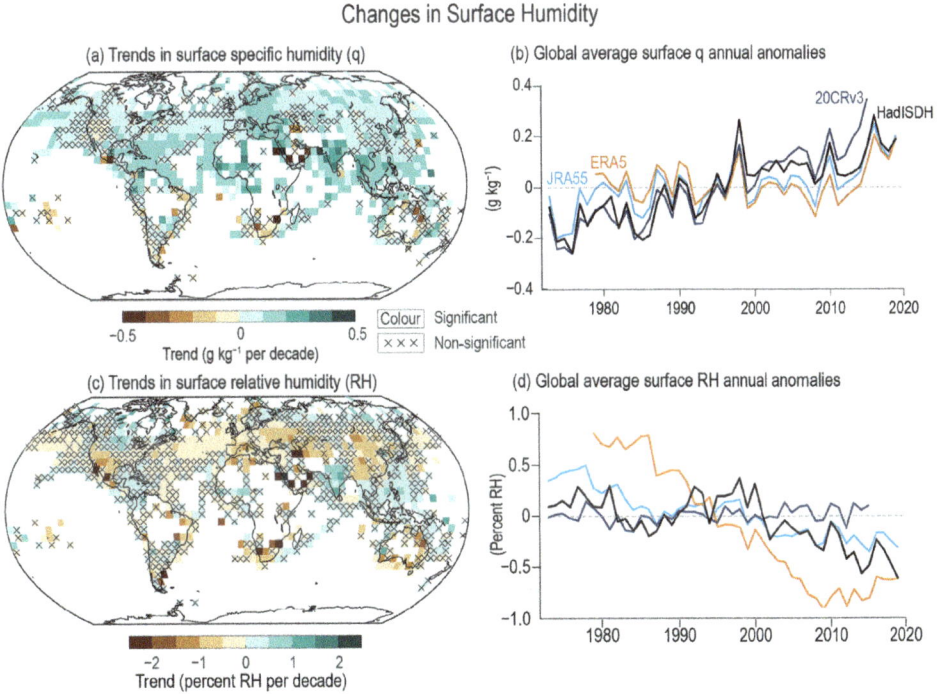

Figure 18.5 Changes in surface humidity. *Source*: see Acknowledgements at the end of the chapter.

Figure 18.6 Time series of global mean total column water vapour annual anomalies (mm) relative to a 1988–2008 base period. *Source*: see Acknowledgements at the end of the chapter.

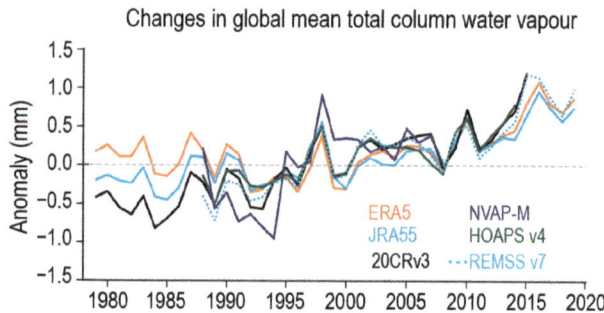

Changes in global mean total column water vapour

global land area since 2000, particularly over mid-latitude regions of the NH, with increases at northern high latitudes.

We now turn to TWVC. As we have already discussed in previous chapters, the Clausius–Clapeyron equation tells us that a warmer atmosphere can hold more water vapour: 7% per °C. But that is just a theoretical possibility; only real data can be considered meaningful. Before the satellite era, our only source of such data was the quasi-global coverage by radiosondes: balloon packages launched twice daily from many weather stations. These data, coupled with other analyses, showed a positive trend from 1910 to 1940.

The late 1970s ushered in the satellite era, with water vapour profile data being one of the key products. These meteorological data are complemented by a number of more specialised data products from radio occultation and dedicated missions. Of course, all satellite systems have finite lifetimes, and occasional calibration issues, leading to a degree of uncertainty in the conclusions that may be drawn. Figure 18.6 {figure 2.14} shows several assessments of the change in global mean total water vapour column (relative to a 1988–2008 base period): note the similarities to Figure 18.5b.

AR6 concludes that 'positive trends in global total column water vapour are *very likely* since 1979 when globally representative direct observations began, although uncertainties associated with changes in the observing system imply *medium confidence* in estimation of the trend magnitudes'.

18.2.2.2 Surface Hydrology

We now combine 3 AR6 topics – global precipitation, precipitation minus evaporation, and streamflow – under this broader heading. Precipitation has been measured for centuries over land, in selected locations. However, as we know, rainfall can be very inhomogeneous, even within a large city, especially when it is the result of convective uplift. Satellite data are now also available to provide a more global perspective, including over the oceans for the first time. The AR6's table 2.6 summarises the results from a small number of analyses, often with widely differing results, indicating the significant uncertainties around this indicator.

Nevertheless, AR6 concludes that

global land precipitation has *likely* increased since the middle of the 20th century (*medium confidence*), with *low confidence* in trends prior to 1950. A faster increase in global land precipitation was observed since the 1980s (*medium confidence*), with large interannual variability and regional heterogeneity. Over

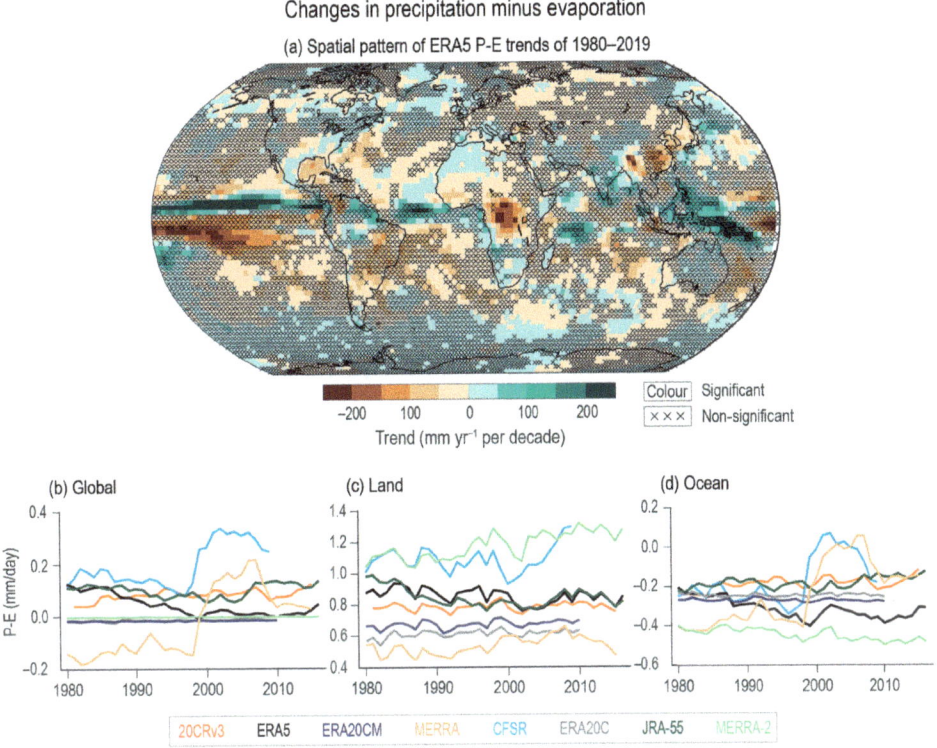

Figure 18.7 Changes in precipitation minus evaporation. (a) Trends in precipitation minus evaporation (P – E) between 1980 and 2019 ('x' marks denote non-significant trends). Time series of (b) global, (c) land-only, and (d) ocean-only average annual P – E (mm day^{-1}). *Source*: see Acknowledgements at the end of the chapter.

the global ocean there is *low confidence* in the estimates in precipitation trends, linked to uncertainties in satellite retrievals, merging procedures and limited in situ observations.

From an agricultural point of view, evaporation is almost as important as precipitation: even more so is the difference; precipitation minus evaporation (P – E). Over the ocean (hardly of agricultural significance), we can turn to measurements of salinity. Since the 1950s, saline surface waters have become saltier, while relatively fresh surface waters have become fresher. To go beyond these observations to the global scale is challenging, due to limited evaporation measurements and inhomogeneities in satellite data sets.

A number of analyses have been attempted, with little consistency. Figure 18.7a {figure 2.16} shows one such global-scale analysis, with most of the globe covered in hatch marks (x), indicating that the data/conclusions cannot be considered significant. There is, however, a clear and significant trend of an increase in P – E over tropical waters, and a decrease in the Pacific Ocean just south of the equator, and over central Africa. The three lower panels show a number of different analyses, split between global, land, and ocean. The lack of consistency is clear.

Precipitation minus evaporation leads to runoff, and thence to streamflow. This is something that is widely monitored – for reasons from irrigation to flood mitigation – although gaps remain over central Asia and Africa. Damming a river will almost certainly lead to an increase in evaporation. What data are available suggest that, since the 1950s, of the rivers which are gauged, slightly more have experienced significantly decreased flows than significantly increased flows (*low confidence*).

18.2.2.3 Human Influence

Again, we will quote from the Executive Summary.

It is *likely* that human influence has contributed to observed large-scale precipitation changes since the mid-20th century. New attribution studies strengthen previous findings of a detectable increase in Northern Hemisphere mid- to high-latitude land precipitation (*high confidence*). Human influence has contributed to strengthening the zonal mean precipitation contrast between the wet tropics and the dry subtropics (*medium* confidence). Anthropogenic aerosols contributed to decreasing global land summer monsoon precipitation from the 1950s to the 1980s (*medium confidence*). There is also *medium confidence* that human influence has contributed to high-latitude increases and mid-latitude decreases in Southern Hemisphere summertime precipitation since 1979 associated with the trend of the Southern Annular Mode towards its positive phase.

18.2.3 Circulation

It is the circulation of the atmosphere, on a range of scales, that is most responsible for local weather and climate. The general circulation, and especially the descending arm of the Hadley cell, creates our desert and semi-desert regions (Chapter 7). The Walker circulation also has profound impacts on countries around the Pacific. Storm tracks at high latitudes have an obvious impact. So any changes to circulation patterns could, potentially, have regional impacts. AR6 has looked at this topic, using the re-analysis products mentioned above. To date, no major changes are apparent, although some small shifts have been detected, which we will briefly discuss.

   After examining a number of studies, AR6 concludes that there has *likely* been a widening of the Hadley circulation since the 1980s, mostly in the Northern Hemisphere. (This, potentially, has implications for food security in parts of Africa.) Trends since 1980 are consistent with a *very likely* strengthening of the Walker circulation that resembles a La Niña-like pattern: the causes are not well understood.

   The total number of extratropical cyclones (i.e. storm systems: not to be confused with tropical cyclones/hurricanes/typhoons) has *likely* increased since the 1980s in the Northern Hemisphere (*low confidence*), but with fewer deep cyclones, particularly in summer. The number of strong extratropical cyclones has *likely* increased in the Southern Hemisphere (*medium confidence*).

18.3 Cryosphere

18.3.1 Sea Ice

Arctic sea ice has been well monitored for around 40 years, allowing IPCC to come to strong conclusions. Over the period 1979 to 2019, the Arctic sea-ice area (SIA) has decreased in all months, with the strongest decrease in summer (*very high confidence*). Decadal means of SIA decreased from the first to the last decade in that period from 6.23 to 3.76 million km^2 in September, and from 14.52 to 13.42 million km^2 in March. Arctic sea ice has also become younger, thinner and faster moving (*very high confidence*). To look back further in time requires drawing on a range of less-precise data sources; however, some conclusions are still possible. Current Arctic sea-ice coverage (annual mean and late summer) is unprecedently low since 1850 (*high confidence*), and with *medium confidence* for late summer for at least the past 1,000 years.

Antarctica is a very different place from the arctic region, as we noted when discussing the Ozone hole, and the changes are much less clear cut. Antarctic sea ice has, in fact, experienced both increases and decreases in SIA during the satellite era, with only minor differences in the decadal means. Record high extents were reached each September from 2012 through 2014. It then dipped rapidly in mid-2016, remaining predominantly below average through 2019.

18.3.1.1 Post AR6

The WMO report of 8 August 2023 also looked at sea-ice extent. 'Antarctic sea ice extent continued to break records for the time of year, with a monthly value 15% below average, by far the lowest July extent since satellite observations began.'

18.3.1.2 Human Influence

From the Executive Summary to AR6 Chapter 3:

It is *very likely* that anthropogenic forcing, mainly due to greenhouse gas increases, was the main driver of Arctic sea ice loss since the late 1970s ... In the Arctic, despite large differences in the mean sea ice state, loss of sea ice extent and thickness during recent decades is reproduced in all CMIP5 and CMIP6 models (*high confidence*).

18.3.2 Snow Cover

The ice–albedo feedback mechanism works at least as well for snow cover as for ice, so we should expect to see a reduction in snow cover as the planet warms. As a result, there has been a decrease in snow-cover duration and persistence, particularly at higher latitudes, due to earlier spring melt. Arctic snow cover has decreased by 2–4 days per decade since the 1970s. For the Northern Hemisphere, maximum snow depth has generally decreased since the 1960s.

In summary, substantial reductions in spring snow cover extent have occurred in the NH [Northern Hemisphere] since 1978 (*very high confidence*) with limited evidence that this decline extends back to the early 20th century. Since 1981 there has been a general decline in NH spring snow water equivalent (*high confidence*).

18.3.2.1 Human Influence

From the Executive Summary to AR6 Chapter 3:

It is *very likely* that human influence contributed to the observed reductions in Northern Hemisphere spring snow cover since 1950. The seasonal cycle of Northern hemisphere snow cover is better reproduced by CMIP6 than by CMIP5 models (*high confidence*).

18.3.3 Land Ice

Glaciers are fascinating frozen rivers, which actually 'flow', but not as we normally understand that word. It was the glaciers in the Swiss Alps that first prompted thinkers such as Louis Agassiz to postulate the idea of an Ice Age. There is *high confidence* that there were times during the Holocene when glaciers were smaller than today. However, it is the past century or so that is of most concern in our currently changing climate. There is *very high confidence* that, with a few exceptions, glaciers worldwide have retreated since the second half of the nineteenth century. This consistency is considered highly unusual for at least the past 2,000 years (*medium confidence*). Glacier mass loss rates have increased since the 1970s (*high confidence*). Current disequilibrium conditions imply that most glaciers are committed to further ice loss.

As we discussed in Chapter 16, water stored in the Greenland and Antarctic ice caps is the key to sea-level variations during the Ice Age cycles. For this reason, both these ice caps have received significant attention over the past few decades. Both gain mass from snow fall, while losing mass around their edges, and via surface melting: both are important. Satellite observations are now able to supply valuable data, which are summarised in Figure 18.8 {figure 2.24}. (AR6 devotes chapter 9 to Ocean, Cryosphere and Sea Level Change, as the details can be very complex.)

The two ice sheets do show some differences in behaviour, again for reasons we understand, which calls for separate assessments. For Greenland, AR6 concludes that its ice sheet reached a peak during the period 1450–1850 before decreasing, and that the rate of loss has increased substantially since the turn of the twenty-first century (*high confidence*). Antarctica is a very much larger ice sheet, and its behaviour is far from uniform, calling for a more extensive database and analysis. AR6 concludes that the Antarctic Ice Sheet has lost mass between 1992 and 2020 (*very high confidence*), as can be seen in Figure 18.8, and there is *medium confidence* that this mass loss has increased. Their contribution to sea level will be covered in Section 18.4.1.

18.3.3.1 Human Influence

From the Executive Summary to AR6 Chapter 3:

Human influence is *very likely* the main driver of the recent global near-universal retreat of glaciers. It is *very likely* that human influence has contributed to the observed surface melting of the Greenland Ice

Figure 18.8 Cumulative Antarctic Ice Sheet (AIS) and Greenland Ice Sheet (GIS) mass changes. Values shown are in gigatons and come from satellite-based measurements for the period 1992–2020. The *very likely* ranges are shaded. *Source*: see Acknowledgements at the end of the chapter.

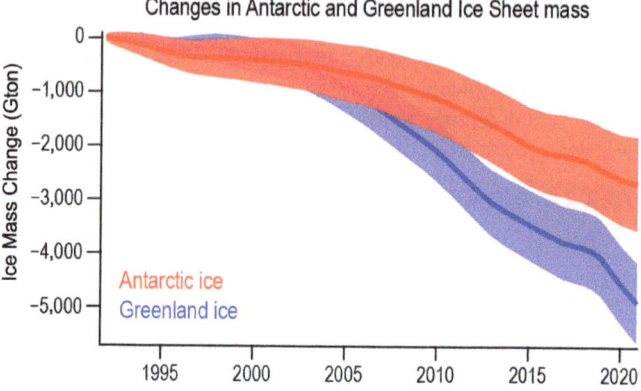

Sheet over the past two decades, and there is *medium confidence* in an anthropogenic contribution to recent overall mass loss from the Greenland Ice Sheet. However, there is only *limited evidence*, with *medium agreement*, of human influence on Antarctic Ice Sheet mass balance through changes in ice discharge.

18.3.4 Permafrost

As we discussed in Section 14.3.3, there are substantial quantities of buried methane at high northern latitudes, trapped under permafrost. What is happening to this permafrost seal? Some measurements are available in certain locations, suggesting increases of ~0.5°C per decade for colder permafrost, but only ~0.17°C per decade for warmer regions. Other evidence includes observations of ground subsidence, indicating permafrost thawing.

Recently, a new potential threat has emerged: viruses and bacteria buried under permafrost. As the permafrost thaws, so do enormous numbers of cells that have been frozen for hundreds of thousands of years. Most will be 'harmless', as their previous hosts are no longer present. But even if only 1% find a host, the possibility of a jump to humans cannot be excluded.

18.4 Oceans

The primary climatic concern in the ocean domain is generally understood to be sea-level rise. While the melting of land ice – glaciers, and the Greenland and Antarctic Ice Sheets – is clearly having some impact, the main driver of sea-level rise in recent decades – and for the next few decades, or at least that is the current consensus – is thermal expansion. This is a simple and direct consequence of the ocean's uptake of a share of the current global energy imbalance that we noted in the previous chapter. However, there is a second major issue to do with oceans and climate modification, and that is their chemistry: pH, deoxygenation, and salinity. Finally, there is a question mark over the ocean's dynamics, and especially the future of the Gulf Stream: i.e. the AMOC.

18.4.1 Heat Content and Sea Level

Measuring oceanic properties at any significant depths is clearly not trivial. Until quite recently the main approach was lowering samplers from ships. However, since 2006 we have had a major improvement in the form of the fleet of 4,000 drifting Argo profiling floats. They are 'parked' at a depth of 1,000 metres, and every 10 days are able to dive to a depth of 2,000 metres. They measure temperature and salinity (plus other parameters), and transmit these data to shore via satellite. These data are freely available to everyone, without restrictions. A warming signal is clear. It is *likely* that the global ocean has warmed since 1871, consistent with the increase in sea-surface temperature. It is *virtually certain* that ocean heat content increased between 1971 and 2018 in the upper 700 m, and *very likely* in the 700–2,000 m layer since 2006.

The uptake of heat by the oceans will clearly lead to thermal expansion; that is, a rise in sea level. We should also point out that it has other effects. One of these is the bleaching of corals around the world, with some suggestions of near-total devastation in the decades to come. A second impact is on tropical cyclones/hurricanes/typhoons, which can only form over surface waters with temperatures above 27°C. Thus we are likely to see such storms, and the damage they cause, penetrate to higher latitudes. In some parts of the world, building standards are set to withstand certain wind speeds. These may have to be expanded.

Sea level has been measured by tide gauges around the world for decades and more. While these data are invaluable, it needs to be understood that local geological effects can also impact these results, relative to the global sea level. For example, northern land masses were under several kilometres of ice comparatively recently, and are slowly 'rebounding' to regain hydro-static balance. More recently we have data from the TOPEX/Poseidon mission. We are now in a position to make clear statements. Global mean sea level (GMSL) is rising, and the rate of rise since the twentieth century is faster than over any preceding century in at least the last three millennia (*high confidence*). Since 1901 GMSL has risen by 0.2 [0.15 to 0.25] m, at an accelerating rate.

In chapter 9 of AR6, table 9.5 provides a summation of the various contributions to the rise in GMSL from various sources, and for five different time periods (both total change, and rate of change). In Table 18.1 we present these numbers for two time periods: 1901–2018 and 1993–2018. The acceleration in recent decades from most contributors is apparent.

AR6 chapter 9 also contains cross-chapter box 9.1: Global Energy Inventory and Sea Level Budget; figure 1. We know that the vast majority of net gain in energy that has resulted from the increase in greenhouse gases has ended up in the oceans. Figure 18.9(a) shows the time series of the breakdown of this energy since 1971, with the bulk clearly entering either the upper ocean (0–700 m) or the middle depths (700–2,000 m), plus a component to the melting of ice. Panel (b) shows the time series of the contributions to global sea level.

In April 2023, the WMO released an updated report on sea-level rise. It concluded that between 2013 and 2022 sea level rose at an average rate of 4.62 mm year^{-1}, a figure distinctly higher than the one given in Table 18.1 (for the period 1993–2018). The cause for this jump was

Table 18.1 **Observed contributions to global mean sea-level (GMSL) change for two different periods. Values are expressed as the total change (mm) in the annual mean or year mid-point over each period along with the equivalent rate (mm year^{-1}). The *very likely* ranges appear in brackets. (*Source*: see Acknowledgements at the end of the chapter.)**

Contribution	1901–2018 total (mm)	1901–2018 mm year^{-1}	1993–2018 total	1993–2018 mm year^{-1}
Thermal expansion	63.2 [47.0 to 79.4]	0.54 [0.40 to 0.68]	32.7 [23.8 to 41.6]	1.31 [0.95 to 1.66]
Land glaciers	67.2 [41.8 to 92.6]	0.57 [0.36 to 0.79]	13.8 [10.0 to 17.6]	0.55 [0.40 to 0.70]
Greenland Ice Sheet	40.4 [27.2 to 53.5]	0.35 [0.23 to 0.46]	10.8 [8.9 to 12.7]	0.43 [0.51 to 0.74]
Antarctic Ice Sheet	6.7 [–4.0 to 17.4]	0.06 [–0.03 to 0.15]	6.1 [4.0 to 8.3]	0.25 [0.16 to 0.33]
Land-water storage	–12.9 [–45.8 to 20.0]	–0.11 [–0.39 to 0.17]	7.8 [3.3 to 12.2]	0.31 [0.13 to 0.49]
Sum of contributions	164.6 [116.9 to 212.4]	1.41 [1.00 to 1.82]	71.2 [2.41 to 3.29]	2.85 [72.1 to 90.2]
Observed GMSL rise	201.9 [150.3 to 253.5]	1.73 [1.28 to 2.17]	81.2 [72.1 to 90.2]	3.25 [2.88 to 3.61]

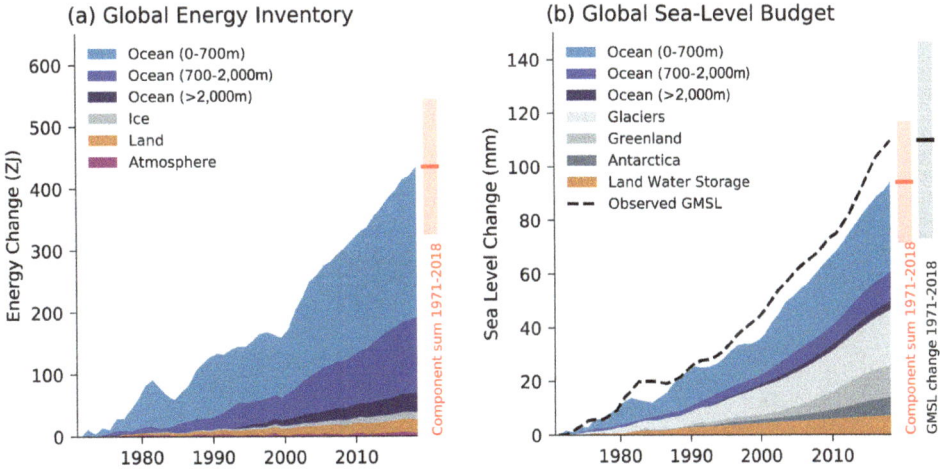

Figure 18.9 Global energy inventory and sea-level budget. (a) Observed changes in the global energy inventory from 1971 to 2018 (shaded time series). Earth System heating for the whole period and associated uncertainty is indicated to the right of the plot (red bar = central estimate; shading = *very likely* range); (b) observed changes in components of the global mean sea level for 1971–2018 (shaded time series). Observed global mean sea-level change from tidal gauge reconstructions (1971–1993) and satellite altimeter measurements (1993–2018) is shown for comparison (dashed line) as a three-year running mean to reduce sampling noise. Closure of the global sea-level budget for the whole period is indicated to the right of the plot (red bar = component sum central estimate; red shading = *very likely* range; black bar = total sea level central estimate; grey shading = *very likely* range). *Source*: see Acknowledgements at the end of the chapter.

put down to both extreme glacier melt and record ocean heat levels, and warned that this trend would continue for millennia. 'We have already lost this melting of glaciers game and sea-level rise game so that's bad news', said WMO secretary-general Petteri Taalas.

18.4.1.1 Human Influence

From the Executive Summary to AR6 Chapter 3:

It is *extremely likely* that human influence was the main driver of the ocean heat content increase since the 1970s, which extends into the deeper ocean (*very high confidence*) ... Updated observations and model simulations show that warming extends throughout the entire water column (*high confidence*), with CMIP6 models simulating 58% of industrial era heat uptake (1850–2014) in the upper layer (0–700 m), 21% in the intermediate layer (700–2000 m) and 22% in the deep layer (>2000 m).

Figure 18.10 {figure 3.29} shows the simulated and observed change in global mean sea level due to thermal expansion, based on natural forcings only (green), greenhouse gases only (grey), aerosols only (blue), and all forcings (brown). The solid lines show the multi-model means, and the shading the 5th to the 95th percentile range. The best estimate of the observations is shown in black. The agreement is clearly more than satisfactory.

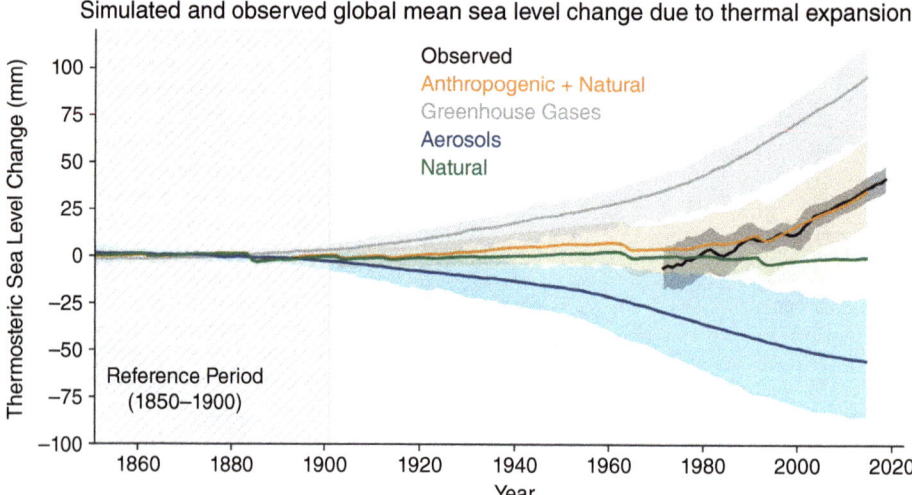

Figure 18.10 Simulated and observed global mean sea-level change due to thermal expansion for CMIP6 models and observations relative to the baseline period 1850–1900. *Source*: see Acknowledgements at the end of the chapter.

From the Executive Summary to AR6 chapter 3:

Combining the attributable contributions from glaciers, ice-sheet surface mass balance and thermal expansion, it is *very likely* that human influence was the main driver of the observed global mean sea level rise since at least 1971. Since AR5, studies have shown that simulations that exclude anthropogenic greenhouse gases are unable to capture sea level rise due to thermal expansion (thermosteric) during the historic period and that model simulations that include all forcings (anthropogenic and natural) most closely match observed estimates. It is *very likely* that human influence was the main driver of the observed global mean thermosteric sea level increase since 1970.

18.4.2 Changing Chemistry

We discussed the implications of the ocean's uptake of CO_2 in Section 3.2, including Figure 3.3 showing how pH is connected to the balance of the various ionic forms of inorganic carbon in the ocean. We know that this process is taking place, and from the available data we are able to conclude that it is *virtually certain* that surface open ocean pH has declined globally over the past 40 years by 0.003–0.026 pH per decade, and a decline in the ocean interior has been observed in all ocean basins over the past 2–3 decades (*high confidence*).

While the solubility of salts increases with temperature, for gases the situation is reversed, as the molecules find it easier to escape. Thus, a warmer ocean should hold less dissolved gases, including both CO_2 and oxygen. Current data are quite limited, but AR6 is able to conclude that there has *very likely* been a net loss of oxygen over all ocean depths since the 1960s in a range of 0.3–2.0%, and that global oxygen levels in the upper 1,000 metres had decreased by 0.5–3.0% during 1970–2010 (*medium confidence*).

In Section 18.2.2 we looked at precipitation minus evaporation over the oceans, which will almost certainly lead to changes in salinity. And we also have some recent Argo data, as well as older but more sporadic sampling. It is *virtually certain* that since 1950, near-surface high-salinity regions have become more saline, while low-salinity regions have become fresher, and it is *very likely* that this extends to the ocean interior along ventilation pathways. It is *very likely* that the Atlantic has become saltier and the Pacific and Southern oceans have freshened. The differences between high-salinity and low-salinity regions are linked to an intensification of the hydrological cycle (*medium confidence*) – as we would expect, of course.

18.4.2.1 Human Influence

Once again, we quote the relevant paragraphs from the Executive Summary to chapter 3.

It is *extremely likely* that human influence has contributed to the observed near-surface and subsurface salinity changes since the mid-20th century. The associated pattern of change corresponds to fresh regions becoming fresher and salty regions becoming saltier (*high confidence*). Changes to the coincident atmospheric water cycle and ocean-atmosphere fluxes (evaporation and precipitation) are the primary drivers of the observed basin-scale salinity changes (*high confidence*) ... The basin-scale changes are consistent across models and intensify through the historic period (*high confidence*).

It is *virtually certain* that the uptake of anthropogenic CO_2 was the main driver of the observed acidification of the global surface open ocean. The observed increase in CO_2 concentration in the subtropical and equatorial North Atlantic since 2000 is *likely* associated in part with an increase in ocean temperature, a response that is consistent with the expected weakening of the ocean carbon sink with warming. Consistent with AR5 there is *medium confidence* that deoxygenation in the upper ocean is due in part to human influence. There is *high confidence* that Earth system models simulate a realistic time evolution of the global mean ocean carbon sink.

18.4.3 Circulation

AR6 looked at several topics under this heading, but we will restrict our coverage to the AMOC. We addressed one major change in the AMOC when we examined the Younger Dryas event in Section 16.5.1. There have been other, generally smaller events associated with abrupt climate changes during the glacial intervals as well (known as Dansgaard–Oeschger and Heinrich events), indicating the complexity of how this circulation is driven. Because of this, the AMOC's history has been the subject of intense investigation by specialists: their work is beyond the scope of this book, although some of their conclusions need to be noted.

After the demise of the Laurentide Ice Sheet ~8,000 years ago, the overall strength of the AMOC has been relatively stable, compared to the previous 100,000 years. Over the past 3,000 years there is evidence that AMOC variability was linked to decreasing production of Labrador Sea Water (a component of North Atlantic Deep-Water formation). Proxy data imply that the AMOC is currently at its weakest point in the past 1,600 years. Marine sediments and other proxies show a decline beginning in the late nineteenth century and over the twentieth century, along with large decadal variability in the latter half of the twentieth century.

Since the 1980s, multiple lines of evidence are available. Ship-based hydrographic estimates of AMOC show no overall decline in strength. Full-depth in-situ measurements report that deep convection, a major driver, has returned to the Labrador Sea, particularly in 2015. In summary, AR6 states that '*confidence* in an overall decline of AMOC during the 20th century is *low*. From the mid-2000s to mid-2010s the directly observed weakening in AMOC (*high confidence*) cannot be distinguished between decadal-scale variability or a long-term trend (*high confidence*)'.

18.4.3.1 Human Influence

From the Executive Summary to AR6 Chapter 3:

While observations show that the Atlantic Meridional Overturning Circulation (AMOC) has weakened from the mid-2000s to the mid-2010s (*high confidence*) and the Southern Ocean upper overturning cell has strengthened since the 1990s (*low confidence*), observational records are too short to determine the relative contributions from internal variability, natural forcing and anthropogenic forcing to these changes (*high confidence*). No changes in Antarctic Circumpolar Current transport or meridional position have been observed. The mean zonal and overturning circulations of the Southern Ocean and the mean overturning circulation of the North Atlantic (the AMOC) are broadly reproduced by CMIP5 and CMIP6 models.

18.5 Biosphere

The changes in the biosphere actually belong more to the province of Working Group II than Working Group I. Nevertheless, WGI does address a number of biospheric issues, and we will examine these in two themes: the response of the biosphere to the increase in atmospheric CO_2; and the response of ecosystems to the resulting (broader) climatic changes.

18.5.1 Carbon Fertilisation

Plants need four things to grow: CO_2, water, sunlight, and nutrients. If we increase the supply of any of these, we might expect increased primary production, unless of course the region is currently at the limits of one of the other three; e.g. water stress. Over the past two centuries, CO_2 levels have increased by 50%, while the impacts that this has had on the energy fluxes equate to about 1°C. So changes in plant growth will be primarily driven by the CO_2 increase (the 'zeroth order effect'; before climatic effects arise): this is carbon fertilisation. Any data we examine must, of course, be from the real (warmer) world.

The CO_2 data from Mauna Loa show a distinct seasonal cycle, while the South Pole data show very little seasonality. This is a reflection of the growth and decay (or harvesting) of plant life, which is much more abundant in the Northern Hemisphere. Analysis of this and related data shows that the amplitude of the CO_2 cycle has increased at Mauna Loa (19.5°N), from ~6 ppm in the 1960s to ~7.5 ppm in the 2010s; with an even bigger increase at Barrow, Alaska (71.3°N) from ~15.5 ppm in the 1980s to ~18.5 ppm in the 2010s. (The increasing temperatures at such latitudes will also aid plant growth, of course.) Thus, 'there is *very high confidence* that the amplitude of the seasonal cycle of atmospheric CO_2 has increased at mid-to-high NH latitudes since the early 1960s. The observed increase is generally consistent with greater greening during the growing season and an increase in the length of the growing season over the high northern latitudes'. Furthermore, 'there is *high confidence* that vegetation greenness (i.e. green leaf area and/or mass) has increased globally since the early 1980s'.

18.5.1.1 Human Influence

From the Executive Summary to AR6 Chapter 3:

The main driver of the observed increase in the amplitude of the seasonal cycle of atmospheric CO_2 is enhanced fertilisation of plant growth by the increasing concentration of atmospheric CO_2 (*medium confidence*). However, there is only *low confidence* that this CO_2 fertilisation has been the main driver of observed greening because land management is the dominating factor in some regions. Earth System models simulate globally averaged land carbon sinks within the range of observation-based estimates (*high confidence*), but global-scale agreement masks large regional scale disagreements.

18.5.2 Phenological Changes

Phenology is a branch of science dealing with the relations between climate and periodic biological phenomena. This includes plant growth and flowering/ripening, bird and other animal migrations, and similar phenomena. Phenological changes may be brought about by environmental changes such as changes in temperature or rainfall, or (in the case of plants), the increase in atmospheric CO_2.

Agricultural authorities have also noticed changes in the growing season of individual crops in their regions, and some valuable long-term records are available, some going back for 600 years. Figure 18.11 {figure 2.32} shows six such time series, over different lengths of time, with Japan's cherry blossom peak and the French grape harvest the longest. What these data reveal is that, while there have been fluctuations over the past 100 years (perhaps a reflection of regional climate variability), there has been a distinct trend towards earlier harvesting, etc., leading to '*high confidence* that the length of the growing season has increased over much of the extratropical NH since at least the mid-20th century'.

The terrestrial biosphere is somewhat easier to study than the marine, and there have been widespread changes in the habitats and ranges of many animal (and plant) species, mainly in response to the increase in temperature. Thus, it should come as no surprise that AR6 was able to conclude 'there is *very high confidence* that many terrestrial species have shifted their geographic ranges poleward and/or upslope over the past century, with increased rates of species turnover. There is *high confidence* that the geographic distribution of climate zones has shifted in many parts of the world'.

While studies of the marine biosphere are more challenging – it is inherently three-dimensional – some conclusions have been drawn. For example, 'there is *high confidence* that the latitudinal and depth limits of the distributions of various organisms in the marine biome are changing'. Chlorophyl distribution is varying between different latitude bands.

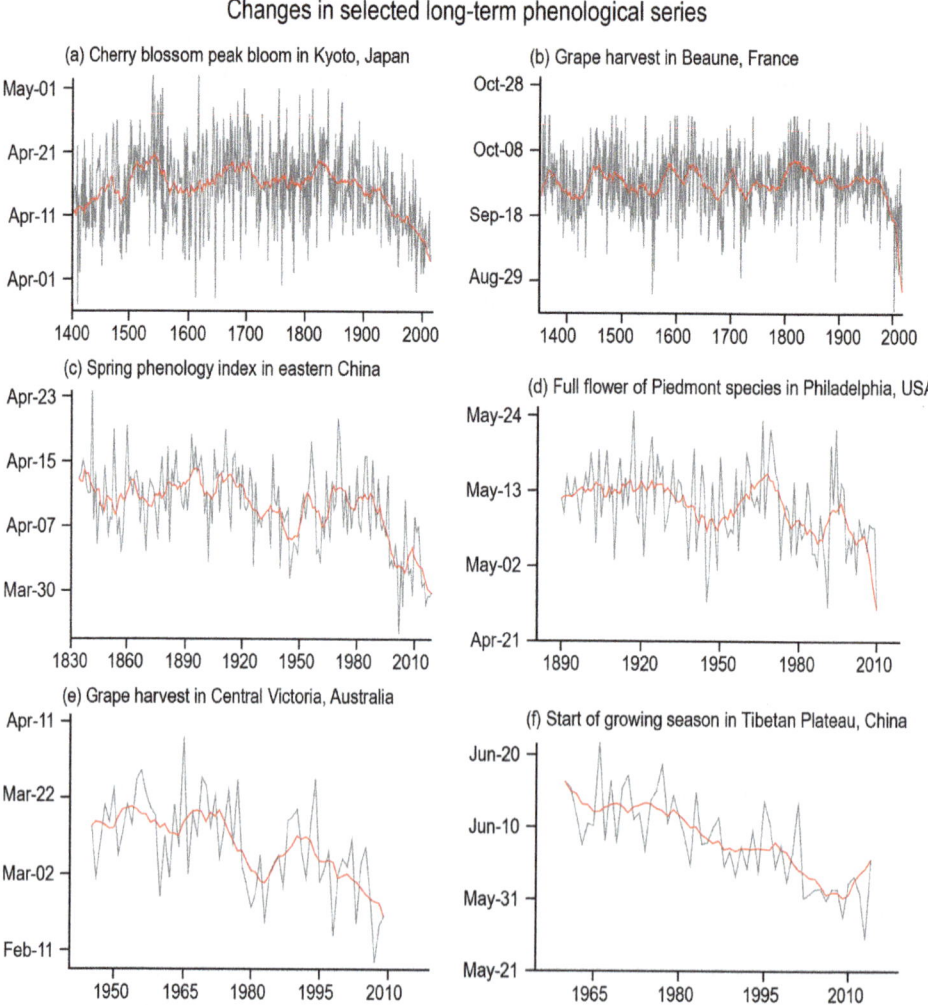

Figure 18.11 Phenological indicators of change in growing season. (a) Cherry blossom peak bloom in Kyoto, Japan; (b) grape harvest in Beaune, France; (c) spring phenology index in eastern China; (d) full flower of Piedmont species in Philadelphia, USA; (e) grape harvest in Central Victoria, Australia; (f) start of growing season in Tibetan Plateau, China. Red lines depict 25-year moving average (top row) or 9-year moving average (middle and bottom rows). *Source*: see Acknowledgements at the end of the chapter.

18.6 Key Working Group II Conclusions

Working Group II covers **Impacts**, Adaptation, and Vulnerability, and can be regarded as the logical follow-on from Working Group I, The Physical Science Basis. While this book is based solely on WGI, we feel that the inclusion of some of the key conclusions from WGII is appropriate, as these refer to the direct and indirect impacts on people and their societies. All have been extracted from the Summary for Policy Makers of the Synthesis Report: we invite you to seek out additional information there.

Approximately 3.3–3.6 billion people live in contexts that are highly vulnerable to climate change. Human and ecosystem vulnerability are interdependent ... Increasing weather and climate extreme events have exposed millions of people to acute food insecurity and reduced water security ... Between 2010 and 2020, human mortality from floods, droughts and storms was 15 times higher in highly vulnerable regions, compared to regions with very low vulnerability (*high confidence*).

Climate change has caused substantial damages, and increasingly irreversible losses, in terrestrial, freshwater, cryospheric, and coastal and open ocean ecosystems (*high confidence*). Hundreds of local losses of species have been driven by increases in the magnitude of heat extremes (*high confidence*) with mass mortality events recorded on land and in the ocean (*very high confidence*). Impacts on some ecosystems are approaching irreversibility such as the impacts of hydrological changes resulting from the retreat of glaciers, or the changes in some mountain (*medium confidence*) and Arctic ecosystems driven by permafrost thaw (*high confidence*).

Climate change has reduced food security and affected water security, hindering efforts to meet Sustainable Development Goals (*high confidence*). Although overall agricultural productivity has increased, climate change has slowed this growth over the past 50 years globally (*medium confidence*) ... Roughly half of the world's population currently experiences severe water scarcity for at least part of the year due to a combination of climatic and non-climatic drivers (*medium confidence*).

Climate change has caused widespread adverse impacts and related losses and damages to nature and people that are unequally distributed across systems, regions and sectors.

Summary

Our world is changing in multiple ways. One of the first challenges the IPCC must address is the reality of those changes, and not just anecdotal evidence, or one particular set of statistics. In this chapter, we have examined a broad range of information on these multiple changes, and not just the obvious: temperature. (In the next chapter we will look at extreme weather events; events which stand out in this maze of data.) While the terms 'climate change' and 'global warming' are often used interchangeably, they are not synonymous.

The increase in surface temperature is well known, and clear. However, global averages are never the whole story, and regional patterns can be even more important: temperatures at high northern latitudes are increasing at more than twice the average rate. Potentially of even greater concern are changes in the hydrological cycle – changes in both precipitation and evaporation – which may have direct impacts on our food supply, either locally or globally.

High northern temperature rises are directly connected to changes in the cryosphere, via the ice–albedo feedback mechanism. Thus we see clear reductions in Arctic sea ice, as well as land ice and snow, and glaciers. Most of the energy imbalance that has been brought about by the build-up of greenhouse gases has ended up in the ocean, causing it to expand, with the obvious result being a rise in sea level. The reduction in land ice has made a growing contribution to this rise. Finally, we have examined changes in the biosphere that reflect both the carbon fertilisation we might expect from the increase in atmospheric CO_2, and changes in growing seasons, mostly as a result of increasing temperatures.

You will remember that this chapter asked two questions, and it is the second which is the key to our future: Why? To answer this question, we turn to our models, and the simulations of our world that they provide, under a range of assumptions. In particular, we run our models with, and without, the changes in atmospheric composition which have occurred over the past ~150 years. Only simulations with all these changes – the known forcings – are able to match the observations: this is attribution. When you read through the subsections headed 'human influence' you will find that, by and large, they are peppered with either *very likely* or *extremely likely* that human activity is the cause.

We will close this chapter by repeating the IPCC's major conclusion: 'It is unequivocal that human influence has warmed the atmosphere, ocean and land surface since pre-industrial times.' There is no stronger language available.

Acknowledgements

Figures 18.1 {2.11}, 18.2 {cross-chapter box 2.3, figure 1}, 18.3 {FAQ3.1 figure 1}, 18.4 {3.9}, 18.5 {2.13}, 18.6 {2.14}, 18.7 {2.16}, 18.8 {2.24}, 18.9 {cross-chapter box 9.1 figure 1}, 18.10 {3.29} 18.11 {2.32} and Table 18.1 {table 9.5}:

Climate Change 2021: The Physical Science Basis. Contribution of Working Group I to the Sixth Assessment Report of the Intergovernmental Panel on Climate Change [Masson-Delmotte, V., P. Zhai, A. Pirani, S. L. Connors, C. Péan, S. Berger, N. Caud, Y. Chen, L. Goldfarb, M. I. Gomis, M. Huang, K. Leitzell, E. Lonnoy, J. B. R. Matthews, T. K. Maycock, T. Waterfield, O. Yelekçi, R. Yu, and B. Zhou (eds.)]. Cambridge University Press, Cambridge, United Kingdom and New York, NY, USA, 2,391 pp. doi:10.1017/9781009157896.

FURTHER READING

There is, of course, a rather large body of literature on climate change, which can easily be divided into the good (both accurate and well written), the bad (accurate, but confusing), and the ugly (denialist nonsense). If you wish to go further than this chapter provides, we hope you will choose your sources carefully. Obviously, IPCC AR6 contains much more than we have had space for, including discussions of the technicalities involved in reaching some of their conclusions and confidence levels. It also contains copious references to the original peer-reviewed science. We have had to make choices as to what we included, at what level of detail, and what we had to leave out.

One reference we are always happy to recommend is *Global Warming, The Complete Briefing* by John Houghton (Cambridge University Press, 2015). [Unfortunately, the final edition only covers AR5.]

If you are interested in more general information on human impacts on the biosphere you will find a number of interesting articles in *The Balance of Nature and Human Impact* edited by Klaus Rohde (Cambridge University Press, 2013).

REVIEW QUESTIONS

1. What are the four 'domains' as employed in AR6?
2. Explain the concept of attribution.
3. What changes have we seen in surface temperature?
4. What has happened to humidity over recent decades?
5. What other changes have been detected in the hydrological cycle?
6. What are the key changes observed in the cryosphere?
7. What is the primary cause of sea-level rise?
8. What other changes have been observed in the ocean?
9. What is carbon fertilisation?
10. What changes have been observed in the biosphere?

EXERCISE

The Carbonator models that we introduced you to in Section 15.6 are primarily designed to simulate the climate of the twenty-first century, which is the subject of Chapter 20. However, the models do run from 1850 to 2100, with historic forcings up until 2005 and scenario forcings thereafter. Now would be a good time to check out the website (Carbonator.org) and run a couple of options. The main cases will all give the same results up until 2005, of course, so you only need to run one of them. However, your interest should be aroused by now, and you might like to start running them all.

19 Recent Weather Extremes

Over the past century, average temperatures have risen a little over 1°C. This may not seem like much: certainly nothing to get terribly worried about, surely? After all, temperatures vary from one day to the next by much more than that, and we take it in our stride – even if we curse Mother Nature for her vagaries. However, over the past couple of decades we have become more and more aware of the rising incidence of what we call extreme events: heatwaves; droughts; wildfires; floods; severe storms. These are the signs of the times; signs, perhaps, that Mother Nature is not happy. Or is this all simply part of the natural unpredictability of the world we live in? Maybe just a media 'discovery'?

In this chapter we will take a look at recent extremes, including multiple examples, based on the IPCC's 6th Assessment Report's chapter 11: 'Weather and Climate Extreme Events in a Changing Climate', along with the recent branch of climate science, event attribution, where we endeavour to assess any human contribution to these events. We illustrate both the nature of extreme events, and our growing understanding, with several detailed case studies.

19.1 Extremes

19.1.1 Definitions

By their very nature, extreme events are almost always local, or at most regional, so both their impacts, and their definitions, can be specific to the location, its underlying weather and climate regime, and socioeconomic circumstances (e.g. resilience). Chapter 11 of AR6 is actually one of the longest, primarily as a result of close to 100 pages devoted to 'Large Tables', giving extensive regional detail. There are 21 tables in all; the first of which looks at 'observed trends, human contributions, and projected changes at 1.5°C, 2°C and 4°C, for temperature extremes in Africa'. Plenty of material there for a research project.

Extremes are almost always linked in some way to climate variability, so that a data record over a sufficient period of time is needed to help define the climatic regime in a given location. This may require both instrumental and proxy data. An event is usually considered extreme if the value of a variable exceeds (or lies below: both can be important) a certain threshold. Such thresholds have been defined in different ways, so that the meaning of an extreme may not be consistent, even if the same words are used.

Let's start with temperature extremes. One metric might count the number of days when the temperature exceeds a threshold, defined (for example) as the 90th (or higher) percentile for that date, over a baseline period. Such an event can occur at any time of the year, of course, including those 'delightful winter days' we enjoy. Another metric might count the number of days above an absolute threshold, such as 35°C, because exceeding such temperatures may have adverse health impacts. High overnight temperatures can be even more dangerous if we can't get a good night's sleep.

Changes in extremes have been examined from two perspectives: changes in frequency for a given magnitude, and changes in magnitude for a particular frequency or **return period**. We have used terms which need to be better defined. You will sometimes hear phrases like 'a 1 in 100-year flood': what does this mean? One hundred years is the so-called return period, but that most definitely does not mean that after you've had one, you are 'safe' for another 100 years! What it means is that there is a 1% chance of such a flood, *per year*.

The sensitivity of changes in extremes to global warming may also differ. Consider, for example, a 'heatwave', which we might define as the number of consecutive days above, say, 35°C. It is quite likely that one or two days either side of that time period had maximum temperatures somewhere between 34.0°C and 35.0°C (this is assuming a somewhat smooth rise and fall in the maximum temperature, which may not be the case). If temperatures in that location were to rise by 1°C as a result of global warming, those 'fringe' days would now be included. Numerically, that heatwave became worse, primarily as a result of how it was defined. Over time, of course, we'll become acclimatised to a warmer world (we'll have little choice!), so should we change the threshold in line with global warming? For the purposes of analysis, no such redefinitions have been considered.

19.1.2 Attribution of Extremes

Attribution science is concerned with the identification of the causes of any changes in the characteristics of the climate system: trends, extreme events, etc. It is inherently statistical in nature, and its outcomes are invariably expressed in such language. Because extreme events do not, usually, follow a Gaussian/normal distribution, there are additional challenges: the technicalities are beyond the scope of this book. In recent years (since AR5), the attribution of extreme events has emerged as a key field in climate research, with an increasing body of literature.

As outlined by Perkins-Kirkpatrick et al., **extreme event attribution** (EEA) is a subfield of climate science where the influence of physical drivers is isolated for specific events. In most cases, the physical driver is, of course, anthropogenic climate change, and we are trying to determine its influence on the frequency or magnitude of an observed extreme. Most such studies involve multiple-model simulations of 'factual' and 'counterfactual' climates; that is, simulations where all known forcings are included, and simulations where anthropogenic forcings are omitted. Simulations are mostly run using atmosphere-only GCMs, with sea-surface temperatures held fixed.

Large sample sizes are the key, using multi-member ensembles of single or multiple models, where the initial conditions (and sometimes the physics) differ slightly. From these we are able

Figure 19.1 Probability per year of reaching a certain temperature versus temperature, for both the present climate and pre-industrial climate. *Source*: see Acknowledgements at the end of the chapter.

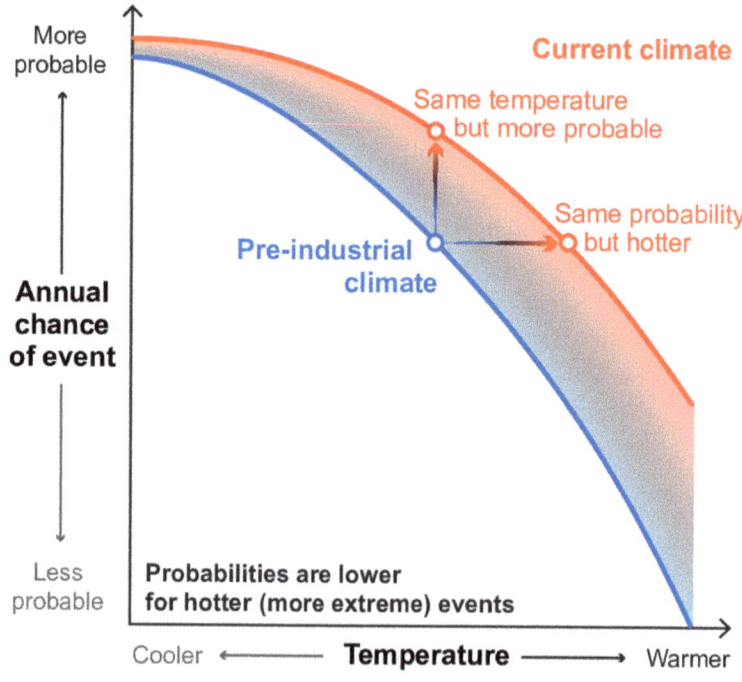

to extract event statistics, such as the probability of exceeding a certain threshold: for example, a certain temperature, or rainfall amount. From these statistics we may define the risk ratio (RR), and the **fraction of attributable risk** (FAR) via

$$RR = \frac{P_{fact}}{P_{cfact}}$$

and

$$FAR = 1 - \frac{1}{RR} = 1 - \frac{P_{cfact}}{P_{fact}}$$

where P_{cfact} is the event frequency (i.e. probability) in the counterfactual simulations, and P_{fact} is the event frequency in the factual simulations.

Figure 19.1 {FAQ 11.3, figure 1} gives a schematic representation of what we might expect in the case of temperature extremes. One curve shows the probability distribution under pre-industrial climatic conditions, while the second shows the distribution in today's warmer climate. We see how a specific above-average temperature will become more probable, while for the same probability, the temperature will be hotter. These two conclusions are, of course, merely different manifestations of the same 'reality'.

The outcome of event attribution depends on the definition of the event, and its temporal and spatial extent, as well as the framing of it, and the uncertainties in both the observations and the modelling. Careful framing is essential. In general, confidence in attribution statements for large-scale heat events, or lengthy extreme precipitation events, have higher confidence than shorter and more localised events such as severe storms.

19.2 Temperature

19.2.1 Observations

Temperature extremes imply both the warm end of the scale and the cold, although virtually all of the interest is at the top end, of course. As the globe as a whole has warmed, it is to be expected that both 'average' maximum and 'average' minimum temperatures must have both increased. Figure 19.2 {figure 11.2} shows the observed changes in (also referred to as anomalies; departures from their 1850–1900 means) global average annual mean temperature (in black), land average annual mean temperature (green), land average hottest daily maximum temperature (TXx; purple), and land average annual coldest daily minimum temperature (TNn; blue).

We see immediately that the coldest temperatures are rising fastest, for an understandable reason. Greenhouse gases trap outgoing long-wave radiation, and re-radiate some of it back to the ground: day and night. By contrast, the cooling effects of anthropogenic aerosols only occur during the day.

Figure 19.3 {figure 11.9} provides maps of the linear trends, over the period from 1960 to 2018, for (a) annual hottest temperature (TXx), (b) annual coldest temperature (TNn), and (c) the number of days exceeding the 90th percentile (TX90p) (based on a reference period of 1961–1990) based on the HadEX3 data set.

A statistical analysis of heatwaves is more challenging than the analysis of individual daily temperature data. Trends in some heatwave measures are observed at the global scale. Globally averaged heatwave intensity, duration, and the number of heatwave days have significantly increased from 1950 to 2011. There are some regional differences in the characteristics of heatwaves, with significant increases in Europe and Australia. In the UK, the lengths of short heatwaves have increased since the 1970s, while the lengths of long heatwaves (more than 10 days) have decreased over some stations in the south-east.

Summarising the observed trends, AR6 concludes

it is *virtually certain* that there has been an increase in the number of warm days and nights, and a decrease in the number of cold days and nights, on the global scale, since 1950. Both the coldest extremes

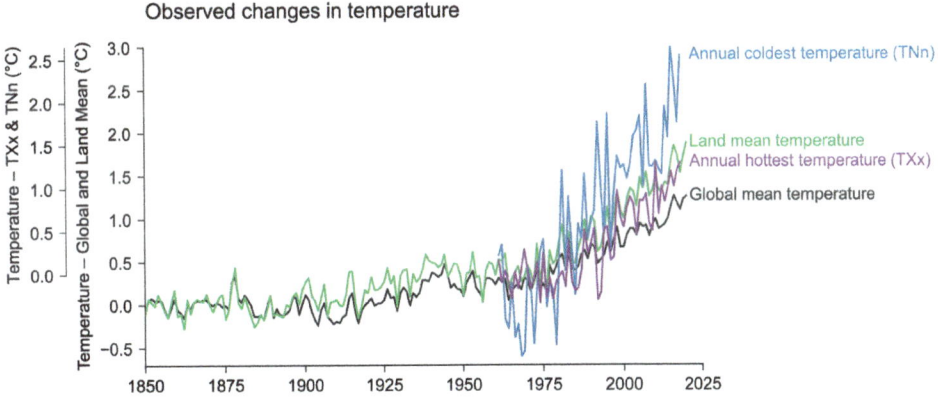

Figure 19.2 Time series of observed temperature anomalies. *Source*: see Acknowledgements at the end of the chapter.

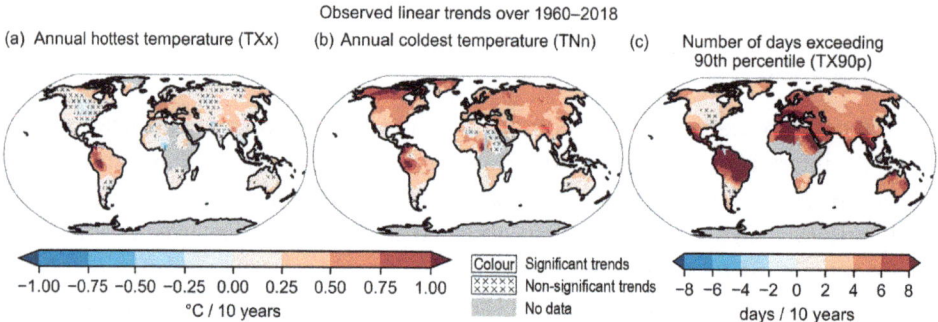

Figure 19.3 Linear trends over 1960–2018 for three temperature extreme indices. *Source*: see Acknowledgements at the end of the chapter.

and the hottest extremes display increasing temperatures. It is *very likely* that these changes have also occurred at the regional scale in Europe, Australasia, Asia, and North America. It is *virtually certain* that there has been an increase in the intensity and duration of heatwaves and the number of heatwave days at the global scale ... Annual minimum temperatures on land have increased about three times more than the global surface temperature since the 1960s, with particularly strong warming in the Arctic (*high confidence*).

The Summary for Policy Makers of the Synthesis Report includes the following:

In all regions increases in extreme heat events have resulted in human mortality and morbidity (*very high confidence*). The occurrence of climate-related food-borne and water-borne diseases (*very high confidence*) and the incidence of vector-borne diseases (*high confidence*) have increased ... Climate and weather extremes are increasingly driving displacement in Africa, Asia, North America (*high confidence*), and Central and South America (*medium confidence*), with small island states in the Caribbean and South Pacific being disproportionately affected relative to their small population sizes (*high confidence*).

19.2.2 Attribution

AR5 concluded that human influence has *very likely* contributed to the observed changes in the intensity and frequency of daily temperature extremes on the global scale in the second half of the twentieth century, and also that 'it is *likely* that human influence has substantially increased the probability of occurrence of heatwaves in some locations'.

Further studies (including regional studies) since then have continued to attribute the observed increases in hot extremes, and the decrease in cold extremes, to human influence, dominated by greenhouse gas emissions, on global, continental, and regional scales. A cooling (i.e. moderating) effect due to aerosols is detectable over Europe and Asia. As much as 75% of moderate daily heat extremes (above the 99.9th percentile) over land are due to anthropogenic forcing. The detected signals are clearly separable from the response due to natural forcing.

A significant recent advance has been the large number of studies focusing on extreme temperature events at monthly to seasonal scales, using various extreme event attribution methods. For example, one study found that human influence has caused a more than 60-fold increase in the probability of the extremely warm 2013 summer in eastern China.

Several studies of recent events have determined a 'risk ratio' approaching infinity, meaning that the probability of the occurrence of such events without anthropogenic influence is close to zero. (This may also be expressed as the fraction of attributable risk, or FAR, approaching 1.) This implies that those events are so far outside the range of simulations with only natural forcings that it is *extremely unlikely* for those events to occur without a human influence.

The relative strength of anthropogenic influences on temperature extremes is regionally variable, in part due to changes in atmospheric circulation, land-surface interactions, and other drivers such as aerosols. In the Mediterranean and western Europe, risk ratios of ~100 have been found, whereas in the USA changes are much less pronounced. This is partly a result of the extreme temperatures of the 1930s, which skews the baseline. Deforestation has contributed about one-third of the total warming of hot extremes in some mid-latitude regions.

From the Executive Summary to AR6 chapter 11:

Human-induced greenhouse gas forcing is the main driver of the observed changes in hot and cold extremes on the global scale (*virtually certain*) and on most continents (*very likely*). The effect of enhanced greenhouse gas concentrations on extreme temperatures is moderated or amplified at the regional scale by processes such as soil moisture or snow/ice-albedo feedbacks, by regional forcing from land-use and land-cover changes, or aerosol concentrations, and decadal and multi-decadal natural variability … Urbanization has *likely* exacerbated changes in temperature extremes in cities, in particular night-time extremes.

In summary, long-term changes in extreme temperatures … have been detected in observations, and attributed to human influence at global and continental scale. It is *extremely likely* that human influence is the main contributor to the observed increase in the intensity and frequency of hot extremes … on the global scale. Some specific recent hot extremes would have been *extremely unlikely* to occur without human influence on the climate system. Changes in aerosol concentrations, as well as land-use changes and practices, have attenuated hot extremes in some regions.

19.2.3 Case Study: European Heatwave, 2003

July–August of 2003 was the hottest summer period in Europe (to that time) since at least 1540, with France hit especially hard. Figure 19.4, based on MODIS data, shows the difference between daytime surface temperatures in July–August 2003 and July–August of 2000/2001/2002 and 2012. The heatwave caused health crises in several countries, and combined with a drought to create crop shortfalls in parts of southern Europe, with the EU's total wheat production down by 10%. Some key rivers were so low as to be unnavigable.

While an accurate death toll is impossible to know, an estimate of about 70,000 is accepted, with a partial breakdown of: France, 15,000–19,000; Portugal, ~2,000; Netherlands, ~1,500; Spain, ~13,000; Germany, ~300; United Kingdom, ~2,000.

In France, for example, parts of the north recorded eight consecutive days with temperatures above 40°C in early August. Most people had little or no experience of this, and did not understand the need for adequate hydration. Summer nights in France generally cool down, but during this heatwave temperatures remained at record high levels, for which many homes were poorly equipped. The United Kingdom recorded its highest ever temperature of 38.5°C at Faversham, Kent, on 10 August, while Greycrook recorded Scotland's highest ever temperature of 32.9°C on 9 August.

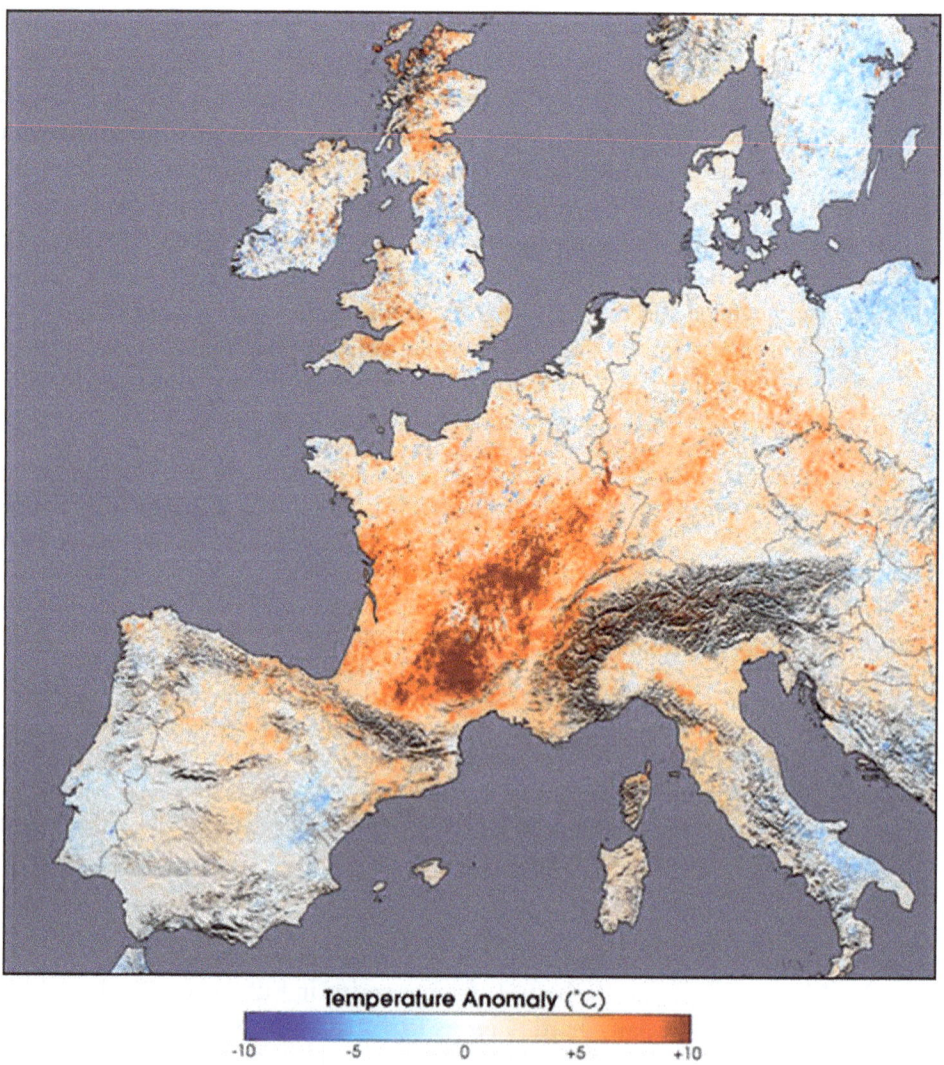

Temperature Anomaly (˚C)

-10   -5   0   +5   +10

Figure 19.4 The difference in average temperature (based on 2000, 2001, 2002, and 2012) from 2003, covering the period from 20 July to 20 August, using MODIS data. *Source*: NASA.

A mid-2023 report in *Nature Medicine* has recently assessed the death toll from the 2022 summer heatwave at more than 61,000. This suggests that two decades of efforts in Europe to adapt to a hotter climate have not been able to keep pace with global warming. (Most of those who died were women, especially those older than 80.)

19.2.3.1 Attribution

This heatwave was one of the first large-scale extreme events to be the focus of attribution studies. Stott et al. used the HadCM3 climate model (with a reasonably coarse resolution of

3.75° longitude by 2.5° latitude), focusing on the relatively large region from 10°W to 40°E, and 30–50°N. Firstly, they investigated the causes of long-term changes in European summer temperatures to determine the changes attributable to anthropogenic drivers, and changes attributable to natural drivers. They ran four simulations, with different initial conditions, each with the same combination of well-mixed greenhouse gases, sulphate aerosols, and changes in tropospheric and stratospheric ozone, as well as changes in solar output and volcanic impacts. The spread of these results encompassed the observed changes, while a simulation using only the natural drivers (solar and volcanic) could not match the data.

Then they estimated the risk of mean July–August temperatures exceeding a particular extreme threshold, based on statistical analysis techniques beyond the scope of this book. Their first (general) conclusion is that it is very likely that past anthropogenic forcing is responsible for a significant fraction of the observed European summer warming. This, of course, is in line with what we have presented in both Chapters 17 and 18. That is saying nothing about any particular heatwave. By pushing such an analysis further, it is possible to estimate by how much human activity has increased the risk of such an event. 'Using a threshold for mean summer temperature that was exceeded in 2003, but in no other year since 1851, we estimate it is very likely (confidence level >90%) that human influence has at least doubled the risk of a heatwave exceeding this threshold maximum.'

19.2.4 Case Study: Russian Heatwave, 2010

In 2010, a series of heatwaves hit the Northern Hemisphere, from May to August, with effects from the United States, Europe, Russia, and across to East Asia, with many new records set. A second, more devastating phase was triggered by a very strong La Niña (one of the strongest ever recorded), with multiple record-breaking temperatures. The period from April to June 2010 was the warmest ever recorded for land areas of the Northern Hemisphere (1.25°C above average).

Some of the worst effects were felt in Russia, and that will be our focus. This heatwave started at the beginning of July with temperatures slowly decreasing at the beginning of August, and finally breaking by the 19th. The persistence of such anomalously high temperatures – more than 5°C above the long-term mean – was mainly due to an atmospheric 'blocking' situation, which is not uncommon for this region.

The heat led to fires throughout the country, the worst drought in nearly 40 years, and the loss of at least 9 million hectares of crops, along with 20% of the grain harvest, and a US$15 billion total loss to the economy. The death toll has been put at 50,000, including ~1,000 who drowned (many, apparently, due to swimming while intoxicated).

19.2.4.1 Attribution

There have been several attribution studies of this heatwave undertaken, with seemingly contradictory results: a more recent study has endeavoured to reconcile their conclusions. In 2011, Dole and co-workers concluded that the heatwave was 'mainly natural in origin'. By contrast, Rahmstorf and Coumou concluded that, with a probability of 80%, 'the 2010 July heat record would not have occurred' without the global-scale warming since 1980, most of which is attributable to the increase in greenhouse gas concentrations.

Otto et al. then undertook to reconcile these studies. (They believed that the primary reason for the different conclusions is that the earlier studies asked somewhat different questions.) They concluded that, while natural climate variability can account for an event of such magnitude, the frequency is likely to have increased as a consequence of greenhouse warming (similar to the 2003 heatwave). They reformulate the Rahmstorf and Coumou conclusion to say that this trend increased the probability by a factor of 5.

19.2.5 Case Study: South-East Australian Fires of 2019/20

Along with the western parts of the United States and the south-western parts of Europe, the south-east of Australia has long been known as one of the most fire-prone parts of the world. For many years, Australia's position in the Southern Hemisphere has allowed the relevant authorities to lend and borrow resources – aerial tankers and skilled personnel – from the USA, as the fire seasons are, of course, reversed. But not any more, as fire seasons are getting longer; and at times, much longer.

The Black Summer of 2019–2020 was the worst on record, driven by unprecedented heat, so we'll start with some numbers: 18 December 2019 was the hottest Australia-wide (i.e. area-averaged) day on record, peaking at 41.88°C. In December 2019 there were 11 days in which the national area-averaged maximum temperature was 40°C or above. Prior to that month there had been only 11 such days since records began in 1910, seven of which occurred in the summer of 2018–2019. Think about that for a moment. These (in our opinion) are stunning – make that scary – numbers! The year 2019 was also the driest on record for Australia as a whole, as was spring (September, October, November), and December.

As we discussed in Chapter 13, Australia's climate is quite variable, mainly driven by ENSO, but also by the Indian Ocean Dipole (IOD) and the Southern Annular Mode (SAM). It is now recognised that the IOD is often the key contributor to our worst droughts. During the latter half of 2019 its mode index was at one of its highest positive levels, drawing heat from the north-west of the continent to the south-east. SAM was also in a persistent negative phase from September to December, implying dry conditions over eastern Australia.

19.2.5.1 Fire Danger

Forest fires are governed by four conditions.

1. Fuel load (biomass). Australia's temperate forests are mainly composed of eucalypts, one of the most fire-prone species in the world. They typically have high fuel loads, comprising leaf and bark litter, dead wood, and living foliage.
2. Fuel dryness. High solar insolation, low relative humidity, and the effects of drought all help dry out the fuel load, allowing the development of large, high-intensity fires. As noted above, this was definitely the case that summer.
3. Ignition. Fires may start by anthropogenic (e.g. arson, fallen power lines, carelessness) or natural (e.g. lightning strikes) ignition. (Some media outlets chose to blame arson which, in reality, accounted for a tiny minority of the fires.)

4. Fire weather. Hot, dry and windy weather, often driven by the passage of a cold front which increases wind speeds, allows fires to spread rapidly. Some embers started new fires several kilometres ahead of the fire front; a frightening scenario for those watching the approaching flames from such distances.

The Forest Fire Danger Index (FFDI) is a common measure of fire weather conditions. It reflects both longer-term rainfall and temperature patterns, and shorter-term weather conditions such as winds. Daily FFDI values can be summed over longer periods of time, and in spring 2019 were the highest on record for Australia as a whole (with the exception of the island state of Tasmania). These conditions continued into December, with the accumulated FFDI that month being twice the long-term average (records back to 1950) over large parts of Australia.

Black Summer was not one fire, but a series of fires, which counted as 16 claims for re-insurance purposes. (All policy holders will be facing higher premiums as a result.) They burned 143,000 km^2, and claimed 34 lives (another ~400 may have died as a result of smoke inhalation), including the crew of a US aerial tanker that crashed due to visibility issues. (We honour their sacrifice.) Economic and property losses have been put at ~A\$80 billion. To place these fires in some context, Australia's most devastating single fire event was on 7 February 2009, when 173 lives were lost north-east of Melbourne in a single weekend.

19.2.5.2 Implications

The causes of Australia's Black Summer are split between climate variability and climate change. It is now clear that the influence of anthropogenic emissions is to make the bad even worse. As we discussed previously, global warming is leading to higher temperatures, and also to reduced relative humidity, all factors which increase fuel dryness. (Ironically, increased rainfall in wet years/seasons, also increases fuel load.) The result is an increase in extreme pyro-convective firestorms, and a massive increase in radiative power. These, of course, are factors which contribute to forest fires wherever in the world they occur, and are almost certain to get worse, as discussed by Abram et al. This clearly calls for increases in both local adaptation measures, and climate change mitigation efforts.

19.2.5.3 East African Floods

A key factor in the fires in Australia was the high positive IOD. Because this mode is a dipole, we should look for contrasting impacts on the western side of the Indian Ocean, in places like Kenya. Between March and May 2020, heavy rainfall led to massive flooding and landslides across the equatorial African region, affecting ~700,000 people, and claiming ~450 lives, half of them in Kenya.

19.2.5.4 Summer of 2023

July 2023 saw devastating fires in many parts of the Northern Hemisphere, and especially the island of Rhodes. The following month saw the devastation of Maui, fanned by winds from an approaching hurricane, causing the fire front and ember attack to arrive so quickly that many residents of Lahaina were unable to escape. Throughout this period, Canada has also experienced unprecedented fires.

19.3   Heavy Precipitation and Floods

19.3.1   Observations

Earlier studies, including AR5, concluded that 'it was *likely* that the number of heavy precipitation events over land had increased in more regions than it had decreased, although there were wide regional and seasonal variations'. AR6 finds that this assessment has been strengthened, with multiple studies finding *robust evidence* of the intensification of extreme precipitation at global and continental scales.

The average annual maximum precipitation amount in a day – Rx1day – has significantly increased since the mid-twentieth century over land. The percentage of observing stations with statistically significant increases in Rx1day is larger than expected by chance, while the percentage of stations with statistically significant decreases is smaller than expected by chance: over the global land as a whole; over North America and Asia; and over global monsoon regions (where data coverage is relatively good). The addition of the past decade of observational data shows a more robust increase in Rx1day over the global land region.

Daily mean precipitation intensities have increased since the mid-twentieth century in a majority of land regions (*high confidence*). The probability of exceeding 50 mm/day increased during 1961–2018. The globally averaged fraction of precipitation from days in the top 5% has also significantly increased. The increase in the magnitude of Rx1day is estimated to be consistent with the Clausius–Clapeyron equation scaling with global mean temperature (7% per °C).

From AR6, section 11.3.4:

In summary, the frequency and intensity of heavy precipitation have *likely* increased at the global scale over a majority of land regions with good observational coverage. Since 1950, the annual maximum amount of precipitation falling in a day, or over five consecutive days, has *likely* increased over land regions with sufficient observational coverage for assessment, with increases in more regions than there are decreases. Heavy precipitation has *likely* increased over North America, Europe, and Asia.

19.3.2   Attribution

AR5 concluded with *medium confidence* that anthropogenic forcing has contributed to a global-scale intensification of heavy precipitation in the second half of the twentieth century. This is based on the evidence of an anthropogenic influence on the hydrological cycle. Since AR5 there has been new and *robust evidence* and improved understanding of a human influence on extreme precipitation. The observed increases in Rx1day and Rx5day over the Northern Hemisphere land area during 1951–2005 can be attributed to anthropogenic forcing, including greenhouse gases and aerosols. This is shown in both CMIP5 and CMIP6 simulations. Human influence was found to have contributed to the increase in frequency and intensity of regional precipitation extremes in North America during 1961–2010.

Some event attribution studies found an anthropogenic influence on the probability and/or magnitude of observed extreme precipitation events. These included European winters; extreme 2014 precipitation over the northern Mediterranean; parts of the USA (for individual events); extreme rainfall in 2014 over Northland New Zealand (north of Auckland); and China.

From AR6, section 11.4.2:

In summary, most of the observed intensification of heavy precipitation over land regions is *likely* due to anthropogenic influence, with greenhouse gas emissions the main contributor. New and *robust evidence* includes the attribution to human influence of the observed increases in annual maximum one-day and five-day precipitation, and the fraction of annual precipitation falling in heavy events.

At the regional level, 'there is *limited evidence* of human influence on extreme precipitation, but new evidence is emerging'.

19.3.3 Floods

Floods are the inundation of normally dry land, and are classified into types – pluvial floods, flash floods, river floods, groundwater floods, surge floods, coastal floods – depending on the space and timescales, and the major processes involved. Because flooded area is often difficult to quantify, many studies have focused on streamflow. For this reason, AR6 has chosen to focus on streamflow, as well as some types of flash floods.

Pluvial and urban floods – flash floods resulting from precipitation intensity exceeding the capacity of natural and artificial drainage systems – are directly linked to extreme precipitation. Hence changes in extreme precipitation are the main proxy for inferring changes in these flood types (assuming no changes in surface conditions). Coastal floods are due to extreme sea levels – tides, storm surges – and are not included here. (They are assessed in chapter 12 of AR6.)

While the amount and intensity of precipitation is clearly the main driver, other key factors include soil moisture (primarily from previous rainfall), and snow in cold regions. Stream morphology, river and catchment engineering, land-cover characteristics, and some climate feedback processes may also be relevant. In many locations, water management changes have, mostly, increased resilience to flooding, masking the impacts of an increase in heavy precipitation, although they do not eliminate very extreme floods. In regions with seasonal snow cover, snow melt is the main cause of extreme river flooding.

Previous studies, including AR5, assessed *low confidence* for observed changes in the magnitude and frequency of floods at the global scale. The vast majority of studies have focused on river floods using streamflow as a proxy. Streamflow measurements are not, of course, distributed evenly across the globe, with gaps in many less-developed regions. This makes the detection of long-term changes difficult. Of more than 3,500 streamflow stations in the USA, central and northern Europe, Africa, Brazil, and Australia, 7.1% showed a significant increase, and 11.9% showed a significant decrease in annual maximum streamflow during 1961–2005 – in direct contrast to the intensification of short-term intense precipitation discussed above.

The only firm conclusion that AR6 was able to arrive at is that the seasonality of floods has changed in cold regions where snow melt dominates the flow regime, in response to global warming (*high confidence*), which we know is amplified at high latitudes and altitudes.

The ice–albedo feedback works wherever there is snow and ice to melt, and that includes the Tibetan Plateau. The glaciers in this region feed a steady flow of water into several of the most important rivers in Asia, acting like a giant reservoir, supplying water to around two billion people. If this were to melt (and the accumulation of soot on the ice will contribute to this), the consequences could be catastrophic.

19.3.4 Case Study: Pakistan 2022

As we were writing these pages, large areas of Pakistan were under water. While it is too early to provide a thorough analysis of an event that is ongoing, some discussion must be considered appropriate. We leave it to readers to seek out more information when it becomes available.

Flooding began in mid-June, following heavier than usual monsoon rains, and melting glaciers that followed severe heatwaves in May and June: all are likely to be climate-related. The provinces of Sindh and Baluchistan received ~9 and 6 times their average rainfall for the month of August. By the end of August, about one-third of the country was under water, affecting 33 million people. Economic losses are likely to exceed US\$50 billion. More than 500,000 people have been displaced.

While the average oceanic temperature rise since 1850 is ~0.7°C, the Indian Ocean has warmed by ~1°C. This provides more moisture for the south Asian monsoon. (Higher than average rainfall was also recorded in both India and Bangladesh.) One preliminary study has suggested that global warming made the flooding ~50% worse, and future floods more likely, although we would caution that further studies are clearly needed. Deforestation is another factor that makes such floods worse, by reducing the biospheric uptake of water.

The phrase 'climate genocide' has been proposed for this and similar events (Pakistan contributes less than 1% of greenhouse gas emissions). This, of course, is highly subjective language, but understandable. Nevertheless, this is a good place to introduce the concept of 'climate justice', as it is so often communities which have contributed the least to climate change who bear its brunt.

19.4 Drought

A drought is a period of time with substantially below normal moisture levels, usually over a large (land) area, which generally results in negative impacts on various components of both the natural and economic sectors. Droughts may be approximately subdivided into

- meteorological drought: a precipitation deficit;
- agricultural drought: reductions in crop yield, often due to a soil moisture deficit;
- ecological drought: plant water stress, with impacts such as tree mortality;
- hydrological drought: a lack of water in reservoirs, lakes and groundwater.

This breakdown is not absolute, and will vary with local circumstances. Thus, there is no universal definition or metric, although many are used in appropriate circumstances: AR6 uses a number of these. Droughts can also range from just weeks (a 'flash drought') to multi-years/decadal rainfall deficits, sometimes known as 'megadroughts'.

Droughts, like most extremes, occur as a result of a combination of factors:

- thermodynamic (greenhouse gas forcings, local-scale heat and moisture fluxes), which are the main drivers of drought changes in a warming climate (*high confidence*);
- dynamic (synoptic patterns such as global circulation, blocking highs, ocean–atmosphere modes such as ENSO), which mostly affect the 'when and where'.

Lack of precipitation – **Precipitation Deficit** (PD) – is the main factor controlling drought onset. 'There is *high confidence* that atmospheric dynamics, on interannual, decadal or longer time scales, is the dominant contributor in the majority of the world's regions.' Regional moisture recycling and land–atmosphere feedbacks play an important role in some cases (*high confidence*).

Atmospheric evaporative demand (AED) quantifies the evapotranspiration (ET) that can happen from a surface where water availability is not limited. AED depends on radiation, relative humidity, and wind speed. The Penman–Monteith equation (not covered in this book) has been recommended by the UNFAO since 1998. The influence of AED on drought depends on the drought type, background climate, and moisture availability. On subseasonal to decadal scales, variations in AED are strongly controlled by circulation variability. Global warming may increase AED by reducing relative humidity.

Soil moisture deficit (SMD) is well correlated with precipitation variability, although ET also plays a key role in depleting moisture from the soil. Soil moisture also plays a role in drought 'self-intensification' under dry conditions. Vegetation cover plays several roles, firstly on surface albedo, but also on the potential for tapping into deeper soil moisture and groundwater. SMD directly affects plant water stress and is a primary cause of plant mortality. (Note that higher atmospheric CO_2 concentration will potentially decrease plant ET, and increase water-use efficiency, because leaves may not need to keep their stomata open as long.)

Hydrological deficits are closely tied to streamflow and surface water deficits, which are complex and strongly dependent on the hydrological system in question, especially in highly regulated river basins. In some regions, snow and glacier melting are key water sources, as noted above.

19.4.1 Observed Trends

Given the complexities involved, both in the types of drought metrics, and the regional and subregional understanding of climate variability, along with differing economic impacts, it is far from easy to draw clear, global conclusions. All such analysis is, by definition, regional, and may be quickly out of date.

Strong precipitation deficits have been recorded in recent decades: the Amazon (2005, 2010), south-western China (2009–2010), south-western North America (2011–2014), Australia (1997–2009), California (2014), the Middle East (2012–2016), Chile (2010–2015), the Horn of Africa (2011), and others, of course, since the cut-off date for AR6's consideration. (As we write, drought in the Horn of Africa is entering its fifth year, with devastating consequences for food supply; exacerbated, this time, by war in Ukraine, a major grain exporter.) Overall trends are mixed. Studies indicate long-term precipitation decreases in parts of Africa and South America, and increases in other regions and subregions.

In several regions, AED increases have intensified recent drought events, enhanced vegetation stress, or contributed to the depletion of soil moisture through enhanced ET (*high confidence*). Given the observed increase in global temperature and decrease in relative humidity over land areas, vapour pressure deficit (saturation vapour pressure minus actual vapour pressure) has increased globally. This trend generally dominates overall AED trends, compared to trends in wind speed and solar radiation.

There are limited long-term in-situ measurements of soil moisture. Microwave-based satellite measurements have been used to analyse trends, although there is only *medium confidence*, due to inhomogeneities in the data. Also, such measurements only capture surface soil moisture and not the root-zone soil moisture tapped by plants. Overall, evidence suggests that several land regions have suffered increased soil moisture drying, or water balance drying, in recent decades. Such trends are generally related to increases in ET rather than decreases in precipitation.

Based on streamflow records, there is evidence of increased hydrological droughts in East Asia and southern Africa. In western, central and northern Europe there is no evidence of changes in the severity of hydrological droughts since 1950. In the Mediterranean region there is *high confidence* in hydrological drought intensification. In North America the results are 'mixed'. In many other regions, the data are insufficient to derive solid conclusions.

Globally, trends in standardised drought indices suggest slightly higher increases in drought frequency and severity in regions by drying, in comparison to a 'simple' precipitation measure. This suggests that AED has contributed more than meteorological droughts in the severity of agricultural and ecological droughts in reducing soil moisture in the dry season, which has increased plant stress and helped trigger more severe forest fires.

From AR6, section 11.4.4:

In summary, there is *high confidence* that AED has increased on average on continents, leading to increased ET, and resulting water stress periods. There is *medium confidence* in increases in precipitation deficits in a few regions of Africa and South America. There is *medium confidence* that agricultural and ecological droughts have increased in several regions on all continents, while there is *medium confidence* in a decrease only in northern Australia.

19.4.2 Attribution

There are only two AR6 regions where there is at least *medium confidence* that human-induced climate change has contributed to changes in meteorological droughts: south-western South America for an increase, and northern Europe for a decrease. In other regions the evidence for contributions to long-term trends is inconclusive. Global warming appears to have led to a decrease in relative humidity over land areas, and hence an increase in vapour pressure deficit. This, as we have seen, is the primary factor in increases in soil moisture deficit. For hydrological droughts, it is difficult to separate the role of climate change from changes in land use and water management on a regional scale.

Attribution studies for some recent meteorological drought events are available. A study of the 2015 central European drought did not find conclusive evidence for a human contribution to the rainfall deficit. A human contribution was found for the 2014 drought in the southern Levant, and the 2015–2017 southern African drought. Event attribution studies also highlight the complex interplay between anthropogenic and non-anthropogenic climatic factors. Anthropogenic warming contributed to the 2014 north-east African drought by increasing east African and west Pacific temperatures, increasing the SST gradient, and reducing rainfall. As our understanding, and methods, evolve, sometimes conclusions change.

From AR6, section 11.6.4:

In summary, human influence has contributed to increases in agricultural and ecological droughts in the dry season in some regions due to increases in ET (*medium confidence*). The increases in ET have been driven by increases in atmospheric evaporative demand induced by increased temperature, decreased relative humidity and increased net radiation over affected land areas (*high confidence*). There is *low confidence* that human influence has affected trends in meteorological droughts in most regions, but *medium confidence* that they have contributed to the severity of some single events. There is *medium confidence* that human-induced climate change has contributed to increasing trends in the probability or intensity of recent agricultural and ecological droughts, leading to an increase of the affected land area. Human-induced climate change has contributed to global-scale change in low streamflow, but human water management and land-use changes are also important drivers (*medium confidence*).

19.4.3 Case Study: South-Western USA Megadrought

The south-western region of North America is delimited by 30° and 40° north, 105° and 125° west, including (much of) the US states of California, Nevada, Utah, Colorado, Arizona, and New Mexico, as well as the Mexican states of Baja California, and Sonora. It includes the Colorado River Basin, as well as the Sonoran, Mojave, and Great Basin deserts. This megadrought (sometimes referred to as the south-western North American megadrought, as parts of Mexico are also affected) has been ongoing since 2000. It is (currently) the driest 22-year period since at least 800 CE (based on tree-ring data).

This is an historically dry region, and has experienced periods of severe drought, and even megadrought, over the past millennium (with impacts on the Pueblo civilisations). The Colorado River is the lifeblood of the region, helping to sustain such cities as Los Angeles, Las Vegas, and Phoenix: 40 million people rely on the Colorado River for water. However, current water levels in Lake Mead, the largest US reservoir, are so low that both hydroelectric generation, and water for human needs, are reduced.

California, in particular, is drought-prone, including 1928–1935, 1947–1950, 1986–1992, plus several more this century (part of the current megadrought). From 2012 to 2015, the Central Valley and south coastal region of California experienced a level of dryness that is unprecedented in the instrumental record (back to 1896), and the driest since the late sixteenth century, based on palaeorecords.

Given the clear record of drought in this region, we should first look to the natural drivers for an explanation. La Niña conditions are usually associated with hotter and drier conditions in California, and beyond, due to the northward movement of the jet stream and the rain that it brings. However, climate change is clearly contributing. While a lack of precipitation is clearly one of the keys to any drought, evaporation is the other. There have been a number of wet years since 2000, but they were not sufficient to eliminate the drought. Water from melting snow is often soaked up by dry soil before it can reach the river system.

We are aware of one attribution study by Williams et al., which focused on the 2012–2014 California drought, not the megadrought as a whole. They concluded that anthropogenic warming has intensified the recent drought as part of a chronic drying trend that is becoming increasingly detectable and is projected to continue throughout this century. As anthropogenic warming continues, natural climate variability will become increasingly unable to compensate for the drying effect of warming. Instead, the soil moisture conditions associated with the current drought will become increasingly common.

19.5 Other Extremes

19.5.1 Severe Storms

As we have stated on several occasions, a warmer atmosphere can hold more water vapour. Hence we should expect that rainfall from all types of storms will be increasing. Evidence suggests that peak rainfall from tropical cyclones is increasing somewhat faster than might be expected due to increased low-level moisture convergence caused by an increase in wind speeds.

Perhaps a more important question relates to tropical cyclone (TC) numbers. 'It is *likely* that the global proportion of Category 3–5 cyclones ... has increased over the past 40 years. It is *very likely* that the average location where TCs reach peak intensity has migrated poleward since the 1940s in the western North Pacific Ocean' (where they are known as typhoons). This, of course, is to be expected. Tropical cyclones need sea-surface temperatures of 26.5°C (80° F) in order to form, and global warming has pushed such warm SSTs to higher latitudes.

It is projected that average peak TC wind speeds, and the proportion of Category 4–5 TCs will *very likely* increase as the globe warms further. 'It is *very likely* that average TC rain rates will increase with warming, and *likely* that they will increase at a greater rate than the Clausius–Clapeyron scaling of 7% per 1°C of warming.' There is *medium confidence* that the total global frequency of TCs will actually decrease, or remain unchanged.

There is *low confidence* in past changes in maximum wind speeds of extratropical cyclones (ETCs), and *medium confidence* that future changes will be small. However, any changes in the location of storm tracks could lead to substantial changes in local extreme wind speeds. There is *high confidence* that average and maximum ETC rainfall rates will increase with global warming.

Severe convective storms (SCSs) are sometimes embedded within synoptic-scale weather systems such as TCs, ETCs and fronts, controlled by large-scale circulation patterns. The occurrence of SCSs and associated events such as tornadoes, hail, and lightning is affected by atmospheric conditions such as convective available potential energy (Section 6.2) and wind shear. It remains uncertain how the balance of these environmental factors may affect severe storm occurrence. There is *medium confidence* that the mean number of tornadoes in the USA has remained more or less constant, but their variability of occurrence has increased over time, especially since 2000, with a decrease in the number of tornado-days per year, but an increase in the number of tornadoes on these days (*high confidence*).

19.5.2 Compound Events

It is very often a combination of events, or compound event, that causes the most damage to ecosystems and infrastructure, disrupting human lives. Compound events may be defined as

1. two or more extreme events occurring simultaneously or successively; or
2. combinations of extreme events with underlying conditions that amplify their impacts; or
3. combinations of events that are not themselves extremes, but lead to an extreme event or impact when combined.

Alternatively, we may choose to define compound events as the combination of multiple climate drivers and/or hazards that contributes to societal or environmental risk. The key here is that such combinations of 'stressors' can more quickly push a system beyond its capacity to cope, as discussed by Zscheischler et al.

Without providing a detailed analysis, AR6 nevertheless concludes that the land area affected by concurrent extremes has increased (*high confidence*); and further that concurrent extreme events at different locations, but possibly affecting similar sectors (e.g. breadbaskets), will become more frequent with increasing global warming, especially above 2°C (*high confidence*).

19.5.2.1 Coastal and Estuarine Regions

Coastal and estuarine regions are prone to many meteorological extreme events, and also concurrent extremes. Floods are a major impact in coastal regions around the world, and their occurrence may be influenced by storm surge, extreme rainfall, and river flow, as well as by sea-level rise, waves and tides, and groundwater in estuaries. The US Gulf and Atlantic coasts are at increasing risk: population growth is also a contributing factor in the rise in property damage, of course.

AR6 concludes that, over the last century, the probability of compound flooding has increased in some locations, including along the US coastline. There is *high confidence* that the occurrence and magnitude of compound flooding in coastal regions will increase in the future due to both sea-level rise and increases in heavy precipitation.

19.5.2.2 Droughts and Heatwaves

Droughts and heatwaves tend to go hand in hand, although this is not a necessity. Clear skies, perhaps the result of ENSO or another oceanic mode, will obviously contribute to both. This combination will clearly lead on to potentially severe fire weather, as we discussed above. The extent of burnt area in western US forests, especially in California, has been linked to climate change via a significant increase in vapour pressure deficit.

AR6 concludes that there is *high confidence* that concurrent heatwaves and droughts have increased over the last century, at the global scale, due to human influence. There is *medium confidence* that fire weather has become more probable in southern Europe, northern Eurasia, the USA, and Australia over the last century. There is *high confidence* that compound hot and dry conditions have become more probable in nearly all land regions as temperatures increase. There is *high confidence* that (severe) fire weather conditions will become more frequent at higher levels of warming in some regions.

Summary

As we learned in the previous chapter, the world is warming, and we are confident that we know why. So it should come as no surprise that we are seeing more heatwaves, however we define them. Natural climate variability also plays a part, at least in the timing and location of individual events. As a result of our growing confidence in attribution science, we can say that 'it is *extremely likely* that human influence is the main contributor to the observed increase in the intensity and frequency of hot extremes on the global scale'.

The European heatwave of 2003 was one of the first extreme events to be extensively studied, allowing Stott et al. to conclude that human influence has at least doubled the risk of such a heatwave. We also took you through two further case studies: the Russian heatwave of 2011, and the South East Australian heat and fires of 2019/2020.

A warmer atmosphere can hold more water vapour, so it is reasonable to expect some increases in heavy precipitation and flooding. What data we have allow us to conclude that the frequency and intensity of heavy precipitation have *likely* increased at the global scale over a majority of land regions where good data are available. The connection between emissions, warming, increased humidity, and heavy precipitation can be considered robust.

It is even harder to draw solid conclusions in the case of floods, due to the range of man-made changes in flood plains and catchments, except that more rapid snow melt is increasing seasonal flooding downstream. We looked at the very recent (2022) major floods in Pakistan. The coastal strip of Australia from Sydney to Brisbane also experienced unprecedent flooding that year, with one river exceeding its previous record height by an astonishing 2 metres.

Droughts are a major threat to human survival. However, they are not easy to quantify, as they depend not only on reduced rainfall over an extended period of time, but also on other components of the hydrological cycle such as streamflow and groundwater. As the climate warms, specific humidity is increasing, while RH appears to be decreasing. This encourages evapotranspiration, creating a soil moisture deficit which appears to be exacerbating droughts. AR6 concludes that such trends are evident across most regions, leading to increased water stress with both agricultural and ecological impacts. The underlying human connection, via the increase in atmospheric temperatures, is noted. The current megadrought across the US South West has not been a period of consistently very low rainfall, but rather a time when such rain as has fallen has not been able to make up for the long-term shortfall.

Further Comments

Both climate change and climate variability are challenges humans and their societies must learn to deal with. However, it is their intersection that provides the most acute aspect of this challenge, because that is where we find most of the extreme events which can lead to so much destruction and misery. Climate scientists are grappling with this challenge, of course, but so must our political leaders, and even the general public.

Greg Mullins is the former Commissioner, Fire and Rescue, New South Wales (our home state). He spent his life fighting fires, and had noticed that in recent years things were changing – for the worse. Fire seasons were getting longer, and fire behaviour harder to predict. In early 2019 it was apparent to him and other Fire Commissioners that the upcoming Australian summer had the potential for disaster. They tried to warm the Federal Government of this danger, but no action was taken. This failure was likely a significant factor in why the government was voted out of office in May 2022. (We have included this story because we believe that such short-sightedness is not confined to one country.)

The World Weather Research Program of the WMO is currently undertaking a project to try to find ways to improve warnings of extreme events: better input data (observations); better numerical weather models; better hazard models (floods/hydrology, fire risk/spread); better communication of results to those likely to be affected. One interesting piece of feedback from people caught in the middle of an extreme event was 'yes, I heard the warning, but I didn't think it would be that bad'. That view, of course, would have been rooted in their experience of past such events, especially floods. We all need to understand that climate change is rewriting the rules as to what is a major flood, for example. Past experience is an imperfect guide.

As well as governments, emergency services, and the general citizenry, there is one other sector of society that is front and centre of extreme events: the insurance industry, or more to the point, the re-insurance industry. They have been watching the rise in claims from such events for several decades, and now employ meteorologists and climate scientists. They also fund research into the better understanding, and quantification, of weather risk.

Acknowledgements

Figures 19.1 {FAQ 11.3, figure 1}, 19.2 {11.2} and 19.3 {11.9}:

Climate Change 2021: The Physical Science Basis. Contribution of Working Group I to the Sixth Assessment Report of the Intergovernmental Panel on Climate Change [Masson-Delmotte, V., P. Zhai, A. Pirani, S. L. Connors, C. Péan, S. Berger, N. Caud, Y. Chen, L. Goldfarb, M. I. Gomis, M. Huang, K. Leitzell, E. Lonnoy, J. B. R. Matthews, T. K. Maycock, T. Waterfield, O. Yelekçi, R. Yu, and B. Zhou (eds.)]. Cambridge University Press, Cambridge, United Kingdom and New York, NY, USA, 2,391 pp. doi:10.1017/9781009157896.

FURTHER READING

Greg Mullins has put his experience of fighting fires, both on the front line, and in command positions, as well as his fears for the future as climate change rages unabated, in an excellent, if scary, book: *Firestorm. Battling Super-Charged Natural Disasters* (Penguin, 2021). The early chapters cover fires as they behaved, and were fought, in the twentieth century. He then focuses on more recent events, in chapters entitled 'Black Summer', 'On the front line of Australia's first giga-fire', and finally, 'We must stop the climate emergency becoming a climate disaster'. We highly recommend it.

- Abram et al., Connections of climate change and variability to large and extreme forest fires in southeast Australia. *Communications Earth and Environment* (2021) **2**, 8.
- Otto et al., Reconciling two approaches to attribution of the 2010 Russian heat wave. *Geophysical Research Letters* (2021) **39**, L04702.
- Perkins-Kirkpatrick et al., On the attribution of the impacts of extreme weather events to anthropogenic climate change. *Environmental Research Letters* (2022) **17**, 024009.
- Stott et al., Human contribution to the European heatwave of 2003. *Nature* (2004) **432**, 610–614.
- Williams et al. Contribution of anthropogenic warming to California drought during 2012–2014. *Geophysical Research Letters* (2015), 42(16), 6819–6828. doi:10.1002/2015GL064924.
- Zscheischler et al., Future climate risk from compound events. *Nature Climate Change* (2018), **8**, 469–477.

REVIEW QUESTIONS

1. What is meant by 'return period'?
2. Explain 'attribution' in the context of extreme events.
3. What does it mean if we talk about a risk ratio approaching infinity?
4. Why are the coldest temperatures (usually the overnight low) rising faster than the warmest temperatures?
5. What has been the role of urbanisation on heat extremes?
6. Why do we expect heavier rainfall in a warming world?
7. List some of the factors that are important in flooding and its impacts.
8. What are the main factors to consider when discussing drought?

EXERCISE

Unlike the material of the previous chapter, which has an essentially continuous time frame, the material in this chapter is episodic, and can be quickly superseded. By the time you get to read these pages, more notable events will, undoubtedly, have occurred, including in your own region. Pick an event and see what you can find about it: details such as rainfall amounts or heatwave duration. Wikipedia is often a good place to start, as its articles are mostly well written and factual, and will also lead you to the key primary sources. Your local weather service will probably have reports on any local extreme events. (For example, the Australian Bureau of Meteorology regularly publishes Special Climate Statements.)

20 Climate(s) of the Twenty-First Century

The title of this chapter – Climate? Climates? – is ambiguous, and intentionally so. Clearly, the twenty-first century will only have one climate – we only inhabit one Earth – although that climate will, undoubtedly, evolve over time. But what will that climate be like? If we knew the answer to that question, our title, indeed the entire chapter, would be much simpler. But we don't, and the reason we don't is that, right now, we have little or no idea what decisions humans, and in particular our leaders in politics, business, finance, technology, and science, will make.

Without that knowledge, how can we possibly address the questions that are at the heart of the reason you are reading this book – your own future? In the absence of the necessary knowledge, we really only have two options: pack up and go home, or make some 'educated' guesses. So that – the educated guesses – will form the first part of this chapter. (Note that the IPCC seeks inputs from a different suite of experts, not from the climate science community, which will then be asked to examine the consequences of those suggestions.)

After that we will take you through the conclusions that the IPCC has been able to draw, based on CMIP6 simulations of those educated guesses. As with Chapter 18, we will focus on the AR6 indicators. We will also look at any implications for policy decisions our leaders may (or may not) make on our behalf.

20.1 Charting a Path to the Future

Throughout its history, one of the key tasks of the IPCC has been to advise the climate modelling community of its consensus view of these educated guesses, in the form of scenarios, and ask the modellers to simulate the consequences. What exactly is a scenario? Put simply, it is an educated guess, based on a *consistent set of plausible assumptions*, as to how the composition of our atmosphere might change in time – greenhouse gases, aerosols, etc. – as well as any land-use changes, which might also impact on the evolving climate.

AR6 discusses scenarios in some detail in {section 1.6.1}: this section draws heavily from it, both paraphrasing, and directly quoting. 'A **scenario** is a description of how the future *may* develop, based on a coherent and internally consistent set of assumptions about key drivers including demography, economic processes, technological innovation, governance, lifestyles, and relationships among these driving forces.'

Scenarios may also be defined (more simply) by just the geophysical drivers such as emissions of greenhouse gases, aerosols and their precursors, and/or land-use changes, and how these may evolve (for example, as populations increase). Scenarios are not predictions; rather, they provide a 'what-if' path to investigate. That is to say, if the world makes certain (conscious or unconscious) choices, what are the likely (climatic) consequences.

Socioeconomic factors are not the province of physical scientists – meteorologists, physicists, chemists, oceanographers – but of social scientists, and their meeting place is Working Group III (WGIII). Hence, a significant input to many scenarios comes from WGIII. Another very relevant input comes from the Paris Agreement to keep temperature rises below 1.5°C, or at least 'well below' 2.0°C. However, we clearly need to examine worlds where we fail to achieve this.

We should point out that, like so much of what the IPCC does, scenarios also evolve, and the ones used in AR6 will, undoubtedly, be updated in the future.

20.1.1 Short Historical Tour

The first transient simulations using a general circulation model were done in 1988. By 'transient' we mean a simulation in which greenhouse gases were allowed to vary in time, as opposed to the previous, simpler case of comparing a doubled-CO_2 world with a baseline. Since then, our models have evolved, our understanding has evolved, our confidence in our science has evolved, but far more significantly, computer power has grown exponentially.

Transient simulations clearly require some assumptions as to how radiatively active agents are changing in (simulation) time; that is, scenarios. These have evolved from one Assessment Report to the next, for a number of reasons. Among them we note two: we learn to ask better questions of the science, and technological and political awareness changes that might not have been foreseen previously.

The first widely used set of IPCC emissions was the **IS92 scenarios**, two of which assumed the stabilisation of CO_2 levels to 350 and 450 ppm; even 450 might well be considered 'optimistic'. By 2000, the IPCC produced the **SRES scenarios**, in four broad 'families': A1, A2, B1, and B2. They were built around socioeconomic storylines, emphasising not just emissions of CO_2 but other GHGs, land-use changes, and aerosols, but no climate policy-induced mitigation.

The next step was the set of **representative concentration pathways** – RCPs – which include time series of emissions and concentrations of the full suite of GHGs, aerosols, and chemically active gases, as well as land use and land cover. The word 'representative' indicates that each RCP is but one of many possible scenarios that would lead to similar radiative forcings. The term 'pathways' emphasises that not only are the long-term concentrations of interest, but also their trajectory over time (through to 2100) to reach a certain outcome. They are labelled by their radiative forcing (in W m^{-2}) in 2100: RCP2.6; RCP4.5; RCP6.0 and RCP8.5. The most 'ambitious', RCP2.6, aims to keep warming below 2.0°C, and includes 'negative emissions beyond 2050, via some form of carbon dioxide removal. At the other extreme, RCP8.5 leads to a warming above 4°C. These RCPs were used in AR5, and are referenced in AR6'.

20.1.2 Shared Socioeconomic Pathways

AR6 uses a core of five illustrative shared socioeconomic pathway scenarios, designated SSP1–1.9, SSP1–2.6, SSP2–4.5, SSP3–7.0, and SSP5–8.5. They span a wide range of plausible societal and climatic futures from potentially below 1.5°C best estimate warming, to over 4.0°C warming by 2100. The last four constitute the so-called 'Tier 1' simulations of the CMIP6 Intercomparison Project, which modelling groups were asked to prioritise. The low-emissions scenario SSP1–1.9 is used in combination with SSP1–2.6 to explore outcomes of approximately 1.5°C and 2.0°C above pre-industrial levels, which are, of course, the Paris Agreement goals.

One key improvement of the SSP scenarios compared to the RCPs is a wider set of assumptions on future air quality mitigation measures, which relate to emissions of short-lived climate forcers (SLCFs). This permits a more detailed investigation of the relative roles of GHG and SLCF emissions on future climate change, and hence on implications for future policy choices.

Both SSPs and RCPs are used in AR6, so we need to first explain the labels.

- RCPY: these are GHG concentrations, aerosol emissions, and land-use pattern time series. Y labels the approximate radiative forcing by 2100. Check back to Section 15.6 where we introduced you to the Carbonator model: four of its options are RCP scenarios.
- Now SSPX–Y. Firstly, X is the SSP family (1–5); that is the socioeconomic storyline that includes (among other things) GDP, population, urbanisation, economic collaboration, and human and technological development projections that describe different future worlds in the absence of climate change and additional climate policy. The quantification of energy, land use, and emissions implications then follow as a second step, using appropriate (WGIII) modelling. The second label, Y, indicates the approximate radiative forcing reached by 2100. This also requires appropriate modelling.

Table 20.1 provides a quick overview of the five SSP scenarios. The information is extracted from Cross-Chapter Box 1.4, table 1. (Note that NDC stands for the Nationally Determined Contributions, which countries commit to under the Paris Agreement.)

20.1.3 Methodology

The primary lines of evidence used by the IPCC in AR6 (as was the case in AR5) are comprehensive climate models: i.e. atmosphere–ocean general circulation models (AOGCMs), and Earth System Models (ESMs) which form CMIP. The latter models differ from the former by including representations of various biogeochemical cycles. There are also models of intermediate complexity.

The models' performances are referenced against their performance over the time period 1850–1900 to the recent past, and also 1995–2014 (see Figure 15.6). Multi-model ensembles are the central focus of future projections, primarily under ScenarioMIP. Other MIPs also investigate future scenarios with a focus on processes and feedbacks: Table 20.2 {table 4.1}.

Various statistical analyses are employed to try to get the best from such models, and ensembles of models. However, internal variability complicates the model 'signals', especially when considering regional climate signals over short timescales.

Table 20.1 Overview of scenarios used in AR6 (*source*: see Acknowledgements at the end of the chapter).

SSPX–Y	Description	Closest RCP scenarios
SSP1–1.9	Holds warming to approximately 1.5°C above 1850–1900 in 2100 after slight overshoot and implied net zero CO_2 emissions around the middle of the century.	No equivalent low RCP scenario exists.
SSP1–2.6	Stays below 2.0°C warming relative to 1850–1900 with implied net zero emissions in the second half of the century.	RCP2.6, although RCP2.6 might be cooler for some settings.
SSP2–4.5	Scenario approximately in line with the upper end of the aggregate NDC emissions levels by 2030. CO_2 emissions remaining around current levels until the middle of the century. New or updated NDCs by the end of 2020 did not significantly change emissions projections up to 2030, but more countries adopted net zero by 2050 in line with SSP1–1.9 or SSP1–2.6.	RCP4.5 and, until 2050, also RCP6.0.
SSP3–7.0	An intermediate-to-high reference scenario resulting from no additional climate policy. CO_2 emissions roughly double from current levels by 2100. Also has high non-CO_2 emissions including high aerosol emissions.	Between RCP6.0 and RCP8.5, although SSP3–7.0 non-CO_2 emissions and aerosols are higher than in any RCPs.
SSP5–8.5	A high-reference scenario with no additional climate policy. CO_2 emissions roughly double from current levels by 2050. (Such emission levels are only obtained under a fossil-fuelled socioeconomic development pathway.)	RCP8.5, although CO_2 emissions under SSP5–8.5 are higher towards the end of the century.

Table 20.2 Model Intercomparison Projects (MIPs) used in AR6 (*source*: see Acknowledgements at the end of the chapter).

MIP/Experiment	Usage	Our section
Deck, 1%, 4×CO_2	Diagnosing climate sensitivity	
CMIP6, Historical	Evaluation, baseline	Chapter 18; reference period
ScenarioMIP	Future projections	This chapter
AerChemMIP	Aerosols and trace gases	20.2
C4MIP	CO emissions-driven simulations	20.2.1
CDRMIP	Carbon dioxide removal	20.4.2
DCPP	Near-term climate change	20.3
GeoMIP	Solar radiation modification	20.5
PDRMIP	Forcing dependence of precipitation	20.2.2
SIMIP	Sea-ice assessment	20.3.1
ZECMIP	Zero emissions commitment	Not covered
CMIP5	RCP scenario assessment	20.4.2

20.2 Atmosphere

AR6 uses the same four 'domains' when it looks to the future, as it did for the recent past (Chapter 18), although not always with the same degree of detail. In this and the next Section we'll look at projections over the course of the twenty-first century – and specifically temperature, precipitation, sea ice, and sea level – before focusing on near-term climate change. Figure 20.1 {figure 4.2} shows the projections for the four indicators just listed, as simulated using the five SSP scenarios, from 1950 to 2100. The pink and blue shadings show the 5%–95% ranges from the model ensembles for the SSP1–2.6 and SSP3–7.0 scenarios, respectively.

20.2.1 Surface Air Temperature

Before proceeding, we need to remind readers of some definitions from Chapter 18. Global mean surface temperature (GMST) is a combination of land-surface air temperatures (LSAT), and sea-surface temperatures (SSTs). Global surface air temperature (GSAT) is a combination of LSAT and marine air temperatures (MATs).

Panel (a) of Figure 20.1 presents the change in global mean surface air temperature (GSAT). The left axis shows the temperature change relative to 1995–2014, while the right axis shows the

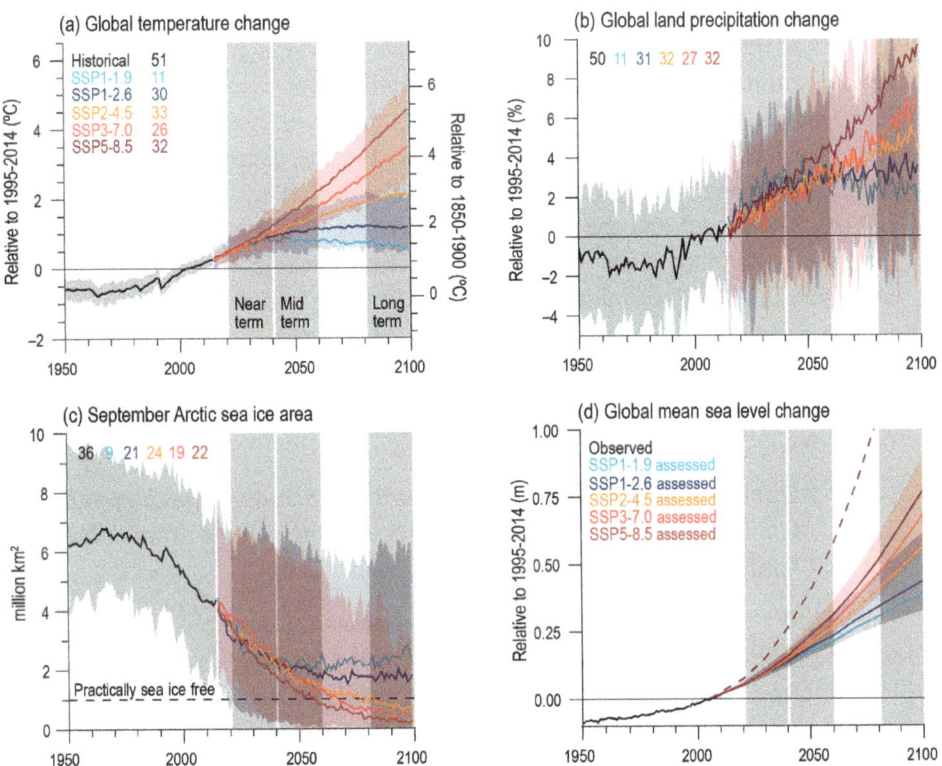

Figure 20.1 Selected indicators of global climate change from CMIP6 historical and scenario simulations. *Source*: see Acknowledgements at the end of the chapter.

Table 20.3 CMIP6 annual mean surface air temperature anomalies (°C) (*source*: see Acknowledgements at the end of the chapter.

Time period and region	SSP1–1.9 (°C)	SSP1–2.6 (°C)	SSP2–4.5 (°C)	SSP3–7.0 (°C)	SSP5–8.5 (°C)
Global: 2021–2040	0.7 (0.3–1.1)	0.7 (0.4–1.1)	0.7 (0.4–1.2)	0.7 (0.5–1.2)	0.8 (0.5–1.3)
Global: 2041–2060	0.8 (0.3–1.5)	1.0 (0.6–1.6)	1.3 (0.8–1.9)	1.4 (0.9–2.3)	1.7 (1.2–2.5)
Global: 2081–2100	0.7 (0.2–1.5)	1.2 (0.6–2.0)	2.0 (1.4–3.0)	3.1 (2.2–4.7)	4.0 (2.7–5.7)
Land: 2081–2100	0.9 (0.3–2.0)	1.5 (0.8–2.6)	2.7 (1.7–4.0)	4.1 (3.0–6.2)	5.3 (3.5–7.6)
Ocean: 2081–2100	0.6 (0.1–1.2)	1.0 (0.5–1.8)	1.8 (1.2–2.7)	2.7 (1.8–4.0)	3.4 (2.3–4.9)
Tropics: 2081–2100	0.5 (0.1–1.1)	1.0 (0.5–1.6)	1.8 (1.2–2.5)	2.7 (2.0–4.0)	3.5 (2.4–4.9)
Arctic: 2081–2100	2.4 (0.5–6.6)	3.3 (0.4–4.7)	5.4 (2.8–10.0)	7.7 (4.5–13.4)	10.0 (6.2–15.2)
Antarctic: 2081–2100	0.5 (0.0–1.1)	1.1 (0.1–2.9)	1.9 (0.6–3.2)	2.8 (1.3–4.5)	3.6 (1.7–5.6)

change relative to 1850–1900. (This has been done to allow for comparison with the way results were presented in AR5. The estimated temperature rise between those two baseline periods is 0.82°C.) In Table 20.3 {table 4.2} we present the best estimate and 5%–95% range for three future time periods – referred to as near-term (2021–2040), mid-term (2041–2060), and long-term (2081–2100) – as simulated for the five SSP scenarios. Note that for SSP1–1.9, temperature actually falls slightly in the latter part of the century, reflecting the drawdown of CO_2 that is a key feature of that scenario.

The CMIP6 models show a general tendency towards greater long-term globally averaged surface warming than did the old CMIP5 models, for nominally comparable scenarios (*very high confidence*). For SSP1–2.9 and SSP2–4.5, the 5%–95% ranges have remained similar to the ranges in RCP2.6 and RCP4.5, respectively, but the distributions have shifted up by about 0.3°C (*high confidence*). For SSP5–8.5 compared to RCP8.5, the 5% bound has hardly changed, but the 95% bound and the range have increased by about 20% and 40%, respectively (*high confidence*). About half the warming has occurred because of more models with higher climate sensitivity in CMIP6 compared to CMIP5; the other half arises from higher effective radiative forcing in nominally comparable scenarios (*medium confidence*).

If we focus on global warming levels (GWLs) of 1.5°C, 2.0°C, and 3.0°C, there is unanimity across all CMIP6 model simulations that GSAT change, relative to 1850–1900, will rise above:

i.   1.5°C following SSP2–4.5, SSP3–7.0, or SSP5–8.5 (on average around 2030);
ii. 2.0°C following either SSP3–7.0 or SSP5–8.5 (on average around 2043);
iii. 3.0°C following SSP5–8.5 (on average around 2062).

Under SSP1–1.9, 55% and 36% of the simulations rise above 1.5°C and 2.0°C, respectively, while for SSP1–2.6, those percentages are 87% and 58%.

AR6 also provides the estimates of temperature rise for several geographic subdivisions for late this century: these are included in Table 20.3. Consistent with CMIP5 and earlier assessments, AR6 models indicate that area-averaged surface air temperature will warm ~50% more over land than over the ocean, and that the Arctic will warm more than 2.5 times faster than the global average. Figure 20.2 {figure 4.12} shows the patterns of warming for two

Seasonal mean temperature change

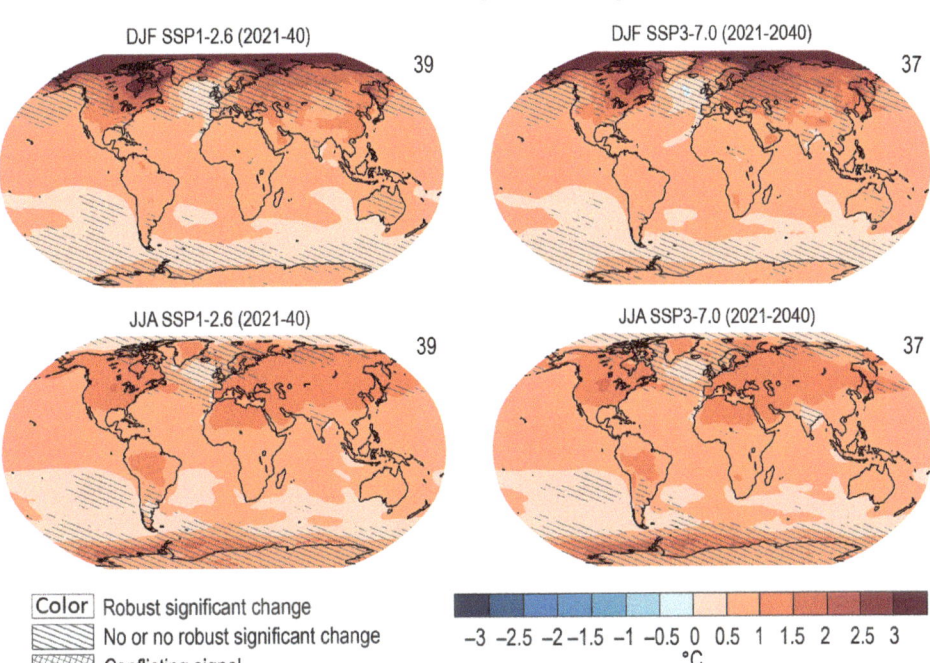

Figure 20.2 Near-term change of seasonal mean surface temperature for two seasons and two scenarios (SSP1–2.6, SSP3–7.0). *Source*: see Acknowledgements at end of the chapter.

seasons – December January February, and June July August – for two of our scenarios – SSP1–2.6 and SSP3–7.0 – for the near-term time period of 2021–2040. Figure 20.3 {figure 4.19} shows the annually averaged patterns for the same two scenarios for the mid-term and long-term periods (50 shades of red).

20.2.2 Precipitation

AR6 has taken a key step on from AR5 in that it now feels able to focus on precipitation over land, rather than globally, which is clearly desirable. Panel (b) of Figure 20.1 shows the changes, as a percentage, relative to 1995–2014. Again, the 5%–95% ranges are shown for SSP1–2.6 and SSP3–7.0. It should come as no surprise that these simulation ranges are much wider than the corresponding temperature ranges: even for 24-hour weather forecasting, rainfall is much more of a challenge than temperature. Based on those results, Table 20.4 {table 4.3} presents the precipitation changes over land for three time periods, under the five SSP scenarios, as well as the 'long-term' global and ocean changes.

Relative to 1995–2014, and across all scenarios, CMIP6 models show greater increases in precipitation over land than over the ocean (*high confidence*). While the different scenarios do differ in their projections – as expected – there is a reasonable degree of agreement that

Table 20.4 CMIP6 precipitation anomalies (%) relative to averages over 1995–2014 for selected future periods, regions, and SSPs (*source*: see Acknowledgements at the end of the chapter).

Region	Period	SSP1–1.9	SSP1–2.6	SSP2–4.5	SSP3–7.0	SSP5–8.5
Land	2021–2040	2.4 (0.7, 4.1)	2.0 (−0.6, 3.6)	1.5 (−0.4, 3.6)	1.2 (−1.0, 3.4)	1.7 (−0.1, 4.1)
	2041–2060	2.7 (0.6, 5.0)	2.8 (−0.4, 5.2)	2.7 (0.3, 5.2)	2.5 (−0.8, 5.1)	3.7 (−0.1, 6.9)
	2081–2100	2.4 (−0.1, 4.7)	3.3 (0.0, 6.6)	4.6 (1.5, 8.3)	5.8 (0.5, 9.6)	8.3 (0.9, 12.9)
Global	2081–2100	2.0 (0.4, 4.2)	2.9 (1.0, 5.2)	4.0 (2.3, 6.7)	4.7 (2.3, 8.2)	6.5 (3.4, 10.9)
Ocean	2081–2100	1.9 (0.6, 4.1)	2.8 (1.1, 5.4)	3.8 (2.0, 6.8)	4.4 (2.1, 7.9)	6.0 (2.9, 10.5)

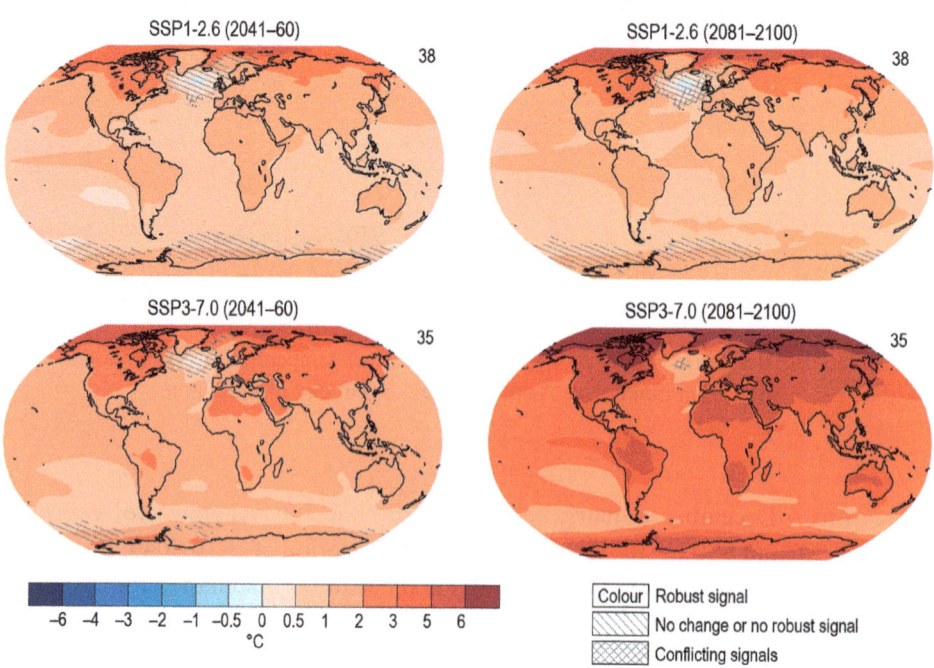

Figure 20.3 Mid- and long-term change of annual mean surface temperature for two scenarios (SSP1–2.6, SSP3–7.0). *Source*: see Acknowledgements at the end of the chapter.

precipitation increases are 'in line' with temperature increases. That is, while the different scenarios take different amounts of time to cross a certain warming threshold, at the time of crossing, the increases in precipitation are reasonably similar.

AR6 does present figures {figure 4.13} showing the geographic changes in precipitation for different time periods and scenarios, similar to the geographic changes in temperature discussed above. As these figures show significant uncertainty (lacking a 'robust' signal), we will not present them. However, some trends do appear to be reasonably clear. The eastern tropical Pacific will see a precipitation increase, along with the region around Sudan/Ethiopia and the

Arabian Gulf, and (in December January February) high northern latitudes {figure 4.24}. By contrast, reduced precipitation is expected in southern Africa, the tropical North Atlantic, the south-east Indian Ocean/south-west of Australia, and the Mediterranean region (especially under SSP3–7.0).

20.3 Cryosphere, Ocean, and Carbon Uptake

We already know that Arctic sea ice is shrinking, and that thermal expansion is leading to a rise in sea levels. How will these two indicators progress in the decades ahead?

20.3.1 Sea Ice

Since AR5, some shortcomings have been identified in the CMIP5 models, particularly around sea-ice thickness: the latest results (for September Arctic sea-ice extent) are presented in panel (c) of Figure 20.1. Table 20.5 {table 4.4} provides the multi-model averages and 5%–95% ranges, for both September (S) and March (M), for all five SSPs and the three time periods. As with the figure, the numbers are sea-ice area, in millions of km^2.

AR6 now concludes that, 'on average the Arctic will be practically ice-free in September by the end of the 21st century under SSP2–4.5, SSP3–7.0 and SSP5–8.5 (*high confidence*)'. Arctic sea-ice area decreases in March, but to a much lesser degree (in percentage terms) than for September (*high confidence*).

20.3.2 Sea Level

Sea level is rising for two fundamental reasons: thermal expansion (a.k.a. thermosteric), and mass loss from glaciers and ice sheets (a.k.a. barystatic). The melting of sea ice does not, of course, make a direct contribution. While AOGCM simulations provide our best estimates of the thermosteric contribution, separate studies are needed to estimate the land-ice contribution.

Panel (d) of Figure 20.1 shows the projections for global mean sea-level rise, as simulated under the five SSP scenarios, with the barystatic contribution added 'offline'. Also included as a

Table 20.5 **CMIP6 Arctic sea-ice area for selected months, time periods, and five SSPs (*source*: see Acknowledgements at the end of the chapter).**

	Time	SSP1–1.9	SSP1–2.6	SSP2–4.5	SSP3–7.0	SSP5–8.5
S	Near	2.6 (1.1–6.5)	2.7 (0.6–6.4)	2.8 (1.1–6.4)	3.1 (1.1–6.4)	2.5 (0.4–5.8)
	Mid	2.2 (0.3–6.5)	2.0 (0.2–6.1)	1.7 (0.1–5.6)	1.7 (0.1–5.7)	1.2 (0.0–5.2)
	Long	2.4 (0.2–6.2)	1.7 (0.0–6.0)	0.8 (0.0–4.6)	0.5 (0.0–3.3)	0.3 (0.0–2.2)
M	Near	14.0 (11.4–18.7)	14.9 (11.9–25.8)	14.9 (11.9–23.5)	15.0 (11.7–27.3)	14.9 (11.9–24.7)
	Mid	13.8 (10.9–18.3)	14.5 (10.9–25.7)	14.3 (11.1–23.3)	14.2 (10.5–27.1)	13.9 (10.2–24.5)
	Long	13.7 (10.9–18.5)	14.2 (10.6–25.7)	13.1 (9.5–22.2)	11.8 (5.4–25.5)	9.7 (9.1–21.6)

dashed curve is a *low confidence* and low likelihood outcome at the high end of SSP5–8.5, and is a reflection of the uncertainties arising from potential ice-sheet instabilities. (By 2100 this curve reaches 1.7 m of GMSL rise, relative to 1995–2014.)

Based on these combined inputs, AR6 concludes that under SSP3–7.0, the *likely* range of GMSL rise, averaged over 2081–2100, relative to 1995–2014 is 0.46–0.74 m. Under SSP1–2.6, the *likely* range over the long term is 0.30–0.54 m. Further, under SSP2–4.5, SSP3–7.0, and SSP5–8.5, the rise in GMSL is projected to accelerate over the twenty-first century. 'In summary, it is *virtually certain* that, under any of the assessed SSPs, there will be a continued rise in GMSL through the 21st century.'

20.3.3 Ice-Sheet Contributions

Melting of the Greenland and Antarctic Ice Sheets, as well as glaciers on land, are a potential source of significant sea-level rise. Ice-sheet dynamics is a very different challenge from the other physical components of the climate system, and is not handled well, if at all, by GCM-type models. Instead, it requires specialist assessment, based on ice-sheet models, which are dependent on the relevant conditions of temperature and precipitation from our suite of Atmosphere–Ocean GCMs.

IPCC AR6 devotes chapter 9 to Oceans, Cryosphere and Sea Level Change. This section is largely drawn from their section 9.4: Ice Sheets. That chapter draws heavily on the Special Report on the Ocean and Cryosphere in a Changing Climate (SROCC). In our Chapter 16 we have already discussed much of the palaeo evidence {section 9.6.2} which helps guide this discussion; see Table 16.1 {table 9.6}.

Ice sheets gain mass from snowfall, and lose mass by surface melt, and also by the effects of ocean warming at the margins. This may even lead to ice-shelf disintegration. The methods and associated modelling of the ice sheets, and especially of their likely future, is well beyond the scope of this book. However, the conclusions are clearly important.

The Greenland Ice Sheet is *likely* to contribute 0.06 (0.01–0.10) m under SSP1–2.6, 0.08 (0.04–0.13) under SSP2–4.5, and 0.13 (0.09–0.18) m under SSP5–8.5 by 2100, relative to 1995–2014. For the Antarctic Ice Sheet there is even greater uncertainty, as the Southern Ocean plays a key role. The corresponding contributions are given as 0.11 (0.03–0.27) m, 0.11 (0.03–0.29) m, and 0.12 (0.03–0.34) m. Glaciers will also make a contribution.

The Exectuive Summary to AR6 concludes:

Both the Greenland Ice Sheet (*virtually certain*) and the Antarctic Ice Sheet (*likely*) will continue to lose mass throughout this century under all considered SSP scenarios. The loss of ice from Greenland will become increasingly dominated by surface melt, as marine margins retreat and the ocean-forced dynamic response of ice-sheet margins diminishes (*high confidence*). In the Antarctic, dynamic losses driven by ocean warming and ice-shelf disintegration will *likely* continue to outpace increasing snowfall this century (*medium confidence*). Beyond 2100, total mass loss from both ice sheets will be greater under high-emissions scenarios than under low-emissions scenarios (*high confidence*).

As we write these closing chapters, an interesting piece of research has just been published in *Nature Climate Change*. (Science never stops just because the IPCC has published its latest 'definitive' update.) In their study, the authors looked at certain sections of the Greenland Ice Sheet, which they refer to as 'Zombie ice'. Zombie, or doomed, ice is ice that is still attached to thicker parts of the ice sheet, but is no longer being fed by the larger glaciers. This occurs when the parent glaciers are receiving less replenishing snow. The doomed ice is melting due to climate change, meaning it will melt and disappear. If (or, rather, when) it all melts it will contribute 27 cm to sea-level rise. Timing is, of course, the central question. While the study's authors were unable to estimate the timing, they did state 'within this century', although with no supporting arguments presented.

20.3.4 Carbon Fluxes

We know that when we increase the atmospheric content of CO_2, some of it will flow into the oceans, and some will be taken up by the biosphere. These fluxes are included in the CMIP6 simulations, and the results presented in Figure 20.4 {figure 4.7}. We note immediately that the spread of model results is much tighter for ocean uptake than land uptake, and the reason is quite simple: one is a physical process, which we understand well; the other is a biological and ecological process, which is far more complex.

AR6 concludes that the cumulative uptake of carbon by the ocean and the land will increase through the twenty-first century for all scenarios except SSP1–1.9 (*very high confidence*).

We also know that the uptake of CO_2 by the ocean will lead to its acidification; that is, a lowering of its pH. Simulation results (for surface waters) for this process are presented in Figure 20.5 {figure 4.8}: again, the spread of model results is tight, as should be expected. AR6 concludes that, with the exception of the two low-emissions SSP1 scenarios, ocean surface pH will decrease monotonically through the twenty-first century.

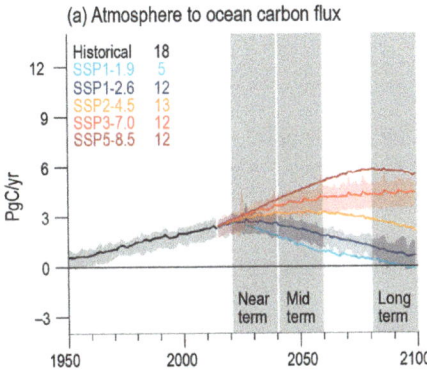
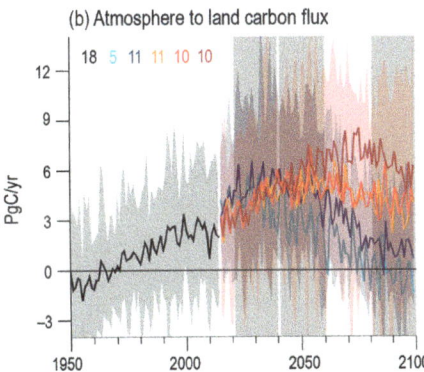

Figure 20.4 CMIP6 carbon uptake in historical and scenario simulations. (a) Atmosphere to ocean carbon flux (PgC year^{-1}). (b) Atmosphere to land carbon flux (PgC year^{-1}). The curves show ensemble averages and the shadings show the 5%–95% ranges for the SSP1–2.6 and SSP3–7.0 ensembles. The numbers inside each panel are the number of model simulations. *Source*: see Acknowledgements at the end of the chapter.

20.3.5 Atlantic Meridional Overturning Circulation

There is one additional element that AR6 looks at, and that is the AMOC, which is central to the climate of (western) Europe. The CMIP6 models show that, over the twenty-first century, the AMOC strength (flow rate), relative to 1995–2014, will decline: Figure 20.6 {figure 4.6}. The units on the axis are Sverdrups: 1 Sv equals 1 million cubic metres per second. AR6 assesses from these results that AMOC weakening over the twenty-first century is *very likely*; and that the rate of weakening is approximately independent of emissions scenario (*high confidence*).

What might be the consequences of a collapse of the AMOC at some time in the future? There is *medium confidence* that it has the potential for such an abrupt change, and that such a change would take centuries to reverse (*high confidence*). While the consequences for the climate of regions around the North Atlantic should be obvious, there is also a suggestion that such a change would have wider impacts on global circulation, including the monsoons.

Figure 20.5 Global average surface ocean pH. The shadings around the SSP1–2.6 and SSP3–7.0 are the 5%–95% ranges across those ensembles. The numbers inside each panel are the number of model simulations. *Source*: see Acknowledgements at the end of the chapter.

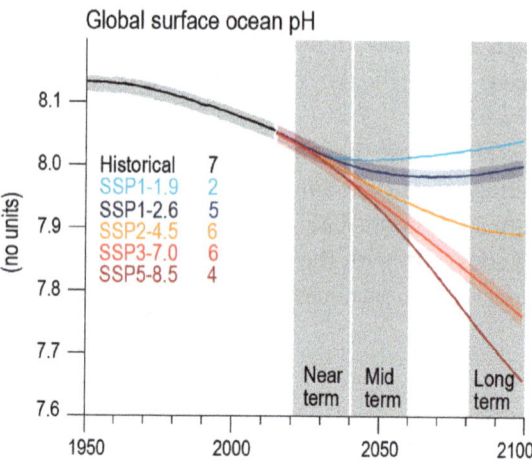

Figure 20.6 CMIP6 annual mean Atlantic Meridional Overturning Circulation (AMOC) strength change in historical and scenario simulations. Changes are relative to the 1995–2014 averages. The shadings show the 5%–95% ranges for the SSP1–2.6 and SSP3–7.0 ensembles. The circles at the right of the panel show the anomalies averages from 2081 to 2100 for each of the available model simulations. The numbers inside the panel are the number of model simulations. *Source*: see Acknowledgements at the end of the chapter.

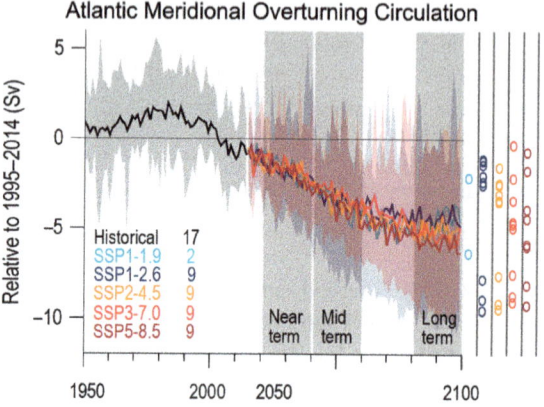

20.3.6 Abrupt and Irreversible Climate Change

AR6 addresses this possibility, although mainly in a section focusing on change beyond 2100. As some of these ideas and concepts have received media attention, we will provide a very brief introduction, and direct you to {section 4.7.2 (page 633)} for more information.

An abrupt climate change is a large-scale abrupt change in the climate system that takes place over a few decades or less, persists for a few decades, and causes substantial impacts to human and/or natural systems. Such a perturbed state is considered irreversible on a given timescale if the recovery timescale from this state due to natural processes takes substantially longer than the timescale of interest. A tipping point may be defined as a critical threshold beyond which a system reorganises, often abruptly and/or irreversibly, and a tipping element as a component of the Earth System that is susceptible to a tipping point.

In {table 4.1}, AR6 looks at 15 components of the Earth System that have been proposed as being susceptible tipping points/abrupt change, under some scenarios. A full analysis of all 15, and the reasons for AR6's conclusions, is beyond the scope of this book, but we will briefly discuss some which we consider the most important. This is a subject you might like to explore.

The West Antarctic Ice Sheet is *likely* to lose mass under all scenarios, and has the potential for abrupt change, which would take decades to millennia to reverse. Global ocean heat uptake, and sea level rise, are considered irreversible for centuries (*very high confidence*). Ocean acidification and deoxygenation are both expected to continue this century. Both are considered to be easily reversible at the surface, but restoring the situation at depth will likely take centuries to millennia.

Both tropical and boreal forests are not expected to undergo abrupt change. However, they are clearly vulnerable to human disturbance, with the Amazon being the best reported. If it were to be seriously degraded it would take many decades to reverse.

20.4 Climate Policy Implications

In the previous two sections we took a look at the climate projections, based on the simple assumption that, for whatever reason (or combination of reasons), the world (i.e. its people) chooses to follow one of the five SSP scenarios (or a closely similar pathway). But there are reasons to believe that humanity will act, via a suite of policy decisions, to change trajectory, at some point in the future. In this section we explore some issues related to such decisions.

We know that the Paris Agreement aims to hold global warming to 1.5°C above pre-industrial, or at least 'well below 2.0°C' above. While almost all countries have signed up to Paris, their current NDCs fall well short of those goals. Of course, part of the Paris Agreement requires all signatories to update (read 'improve') their NDCs every five years, so we should not lose hope. Nevertheless, 1.5°C does look like wishful thinking, while below 2.0°C should be considered an achievable goal: if we can maintain the political/economic/technological momentum.

One task of the IPCC, it might be argued, is to help maintain, and even boost, that momentum. To that end, AR6 devotes section 4.6 to examining the world at different levels of warming by 1.5°C, 2.0°C, 3.0°C, and 4.0°C – regardless of how, or when, such a level is reached. It also looks at the impacts of various decisions we might make along the way: mitigation on the one hand, and geoengineering on the other. In this section and the next we will provide a short overview of this set of issues.

20.4.1 Overshoot and Drawdown

Some of the scenarios that aim to keep global warming to below 1.5°C (or even 2.0°C) by 2100 involve an overshoot: that is, the effective radiative forcing temporarily exceeds a value which would be consistent with 1.5°C (or 2.0°C), before being reduced by some assumed mechanism/process involving net negative CO_2 emissions. While that may be the best we can achieve, we need to be cognisant of any specific effects resulting from the overshoot. The key question is: can any harmful changes that may occur above 1.5°C/2.0°C be reversed?

While some indicators are reversible after overshoot, some do not seem to be. Sea-level rise depends, in part, on the movement of heat to deeper waters, and this may not be fully reversible. Similarly, the melting of land ice would not be reversible, due to ice–albedo feedback effects. Any melting of the permafrost that allowed for the release of greenhouse gases would certainly not be reversible.

Any attempt to bring temperatures back down after an overshoot almost invariably involves the removal of CO_2 from the system: the CDR side of geoengineering. How feasible might this be? CDR options include afforestation, soil carbon sequestration, bioenergy with carbon capture and storage, wet land restoration, ocean fertilisation, ocean alkalisation, enhanced terrestrial weathering, and direct air capture and storage. These options are also referred to as 'negative CO_2 emissions technologies'.

Deployment of CDR will only lead to a reduction in atmospheric CO_2 if the uptake by sinks exceeds net CO_2 emissions, so there could be a (potentially long) delay between the initiation of CDR and net CO_2 emissions becoming negative. The climatic evolution after that would depend on the combination of anthropogenic emissions, CDR, and natural CO_2 sinks. The effect on warming (or cooling) would be proportional to the total CO_2 removed by CDR.

The required scale of CDR can vary from 1–2 $GtCO_2$ per year from 2050 to as much as 20 $GtCO_2$ per year, depending our emissions pathway. In SSP1–1.9, net emissions turn negative around 2050, and around 2070 in SSP1–2.6. (There are variants of other pathways that also include such options.) This implies that CDR will be pivotal in limiting warming to 1.5°C, or even 2.0°C. However, two extensive reviews conclude that it is implausible that any CDR technique can be implemented at the scale needed before 2050. (This, of course, is not strictly a climate science issue, and new research may change such a conclusion.)

Careful analyses of CDR options generally require an Earth System Model, as there are many potential interactions. Replacing croplands by forests decreases the albedo, partly offsetting the gains. For a high-emissions pathway, afforestation, ocean fertilisation, and surface ocean alkalisation are relatively ineffective. Thus, a lowering of anthropogenic emissions is a must.

AR6 {section 4.6.3} concludes that:

there is *high confidence* that, due to the near-linear relationship between cumulative carbon emissions and GSAT change, cooling or avoided warming due to a CDR option would depend on the cumulative amount of CO_2 removed by that option. The climate system response to the deployment of CDR is expected to be delayed by years (e.g., temperature, precipitation, sea ice extent) to centuries (e.g., sea level and AMOC) (*high confidence*).

20.4.2 Mitigation

Working Group II of the IPCC covers, among other topics, Adaptation: the necessary steps that societies and communities will need to take to cope with the impacts of climate change to date, and the changes locked in for the near-term future. While this is clearly essential, it is even more important that humanity takes all feasible steps to halt global warming. These steps fall under the heading Mitigation, and are the province of Working Group III. The material covered in that report looks at many options, and is deserving of a book/course – or at least a chapter – of its own. We do not have the space to do justice to this most important topic, so instead we will simply select the most important of the conclusions of WGIII, from the Summary for Policy Makers of the Synthesis Report. For a serious dive into this topic, we would suggest that you, in fact, bypass the Summary and read the full Synthesis Report.

The Summary starts with what should be obvious by now. 'All global modelled pathways that limit warming to 1.5°C with no or limited overshoot, and those that limit warming to 2°C, involve rapid deep, and in most cases, immediate greenhouse emissions reductions in all sectors this decade.' The details are then developed.

Rapid and far-reaching transitions across all sectors and systems are necessary to achieve deep and sustained emissions reductions and secure a liveable and sustainable future for all. These system transitions involve a significant upscaling of a wide portfolio of mitigation and adaptation options. Feasible, effective, and low-cost options for mitigation and adaptation are already available, with differences across systems and regions (*high confidence*).

WGIII devotes chapters to many of the economic sectors responsible for GHG emissions, much as we did (briefly) in Section 3.2. The SPM provides the key conclusions for these, under the headings: Energy Systems; Industry and Transport; Cities, Settlements and Infrastructure; Land, Ocean, Food, and Water; Health and Nutrition; and Society, Livelihoods, and Economies. Related issues addressed include Equity, Governance, and Finance.

How quickly might we see the results, and start to pat ourselves on the back? Internal variability in the climate system means that the response to mitigation is largely masked in the near term, especially on regional scales (*high confidence*), with the benefits only emerging later in the century. This 'fact of life' needs to be understood by policy makers.

20.5 Solar Radiation Management (SRM)

Most SRM approaches, such as stratospheric aerosol injection (SAI), marine cloud brightening (MCB), and surface albedo enhancements, aim to cool the Earth by reflecting more solar radiation back to space. By contrast, cirrus cloud thinning (CCT) aims to cool the planet by increasing long-wave emission to space: sort of a 'double negative' effect. We include it here, as does AR6 {section 6.6.3.3} for consistency. In this section we will look at the potential climatic effects of SRM, based on that section. Other chapters/sections of AR6 look at some of the potential dangers of SRM, such as the biogeochemical implications of SRM, and water cycle responses. (We also addressed some related issues in Section 9.5.2.) While these are clearly important, we do not have space to go to that level of detail.

SRM is primarily considered as a potential supplement to deep mitigation, such as within overshoot scenarios. AR5 assessed both the climate response to and the risks/side effects of SRM, and concluded with *high confidence* that, if practicable, SRM could substantially offset a global temperature rise, and partially offset some other impacts of global warming; but the compensation for the effects of GHGs would be imprecise. Regional impacts would vary.

Earlier results from relatively simple/idealised model simulations have recently been supplemented by the coordinated work of the Geoengineering Model Intercomparison Project (GeoMIP). Studies show that the reflection of an extra 2% of solar radiation – or an increase in the planetary albedo from 0.31 to 0.32 – would suffice to offset the warming from doubled CO_2, although the details differ for the different SRM methods.

20.5.1 Stratospheric Aerosol Injection (SAI)

Most SRM research has focused on stratospheric aerosol injection, mostly the injection of sulphate particles or their precursor gases such as SO_2. We covered some of this in Section 9.4. Other options proposed include $CaCO_3$, TiO_2, Al_2O_3, and engineered nanoparticles. Volcanic eruptions provide an imperfect analogy to sulphate aerosol injection, including some quantification of the effects on surface temperature. The cooling potential of sulphates depends on factors such as amount injected, microphysics, the spatial and temporal pattern of injection, the response of stratospheric chemistry and dynamics, and effects on high cirrus clouds. A negative forcing of anywhere between 1 and 8 W m^{-2} could be achieved.

There is a large uncertainty in the stratospheric response to SAI, and the change in both the dynamics and chemistry of the stratosphere would depend on the amount, size, type, location, and timing of injection. There is *high confidence* that aerosol-induced stratospheric heating will play an important role in surface climate change by altering the effective radiative forcing, lower stratospheric stability, quasi-biennial oscillation (QBO; which we have not covered in this book), and North Atlantic Oscillation. Simulations indicate stronger polar jets and weaker storm tracks, and a poleward shift of the tropospheric mid-latitude jets in response to injection in the tropics. Off-equatorial injection is *likely* to result in reduced change in stratospheric heating, circulation, and QBO.

Stratospheric ozone response to sulphate injection is uncertain. It is *likely* that sulphate injection would delay the recovery of the Antarctic ozone hole, with implications for UV radiation and surface ozone. Injection of non-sulphate aerosols is *likely* to result in less stratospheric heating and ozone loss. Another side effect of SAI is an increase in sulphate deposition at the surface. However, this is balanced by a projected reduction in anthropogenic SO_2 emissions.

20.5.2 Marine Cloud Brightening (MCB)

Marine cloud brightening involves injecting small aerosols such as sea salt into the base of marine stratocumulus clouds, where they act as cloud condensation nuclei (CCN). In the absence of other changes, an increase in CCN would produce higher cloud droplet number concentrations, with reduced droplet sizes, increasing the cloud albedo. There are competing effects around an increased liquid water path versus suppressed precipitation and even enhanced cloud water evaporation. The well-documented 'ship tracks' phenomenon (Section 6.3.1, Figure 6.5) is a good indicator. A recent study found a substantial increase in cloud reflectivity in the south-east Atlantic basin, suggesting that a regional-scale test of MCB in a stratocumulus-dominated region could be successful.

Modelling studies suggest that MCB has the potential to achieve a negative forcing of about $1–5$ W m^{-2}, depending on deployment area and seeding strategies. Regional applications have also been suggested to offset the severe impacts from tropical cyclones whose genesis requires high SSTs, and also for protecting coral reefs from the bleaching effects of high SSTs. Such localised approaches involve large uncertainties relating to responses and consequences.

Relative to a high-GHG climate, it is *likely* that MCB would increase precipitation over tropical land due to the inhomogeneous forcing pattern over ocean and land (*medium confidence*). The overall climate response to MCB remains uncertain due to the high level of uncertainty associated with cloud microphysics and aerosol–cloud–radiation interactions.

20.5.3 Cirrus Cloud Thinning (CCT)

Cirrus clouds reflect very little sunlight, but are efficient absorbers of long-wave radiation. When that energy is re-radiated it is at a much lower temperature than the surface, and hence the radiation escaping to space is reduced. In other words, radiation is trapped by cirrus clouds, which has an overall warming effect on the climate. If we were to seed such clouds with an appropriate concentration of ice-nucleating particles this should enhance the Bergeron–Findeisen (cold cloud) process (Section 6.3), and 'thin' such clouds.

Under present-day conditions, cirrus clouds exert a net positive radiative forcing of ~5 W m^{-2}, so this is the maximum cooling potential of CCT if all cirrus were to be removed. However, modelling studies suggest a much smaller potential of $1–2$ W m^{-2}, even for optimal seeding. Complex microphysical effects may well reduce that even further. Thus, there is *low confidence* in the cooling effect of CCT, due to limited understanding of cirrus microphysics.

20.5.4 Further Considerations

Modelling studies have consistently shown that SRM has the potential to offset some effects of increasing GHGs on global and regional climate (*high confidence*), but there would be some questionable effects at the regional and seasonal scale (*high confidence*). Patterns of climate change would vary for different SRM options. In the absence of mitigation, ocean acidification would continue to increase. There is *high confidence* that a sudden and sustained termination of high-level SRM, against a high-GHG background, would cause a rapid increase in temperature at a rate that far exceeds that projected for climate change without SRM.

Acknowledgements

For Figures 20.1 {4.2}, 20.2 {4.12}, 20.3 {4.19}, 20.4 {4.7}, 20.5 {4.8}, 20.6 {4.6}; Tables 20.1, 20.2 {4.1}, 20.3 {4.2}, 20.4 {4.3}, 20.5 {4.4}:

Climate Change 2021: The Physical Science Basis. Contribution of Working Group I to the Sixth Assessment Report of the Intergovernmental Panel on Climate Change [Masson-Delmotte, V., P. Zhai, A. Pirani, S. L. Connors, C. Péan, S. Berger, N. Caud, Y. Chen, L. Goldfarb, M. I. Gomis, M. Huang, K. Leitzell, E. Lonnoy, J. B. R. Matthews, T. K. Maycock, T. Waterfield, O. Yelekçi, R. Yu, and B. Zhou (eds.)]. Cambridge University Press, Cambridge, United Kingdom and New York, NY, USA, 2,391 pp. doi:10.1017/9781009157896.

FURTHER READING

There are many valuable books you might now wish to read, to further your understanding of the issues raised in this chapter, and, indeed, in this book. Two we suggest are:

- *The New Climate War* by Michael Mann (Public Affairs Books, 2021).
- *Introduction to Modern Climate Change* by Andrew Dessler (Cambridge University Press, 2021).

Final Remarks

So, we have come to the end of the book, and the end of a journey. What have we learnt?

The Basic Science of Climate

We started out looking at the important *Chemistry* of our atmosphere: some basic and important photochemistry, including the OH radical, nature's garbage collector; the lessons from air pollution and attempts to contain it; the sources and sinks and cycles of the all-important greenhouse gases, along with the perturbations to those cycles that have resulted from human activity; and the importance of atmospheric aerosols, from their formation through to their impacts, especially on the flow of solar radiation.

We then switched to *Physics*: the thermal properties and basic laws of the atmosphere, without, and then with, the inclusion of water vapour (including saturation and the Clausius–Clapeyron equation, latent heat, cloud formation, and microphysics); and the forces at work that drive atmospheric circulation (with the underlying thermal drivers seen as paramount).

Central to climate, and especially to our changing climate, is the flow of *Radiation*. While this is clearly a branch of physics, it is of such importance we have placed it on its own pedestal. The laws of thermal radiation determine the overall temperature of our planet as a member of the solar system. However, surface temperatures are only liveable due to the presence of certain radiatively active, or greenhouse gases: primarily water vapour, carbon dioxide, methane, and nitrous oxide. Why these gases, and not the more abundant nitrogen or oxygen? To answer this question, we took you on a brief journey through molecular quantum physics. At the end of that journey we also asked, and answered, another key question: how are the planet's energy fluxes being altered by the steady increase in these greenhouse gases?

Applications

We then pulled these various threads together, to see how well we understand climate. Our planet is large, and the climate system is complex: the full Earth System even more so. We need to add the oceans, and especially how their 'modes' contribute to climate variability (we still have much to learn here). We also need the cryosphere and the land surface. Central to the behaviour of any complex system is the suite of interactions and resulting feedbacks, and the climate is no exception. To help you understand how these work, we introduced you to a 'simple' Energy-Balance model.

If we wish to really understand climate, and especially climate change, we have no alternative but to build detailed numerical models, even while we acknowledge their imperfections. In fact, it is this recognition, and how we address the challenge, which makes us scientists. The first test we put our knowledge to was the climate, and its changes, at various stages of planetary history; a fascinating story, which we only partially comprehend. With due allowance for the limitations in our knowledge of the past, we believe we passed that test. One of the most active areas of recent palaeoclimate research focuses on sea level at various times in the (not too distant) past, and the lessons we might learn for the future.

Finally, in the last four chapters, we have addressed the most important questions many (we might hope, all) of our readers want answered. What has human industrial and agricultural activity done to our environment over the past century or so? What has been happening to our climate over the past century? Are these changes interconnected? How confident are we of that answer? Recently, the media has noted an increase in 'extreme weather' events. Is this just anecdotal – a media discovery – or is this a real trend? If so, can it be explained? And finally, the ultimate question: what does the future hold? What have we been able to conclude?

Our Changing Climate

In Chapter 17 we looked at how the composition of our atmosphere is changing (and also the impacts of any land-use change). From that we were able to compute the changes in the fluxes of energy both into, and out of, the earth–atmosphere system. We concluded that the increases in greenhouse gases have contributed a flux change of ~3.1 W m^{-2}: positive, and thus warming the planet. But we also found that the increase in anthropogenic aerosols (and their precursor gases) has cancelled out more than one-third of this, mainly due to the modification of cloud microphysics, and thus reflectance/albedo. This means that a significant fraction of the GHG warming is currently being masked: have we been deceiving ourselves?

In Chapter 18 we asked two questions. Firstly, how is our climate changing (over the past century or so)? Secondly, why is it changing? And more to the point, can we be confident in the second answer? The available data show clearly that global average surface temperature has increased by around 1 degree, and this warming is far greater at high northern latitudes. This is easily explained as a consequence of the melting of ice and snow, and the resulting ice–albedo feedback. The data also show an understandable rise in sea level, due to the combination of thermal expansion and the melting of land-based ice. To answer the second, more important question, we run climate simulations both without, and with, all of the drivers identified in Chapter 17: a process known as attribution. In all cases, the IPCC – meaning the science community – sees a clear human fingerprint. For example, 'It is unequivocal that human influence has warmed the atmosphere, ocean and land surface since pre-industrial times.'

While Chapter 18 took a broad look at our changing climate, Chapter 19 focused on a number of extreme events that have recently become apparent. Rising global average temperatures will, understandably, lead to an increase in heatwaves, although details such as location and timing might also be dictated by climate variability factors. A warmer atmosphere can hold more water vapour, providing the potential for heavier rainfall and flooding. Droughts are driven by rainfall deficit, but also by increased re-evaporation when it does rain. In recent years, attribution studies have begun to focus on some of these extremes, such as the 2003 European heatwave, allowing science to make quantitative statements about the role of human activity (e.g. greenhouse gas emissions) in either the magnitude, or probability, of some of these events.

What Does Your Future Hold?

In the final chapter we tried to answer the key question we are sure you would like answered: how will climate change in the years and decades ahead? Of course, we can't know the answer, as it depends very much on choices and decisions yet to be made. So the science community, through the collective endeavours of the IPCC (and not just Working Group I), has chosen a small number of options – scenarios – to reflect plausible paths which humanity might take, and then simulated the climate of the twenty-first century (and beyond) under each.

In AR6, the IPCC has provided a specific focus on three time periods: near-term (2021–2040), mid-term (2041–2060), and long-term (2081–2100). We, your authors, might just make it into the beginning of the mid-term. But, you, our readers, are likely to make it into the long-term: hence the strikethrough in the section heading above. You will inherit a world, and its problems, that previous generations have not managed as well as they should. We can only hope you will do better: we believe you can.

There is little point in trying to summarise this chapter for you, as there remain too many open questions. It will get hotter. How much hotter depends directly on the commitments nations make under the Paris Agreement, and how well they live up to them. Good intentions are the place to start, but they are not enough. Right now, we are on a path to 2.5°C, not 1.5°C, or even 2.0°C. And as we have seen all too clearly, unforeseen world events can mean that even the best of intentions might be put on hold.

We have looked at some of the likely, even 'certain', implications of such warming. It will be warmer virtually everywhere, but especially at high northern latitudes. What are the further implications? How much more will ice melt, and sea level rise, especially if such warming cannot be brought back down? Even if we can, eventually, remove some CO_2 from the atmosphere (i.e. 'overshoot and drawdown'), these changes may not be reversible. The melting of permafrost, and subsequent release of methane, is certainly not reversible. What other tipping points might be crossed? In Chapter 19 we looked at a selection of the recent extreme events we have experienced with a warming of ~1.0°C: right now, we are heading for at least twice that number. How much such mayhem can societies bear?

But Should I Believe?

Of course, a question you must each ask yourself is, how much of what the models and their prognostications are saying am I prepared to accept? This is a valid question, and one only you can answer. The late Stephen Schneider, a leading climate scientist and populariser/advocate, once described climate models as 'dirty crystal balls', then asked 'how long do we go on polishing the glass before we decide to act on what we think we see inside?'

In *Failure Is Not An Option* (a very readable book), Gene Kranz relates his personal experience of the Apollo-13 rescue (he was a mission controller), when his teams had to make life-and-death decisions with far too little time to check everything. He told his team 'If you don't have an answer, we need your best judgement': give me your best shot, and I will back you. '**A 100 per cent correct answer, too late to be of use, was worthless.**' This truly profound statement is as true today as it was in 1970.

There are two points we'd like to focus on before we sign off. It is virtually certain, given all we know about climate science, economics, technology, and politics, plus human nature, that we will not be able to hold global temperature rise below 1.5°C, and even 2.0°C is doubtful. So AR6 has looked at the question of overshoot and drawdown (Section 20.4.2): that is, how to get back to a 'reasonable' climate after going above the desired limit. This will require the removal of carbon dioxide from the atmosphere, something we currently struggle to do. This is clearly an area that calls for more research, and more researchers. The alternative of solar radiation management (Section 20.5) is also doable, and is an active field of research – in particular stratospheric aerosol injection – but is not without its drawbacks.

Existential Issues

Finally, a few words on sea-level rise. This is one aspect of the Earth System that responds to forcings on comparatively long timescales, and is one of the key motivators of current work in palaeoclimatology as we saw in Chapter 16. So, even after we stabilise both CO_2 and global temperatures, the oceans will continue to expand – rise – in the decades and centuries to come.

The effects of sea-level rise will, of course, not be felt uniformly. At the 2022 annual United Nations General Assembly, most world leaders chose to concentrate on conflict in Ukraine, and its impacts on food and energy supply: who would blame them? However, small island nations, led by Vanuatu, focused on their existential future. President Nikenike Vurobaravu told the Assembly:

Every day, we are experiencing more debilitating consequences of the climate crisis. Fundamental human rights are being violated and we are measuring climate change not in degrees Celsius or tonnes of carbon, but in human lives. We call for the development of a Fossil Fuel Non-Proliferation Treaty to phase down coal, oil and gas production in line with 1.5°C, and enable a global just transition for every worker, community and nation with fossil fuel dependence. (Quoted in the *Sydney Morning Herald*, 24 September 2022.)

While small island nations face an existential threat, people living near river deltas – the Nile, Mekong, Mississippi, Indus, and especially Ganges/Brahmaputra – also face the prospect of having to relocate, not just their homes, but their lives. And what of Venice, and other coastal cities?

Between the Australian mainland, and Papua New Guinea, lie the Torres Strait Islands, which are part of Australia. (Australia's First Nations people are known collectively as Aboriginal and Torres Strait Islanders.) Their homes are under threat from rising seas, including by salt-water incursion. Recently they took their case to the United Nations Human Rights Committee, which ruled that the Australian Government had violated their rights as Torres Strait Islanders. Only a moral victory, of course, but one with wider implications.

We have not included any Exercises for this chapter. Instead, we'd rather you spent a little time reflecting, not only on what you have read, but what may have happened since we wrote these last paragraphs. Do you see signs of improvement? Or have matters become worse? Do you have any ideas on how you might make even a small positive contribution to the future of the world, either in your daily lives, or by your career choices? We wish you well.

Glossary

Absorption A process whereby the energy of incident electromagnetic radiation is converted to internal energy (e.g. heat) in a particle, or a gas

Absorption cross-section A measure (in m^2) of the amount of radiation absorbed by an object

Acid rain Precipitation with a pH less than ~5.5

Adiabatic A process that takes place without the exchange of heat (energy)

Aerosol Suspended atmospheric particulate

Air parcel Imaginary small amount of air, especially for considering vertical stability

Albedo The fraction of incoming light reflected by a body or a surface

Anomaly Departure of a climatic parameter from the climatological mean, or some reference value

Attribution The scientific process of deciding the cause(s) of any change in climate.

Austral Southern: hemisphere, or high latitudes

Boiling point Temperature at which saturation vapour pressure equals ambient pressure

Boreal Northern: often confined to the higher latitudes

Climate sensitivity The increase in temperature that would result from a doubling of CO_2 in the atmosphere

Convection Overturning motions that transfer heat (and other relevant properties) vertically

Coriolis force A fictitious force introduced into the equations of motion to account for the Earth's rotation

Cryosphere The part of the climate system dominated by frozen water (ice, snow, permafrost)

CSIRO The Commonwealth Scientific and Industrial Research Organisation, Australia

Dobson unit One 'milli-atmosphere-centimetre' of ozone. It is equal to 2.69×10^{16} ozone molecules per square centimetre of surface area

Extinction cross-section Sum of absorption cross-section + scattering cross-section

Evapotranspiration Water vaporisation via direct **evaporation** of surface water, plus release of water vapour through leaf pores (**transpiration**)

Feedback Environmental process whereby a change in a variable of the system, through interactions within the system, either reinforces the original process (*positive feedback*) or suppresses the process (*negative feedback*)

Free radical A molecule containing an unpaired electron

Gradient The change in a variable (e.g. temperature) per unit of distance

Greenhouse gas A radiatively active gas

Halogen Fluorine, chlorine, or bromine

Hydrometeor Water in liquid or solid form which precipitates

Insolation (*in*coming *sol*ar radi*ation*) Solar radiation/energy at the Earth's surface (or another level; e.g. top of the atmosphere)

Interannual Having a timescale of years

Interdecadal Having a timescale of decades

Lapse rate The rate of decrease of temperature with altitude

Latent heat Amount of energy (J) required to convert 1 kg of a substance from one phase to another (solid to liquid; liquid to vapour; solid to vapour)

Long-wave radiation Electromagnetic radiation with wavelengths greater than 4.0 μm

Luminosity Total radiant power emitted by the Sun (or any star)

Mass extinction efficiency Extinction cross-section divided by mass

Meridional In the south–north direction

OLR Outgoing long-wave radiation (leaving the top of the atmosphere)

Permafrost A layer of soil or bedrock at a depth below the surface in which the temperature has been below freezing for more than a few thousand years

Photon The 'quantum' of radiant energy

Primary aerosol Particle directly emitted to the atmosphere

Radiatively active gas Any gas that absorbs long-wave radiation

Re-analysis A full, four-dimensional picture of the wind, pressure and temperature fields as a function of time

Relative humidity Actual vapour pressure divided by saturation vapour pressure

Salinity Concentration of salt in water

Scattering cross-section A measure (in m^2) of the amount of radiation scattered by an object

Secondary aerosol Particle formed by gas-to-particle conversion

Short-wave radiation Electromagnetic radiation with wavelengths less than 4.0 μm

Single scattering albedo Scattering cross-section divided by extinction cross-section

Solar constant Amount of solar radiation incident at the top of the atmosphere (at mean Earth–Sun distance)

Specific heat Amount of energy (J) required to raise the temperature of 1 kg of a substance by 1°C

Specific humidity, q Mass of water vapour divided by the mass of moist air

Specific volume, α Volume per unit mass (the inverse of density)

Spectrum The range of wavelengths of electromagnetic radiation relevant to a particular process (also spectral range)

Terrestrial radiation Long-wave radiation emitted by the Earth and/or atmosphere

Thermal circulation Atmospheric circulation driven by temperature differences

Thermocline Subsurface ocean layer at ~100+ m depth which separates the relatively warm upper ocean from the cold ocean depths

Ultraviolet radiation Electromagnetic radiation with wavelengths less than 0.4 μm

Visible radiation Electromagnetic radiation with wavelengths between 0.4 μm and 0.7 μm

Vapour pressure deficit Saturation vapour pressure minus the actual vapour pressure

Zonal In the west–east direction

Index

For EU product safety concerns, contact us at Calle de José Abascal, 56–1°, 28003 Madrid, Spain or eugpsr@cambridge.org.

www.ingramcontent.com/pod-product-compliance
Ingram Content Group UK Ltd.
Pitfield, Milton Keynes, MK11 3LW, UK
UKHW052023150126

466966UK00015B/168

9 781009 372336